Engineering Design Process
Third Edition, International Edition

· ·

Yousef Haik
Hamad Bin Khalifa University

Sangarappillai Sivaloganathan
United Arab Emirates University

Tamer Shahin
Nuviun Corporation

CENGAGE
Learning™

Australia • Brazil • Japan • Korea • Mexico • Singapore • Spain • United Kingdom • United States

Engineering Design Process,
Third Edition, International Edition
Yousef Haik, Sangarappillai Sivaloganathan, Tamer Shahin

Product Director, Global Engineering:
 Timothy L. Anderson

Associate Media Content Developer:
 Ashley Kaupert

Product Assistant: Teresa Versaggi

Marketing Manager: Kristin Stine

Director, Higher Education Production:
 Sharon L. Smith

Content Project Manager: D. Jean Buttrom

Production Service: SPi Global

Compositor: SPi Global

Senior Art Director: Michelle Kunkler

Internal Designer: Lou Ann Thesing

Cover Designer: Tin Box Studio

Cover Image: Konstick/Dreamstime.com

Internal Graphic Images:
 iStock.com/naqiewei

Intellectual Property
 Analyst: Christine Myaskovsky
 Project Manager: Sarah Shainwald

Text and Image Permissions Researcher:
 Kristiina Paul

Manufacturing Planner: Doug Wilke

Library of Congress Control Number: 2016952377

ISBN: 978-1-305-25330-8

Cengage Learning

Cengage Learning International Offices

Asia
www.cengageasia.com
tel: (65) 6410 1200

Australia/New Zealand
www.cengage.com.au
tel: (61) 3 9685 4111

Brazil
www.cengage.com.br
tel: (55) 11 3665 9900

India
www.cengage.co.in
tel: (91) 11 4364 1111

Latin America
www.cengage.com.mx
tel: (52) 55 1500 6000

UK/Europe/Middle East/Africa
www.cengage.co.uk
tel: (44) 0 1264 332 424

Represented in Canada by
Nelson Education, Ltd.
tel: (416) 752 9100 / (800) 668 0671
www.nelson.com

Cengage Learning is a leading provider of customized learning solutions with office locations around the globe, including Singapore, the United Kingdom, Australia, Mexico, brazil, and Japan. Locate your local office at: **www.cengage.com/global**

For product information: **www.cengage.com/international**
Visit your local office: **www.cengage.com/global**
Visit our corporate website: **www.cengage.com**

Unless otherwise noted, all items © Cengage Learning.

Printed in the United States of America

Print Number: 01 Print Year: 2016

TO OUR PARENTS, WIVES,
AND CHILDREN.

TO FUTURE DESIGNERS.

CONTENTS

Preface   ix

Design is the practice of turning "dreams of wonderful products" into reality. It converts an abstract idea to a tangible physical product. The complexity of any design stems from the abstraction of the idea, the unknown aspects of the "wonderful product dreamt", and the identification of the best combination of creativity, science, and technology used to realize it. Thus, the challenges facing the designer are (1) defining the problem well, and (2) identifying the optimal solution.

In traditional practice, the designer handles these challenges in private and this ability is treated as natural talent. *Engineering Design Process* presents the model of systematic design, which challenges the assumptions of traditional design practice and makes the design process transparent so that any motivated person can engage in design and produce dream products.

We have taken special consideration of the needs of undergraduate students taking a design course. We describe an orderly process of collecting, recording, and analyzing all facts to enable students to realize physical products that meet societal needs. The book makes it easy for students at various levels to follow the design process as different methods at each stage of the design process are introduced. The technical analysis component of the design was kept at an introductory level so that students with technical capabilities at the level of college physics and calculus can easily understand the analysis component. The book can be used by first- and second-year students starting out in design while still being a useful reference for students working on capstone projects and entering professional practice.

NEW TO THIS EDITION

This third edition introduces the systematic design process in five stages:

1. Requirements
2. Product concept
3. Solution concept
4. Embodiment design
5. Detailed design

This design model tackles the two primary challenges to designers: (1) ill-defined design problems, and (2) unknown solution spaces.

The third edition presents methods in each of the design stages designed to lead the reader towards the optimal design solution. New methods are added in all of the five stages. New relatable, easy to comprehend examples have been presented.

The five stages are presented in a coherent fashion to help students navigate the design model systematically with supporting exercises and labs. A new lab in Chapter 1 introduces beginner designers to the design process and helps show how to apply the methodology in engineering and other design problems.

TRUSTED FEATURES

This book incorporates a consistent approach to teaching the engineering design process, making it easily comprehensible to all engineering students. Students learn how to regularly and carefully follow each important step, including identification of a need and setting goals, market analysis, specifications and constraints, function analysis, generating concepts, evaluating alternatives, embodiment, analysis, experiment, and marketing.

Particular care is taken to emphasize the issues students need to consider before proceeding with the design process. These key prerequisite considerations include scheduling, human factors, safety considerations, and presentation style.

Examples, activities, and labs ensure students have the knowledge and skills necessary to succeed as practicing engineers. Practical, illustrative examples throughout the text clarify and visually reinforce the key steps in engineering design and show how the material is applied. Design labs are integrated into each chapter. In addition to giving students important practice in teamwork as part of the design process, these projects help students practice material selection, ergonomics, finite element method analysis, geometric tolerance, and scheduling.

ORGANIZATION

The book consists of 15 chapters divided into seven parts. We recommend reading each part completely and thoroughly before embarking on that stage. This will help the reader to decide the tools, methods, and approaches to be used at each stage of the design project.

- Chapter 1 presents an overview of the design process with an introductory discussion on the five stages design model and some of the design methods for each of the stages. The Chapter highlights the design challenges, broadly reviews the various aspects associated with design, illustrates conventional design, summarily describes systematic design process, explains how the function providers of a complex machine can be conceptualised as integration of simple and individual function providers. An illustrative example "a tale of developing a sandwich" is introduced in Lab 1 to orient beginning design students and to demonstrate the versatility of this text's design model. This is a 'must read chapter' for anyone using this book.
- **Part 1: General:** Covers the general skill set and norms used by a successful professional engineer. Students should master them before embarking on the use of systematic design process. Chapter 2 covers the basic competencies students need to succeed in the design process, including working in teams, scheduling, research skills, technical writing, and presentation skills. Chapter 3 discusses ethics and professionalism, including the Code of Ethics developed by the National Society of Professional Engineers.
- **Part 2: Requirements:** The genesis of a product starts with the design brief stating "what the product is" and the list of customer requirement stating "what the product would or should do" as expected by the customer. This part covers these in detail. Chapter 4 describes how market information is gathered by senior management to establish the design brief. The design brief is the first document that defines what the product is and the constituent elements are also described in Chapter 4. Chapter 5 describes how to determine customer requirements, then standardize and prioritize them. The set of requirements thus established describe what the product would or should do.
- **Part 3: Product Concept:** This part describes the design model consisting of functional description and specifications. Chapter 6 covers the process of establishing function structures, starting with describing the product with its physical and functional elements, creating a function tree for a product, and developing function structure in terms of flow for a product. Chapter 7 explains how to identify and define specifications based on customer requirements.
- **Part 4: Solution Concept:** This part describes the conceptual design methods, evaluation of alternatives, and assembling and assessing prototypes. Chapter 8 covers concept design, showing how to create morphological charts, use systematic methods to generate function-based designs, conceptual designs, and analyze and evaluate conceptual design approaches. Chapter 9 discusses the various methods for evaluating concepts developed in the previous stages, including using decisions matrices and Pugh's concept selection.
- **Part 5: Embodiment Design:** The design starts to take shape in this stage. Chapter 10 provides an overview of prototyping concepts and practices, including design for "X." Chapter 11 covers the steps needed for the embodiment design stage to bring together the stages of concept design and detail design.
- **Part 6: Detailed Design:** The last step in the synthesis is to make the design complete with every detail and to prove the design with the necessary analyses. Chapters 12 and 13 introduce methods to generate detailed design and design analysis.

- **Part 7: Closure:** This section brings together all of the concepts introduced in the book. Chapter 14 presents a detailed case study of designing a single stage scissor platform and Chapter 15 presents a list of project descriptions that can serve as an entry point to instructors' assignments.

The design labs are integrated within the chapters. The purpose of these labs is to create design activities that help students, especially first and second year students, to adjust to working in teams. The first few labs are geared toward team building. Instructors may want to include other activities in their design classes.

SUPPLEMENTS

An instructor's solutions manual and Lecture Note PowerPoint Slides are available via a secure, password-protected Instructor Resource Center at https://login.cengage.com/.

MINDTAP ONLINE COURSE

Engineering Design Process is also available through **MindTap**, Cengage Learning's digital course platform. The carefully crafted pedagogy and activities in this textbook are made even more effective by an interactive, customizable eBook, automatically graded assessments, and a robust suite of study tools.

As an instructor using MindTap, you have at your fingertips the full text and a unique set of tools, all in an interface designed to save you time. MindTap makes it easy for instructors to build and customize their course, so you can focus on the most relevant material while also lowering costs for your students. Stay connected and informed through real-time student tracking that provides the opportunity to adjust your course as needed based on analytics of interactivity and performance. **End-of-chapter assessments** test students' knowledge of topics in each chapter. Students can submit design portfolios of projects they work on in the course using the **PathBrite App**. Curated **Engineering Design Resources** are available in each chapter to give students videos and links to complement the material in the book.

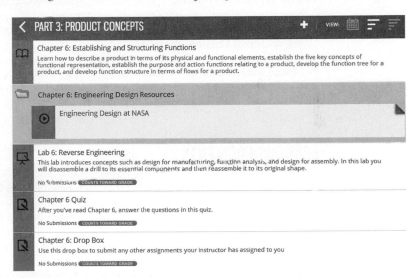

How does MindTap benefit instructors?

- You can build and personalize your course by integrating your own content into the **MindTap Reader** (like lecture notes or problem sets to download) or pull from sources such as RSS feeds, YouTube videos, websites, and more. Control what content students see with a built-in learning path that can be customized to your syllabus.
- MindTap saves you time by providing you and your students with **automatically graded assignments and quizzes**. These problems include immediate, specific feedback, so students know exactly where they need more practice.

- The **Message Center** helps you to quickly and easily contact students directly from MindTap. Messages are communicated directly to each student via the communication medium (email, social media, or even text message) designated by the student.
- **StudyHub** is a valuable studying tool that allows you to deliver important information and empowers your students to personalize their experience. Instructors can choose to annotate the text with **notes** and **highlights**, share content from the MindTap Reader, and create **flashcards** to help their students focus and succeed.
- The **Progress App** lets you know exactly how your students are doing (and where they might be struggling) with live analytics. You can see overall class engagement and drill down into individual student performance, enabling you to adjust your course to maximize student success.

How does MindTap benefit your students?

- The **MindTap Reader** adds the abilities to have the content read aloud, to print from the reader, and to take notes and highlights while also capturing them within the linked **StudyHub App**.
- The **MindTap Mobile App** keeps students connected with alerts and notifications while also providing them with on-the-go study tools like Flashcards and quizzing, helping them manage their time efficiently.
- **Flashcards** are pre-populated to provide a jump start on studying, and students and instructors can also create customized cards as they move through the course.
- The **Progress App** allows students to monitor their individual grades, as well as their level compared to the class average. This not only helps them stay on track in the course but also motivates them to do more, and ultimately to do better.
- The unique **StudyHub** is a powerful single-destination studying tool that empowers students to personalize their experience. They can quickly and easily access all notes and highlights marked in the MindTap Reader, locate bookmarked pages, review notes and Flashcards shared by their instructor, and create custom study guides.

For more information about MindTap for Engineering, or to schedule a demonstration, please call (800) 354-9706 or email higheredcs@cengage.com. For those instructors outside the United States, please visit http://www.cengage.com/contact/ to locate your regional office.

ACKNOWLEDGMENTS

The authors wish to thank Patrick Tebbe, Minnesota State University, Mankato, and Patricia Buford, Arkansas Tech University for their helpful suggestions.

The authors also wish to acknowledge and thank our Global Engineering team at Cengage Learning for their dedication to this new book: Timothy Anderson, Product Director; Ashley Kaupert, Associate Media Content Developer; D. Jean Buttrom, Content Project Manager; Kristin Stine, Marketing Manager; Elizabeth Brown and Brittany Burden, Learning Solutions Specialists; Teresa Versaggi, Product Assistant; and Rose Kernan of RPK Editorial Services, Inc. They have skilfully guided every aspect of this text's development and production to successful completion.

Students have been and are the inspiration behind this book. A lot of their work has initiated the thought process in several examples. Their contributions are very much appreciated. The authors are grateful to all colleagues and students who helped in producing this book. Students and instructors are encouraged to submit their comments and suggestions to the authors by emailing globalengineering@cengage.com.

Yousef Haik,
Sangarappillai Sivaloganathan,
and Tamer Shahin

INTRODUCTION

"The life and soul of science is its practical application."

~Lord Kelvin

Systematic design process is a sequence of stages with a design model and a combination of design methods, which are tools and techniques usable at different stages of the design process. We start by providing an overview of the design process by explaining the nature and definition of design, the design solution space, and the two challenges of design. These are (1) defining the problem right and (2) identifying and developing the unknown optimal solution from the unknown solution space. We will also explore the traditional design process, which can lead to a satisfactory solution. The systematic design process is introduced at this point together with a brief explanation of the systematic design process adopted in this book. Following that is a discussion of the systematic design process as advocated by two other authors. The chapter concludes with an explanation of the book's structure.

1.1 OBJECTIVES

By the end of this chapter, you should be able to

1. Explain and define design.
2. Appreciate the importance and challenges of design.
3. Understand the need for a formalized systematic design process.
4. Name and describe the stages for the systematic design process.
5. Distinguish between different systematic design models.

1.2 NATURE AND DEFINITION OF DESIGN

A design problem is the characterization of a societal *need*. It describes a problem in society for which solutions are needed. A design problem characterizing a societal need typically has several possible solutions, and all possible solutions constitute the design solution space. Some of these solutions may be (1) high-tech while others are not, (2) expensive while others are not, (3) efficient while others are not, (4) manual while others are not, (5) developed by people who have limited theoretical knowledge but do have a wealth of experience (e.g., craftsmen) or not, and (6) acceptable solutions according to some criteria or not. Acceptable solutions meet several requirements, whereas the optimal and near-optimal solutions meet most of the requirements. They are subsets of the acceptable set of solutions. In a Venn diagram the situation can be mapped as shown in Figure 1.1.

Two characteristics of a design problem are that (1) the problem is ill defined and (2) the solution space is unknown. *The challenge is to define the problem well and then identify the optimal or near-optimal solution from the unknown solution space in an efficient way.*

For example, a need statement may read "researchers often refer to more than one book while reading and taking notes. They **need** *a mechanism or appliance to keep these books open by the side for prolonged periods when they refer and cross-refer.*" This is a design problem that needs a solution. The loose nature of this need permits several design solutions that could very well satisfy the need and could be fascinating and exciting as well.

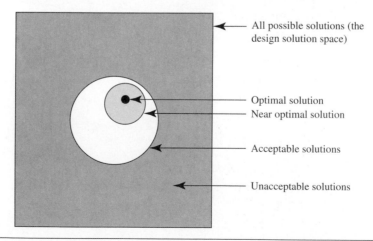

All possible solutions (the design solution space)

Optimal solution
Near optimal solution

Acceptable solutions

Unacceptable solutions

Figure 1.1: Design Solution Space

Identification of design problems is one of the crucial parts of product design and development. All design problems have certain common characteristics:

1. There are several possible solutions to the design problem.
2. The problem solver must formulate the potential solution.
3. Often proposed solutions provide better insight into the problem.
4. Solutions are not right or wrong; instead they range from better to not-so-good.
5. The designer needs to choose from the alternatives according to some implicit or explicit criteria. In this book an *objective-based method* is prescribed as the basis for the criteria and their weighting.

Design refers either to the product that has been designed or the process that has been followed to produce the design. This book focuses on the design process.

In the 1960s a committee from the Science and Engineering Research Council of the UK [1] described engineering design as "the use of scientific principles, technical information and imagination in the definition of a mechanical structure, machine or system to perform pre-specified functions with the maximum economy and efficiency." A task force for the U.S. National Science Foundation defined the new discipline called *Design Theory and Methodology* [2]. In this discipline, design theory refers to "systematic statements of principles and relationships, which explains the design process and provides a useful procedural way for design. In the same way the methodology refers to the collection of procedures, tools and techniques, which the designer may use in applying the design theory to the process of design."

The task force stated that the new discipline has two constituent parts called (1) conceptual design and innovation and (2) quantitative and systematic methods. Another formal definition of engineering design can be found in the curriculum guidelines of the Accreditation Board for Engineering and Technology (ABET). The ABET definition [3] states that "engineering design is the process of devising a system, component, or process to meet desired needs. It is a decision-making process (often iterative), in which the basic sciences, mathematics, and engineering sciences are applied to optimally convert resources to meet a stated objective."

In another approach, adapted from Gero [4], an analytical model for a design can be given as a function in the following way:

$$D = f(F, B, S, K, C)$$

Where

D = Design of the product

F = Functions intended to be performed by the product

B = Behavior of the structural elements that provide the intended functions

S = Structural attributes

K = Knowledge used in the design

C = Context of the product

The fundamental elements of the design process include the establishment of objectives and criteria, synthesis, analysis, construction, testing, and evaluation. It can be viewed from several different perspectives as explained in sections 1.3.1 to 1.3.3. From these definitions it is evident that design is both a scientific and a creative process. Albert Einstein's assertion "imagination is more important than knowledge, for knowledge is finite whereas imagination is infinite" reaches its full embodiment in design.

In this book, we will describe an orderly design process of collecting, recording, and analyzing all facts to enable beginner designers to define the problem sufficiently well as well as steps for identifying a near-optimal or optimal solution and its complete definition.

1.3 THE CHALLENGES OF DESIGN

The challenges of design are two-fold: (1) defining the problem and (2) identifying the solution. When any one of these is not done well, the design can fail to produce the expected results. The systematic design process was introduced to help guide designers to achieve their goals without hindering creativity. Classifying the design task as a set of categories helps to eliminate a significant part of the solution space and focuses on the small part that contains the optimal and near-optimal solutions. The following sections discuss this in more detail.

1.3.1 Classification of Design According to the Level of Difficulty

Based on the degree of difficulty, design can be classified as *adaptive design, development design*, and *new design*.

Adaptive design: In the great majority of instances, the designers' work will be concerned with the adaptation of existing designs. There are branches of manufacturing in which development has practically ceased; there is hardly anything left for the designer to do except make minor modifications, usually in the dimensions of the product. Design activity of this kind demands no special knowledge or skill, and a designer with ordinary technical training can easily solve the problems presented. One such example is the elevator, which has remained the same technically and conceptually for some time. Another example is a washing machine. This has been based on the same conceptual design for the past several years and varies in only a few parameters, such as its dimensions, materials, and detailed power specifications.

Developmental design: Considerably more scientific training and design ability are needed for developmental design. The designer starts from an existing design, but the final outcome may differ markedly from the initial product. Examples of developmental design include moving from a manual gearbox in a car to an automatic one and moving from the traditional tube-based television screen to the modern plasma and LCD versions.

New design: Only a small number of designs are new designs. This is possibly the most difficult level because generating a new concept involves mastering all the skills needed to work on the previous two categories in addition to creativity and imagination, insight, and foresight. Examples of this are the design of the first automobile, airplane, or even the wheel (a long time ago). Take a moment to think of entirely new designs that have been introduced during the past decade.

1.3.2 Modular Design

Modular design is a technique whereby units or modules that perform distinct functions and are easy to assemble are designed and developed so that different combinations of them could be assembled to develop distinctly different products. A good example of modular design is a personal computer. The modules involved are the memory chips, hard discs of various sizes, CD drives, and graphics cards along with many other modules. The basic product can have various combinations of modules resulting in different configurations with different functionalities. Classifying the design task as modular or otherwise at the beginning can narrow down the research area considerably. Further it enables the use of any *design methods* (special techniques) in modular design.

1.3.3 Platform Design

In simple terms, a product platform consists of the basic design and components that are used in several products or in a family of products. It is the structural or technological form from which various products can emerge without the expense of developing a new process or technology. Examples of platform design include power tools, cars, and printers. To elaborate, a family of cars in platform design share the same platform but differ in some features; for example, although the motor and the structure are the same for two different car products, some features such as seating may distinguish two car products.

A famous example of how the platform approach to development design could be a key success is the development of the 35-mm single-use camera. In 1987 Fuji introduced the QuickSnap 35-mm camera in the U.S. market and was able to dominate the market for a short time. Kodak was able to take the market back in 1994 by developing three different cameras that share common platform. Platform design is a key approach that allows tailoring products to meet individual customers' needs and also allows differentiated products to be delivered to the market without consuming excessive resources. Robertson and Ulrich [5] define a platform as "the collection of assets that are shared by a set of products." These assets may include components, knowledge, and production processes. Effective platform planning balances the market value of product differentiation against the economies achieved through commonality and thus reduces the research area in the design solution space.

1.4 CONVENTIONAL DESIGN PROCESS

The conventional design process can be best introduced by describing a meeting with an older traditional designer in the 1980s. Two students (one of the authors was one) and their professor visited the managing director and founder/owner of a company that designed and supplied products to the Ministry of Defense for the United Kingdom's government. The professor was a good friend of the designer. The designer enthusiastically received and offered tea to the visitors. While having tea, the designer got a call to attend to an urgent problem at the factory. The three visitors were invited to wander around the office to pass the time while the designer rushed to the factory. The office was fairly large and had four drawing boards with drawing papers with half-developed drawings clipped on them. No one was working on them. There were a calculator, a white board, and marker pens. The designer returned, and the professor told his friend that the students were curious to know why there were four drawing boards that had no one working on them. Oh that is my "brainstorming room" replied the designer. He then took the three visitors and started explaining. On the first board, he had developed the idea for a product for the defense ministry, and halfway into it, he realized that it would not work within the given constraints. He then moved to the second board and started working fresh. He soon progressed to a level beyond the previous one. Again he realized that it would not work. He then moved to the third board and made substantial progress there. However, now he realized that he could incorporate some parts of his first idea into the third idea. The board and the marker pens would help to write down an occasional calculation or a spark of an idea. The office was in effect a mini-library with books and handbooks. The design process this designer employed can be summarized as follows:

- The designer understands the problem well and can start work on a solution in earnest.
- No other help is made available other than what is provided by textbooks and handbooks.

- The solution to the problem lies within the designer.
- Immersion and long hours of thinking are the working methods.
- Different solution concepts are considered one after the other.
- Better insights can be achieved from unsuccessful concepts.
- The process continues until a working solution to the problem has been reached.

Similar accounts of activities can be seen in the annals of engineering designers. An account about James Brindley (the famous British canal builder) reads "When any extraordinary difficulty occurred, having little or no assistance from books or the labors of other men, his resources lay within himself. In order therefore to be quiet and uninterrupted while he was in search for necessary expedients, he generally retired to his bed; and he has been there for one, two or three days, till he had attained the object in view. He would then get up and execute his design, without any drawing or model. Indeed, it was never his custom to make either unless he was obliged to do so by his employers." [6]

Historically, design started with the craftsmen who made artifacts, which they improved by trial and error. Over time the products reached perfection. Technological advancement came slowly, and the product still remained current with the existing technology. Designs were the embodiment of the saying "A good carpenter will cut the wood twice; once in his head and once physically." There was no drawing. As the products became increasingly complex and required the application of science and mathematics, product development went into the hands of a new individual called the *design draftsman*. That changed a few decades ago when research in design and the introduction of computers into the design office moved product development into the hands of design engineers who practice the *systematic design process*.

1.5 INTRODUCTION TO SYSTEMATIC DESIGN

Engineering students during their training are presented with a vast amount of theoretical material and information. They face a challenge only when they are faced with the task of logically applying what they have learned to a specific outcome. As long as their work is based on familiar models or previous designs, the knowledge they possess is perfectly adequate for finding a solution along conventional lines. As soon as they are required to develop something already in existence to a more advanced stage or to create something entirely new without a previous design, they will fail miserably unless they have reached a higher level of understanding. Without a set of guidelines, they are at a loss for a starting point and a clear finishing goal line. The design process was formalized to enable both students and professional designers to follow a systematic approach to design and to help them guide their creativity and technical problem-solving skills to reach a satisfactory outcome.

The systematic approach to the design process divides the process into constituent components and utilizes the components to achieve the desired result. The process can be based on either (1) the intermediate stages (milestones) the design process goes through or (2) the activities that are carried out during the design process. When a systematic approach is adopted, no specific parts are overlooked and, more importantly, definite starting point and finishing point are defined.

1.5.1   Design Stage Model

The *design stage model* uses the stage at which the design is at different time points during the product development to describe the design process. Different researchers have identified different landmark stages, as few as four to as many as nine. They all involve following the same basic stages, which are the stages used in this book:

- Requirements
- Product concept
- Solution concept
- Embodiment design
- Detail design

1.5.2   Design Activity Model

In the design activity model, we describe the process as a sequence of activities that occur during the design process. These can be more detailed for each of the stages and also can incorporate several steps as the complexity of the product increases. The requirements stage can incorporate activities such as (1) recording customer verbatim (statements/expectations), (2) converting customer verbatim to customer requirements, (3) defining metrics and units that provide each of the requirements and establishing the target specifications for a reasonably complex product. As the complexity of the product increases, it becomes fundamentally important to define the activities in each stage. Designers carry out an evaluation at the end of each stage before proceeding to the next stage. Such design models are called the *stage gate models*.

1.5.3   Design Methods

Cross [7] defines design methods as tools and techniques that can be used at different stages of the design process. The development of several design methods has occurred during the past few decades. Typical examples include brainstorming, collaborative sketching, decision matrix for concept selection, Pugh's matrix for concept evolution, and house of quality for eliciting customer requirements, among others. It has been said that there are several hundreds of design methods to choose from.

1.5.4   Scaffolding the Design Process

Scaffolding provides a framework that facilitates access to previously inaccessible parts of a building under construction and acts as a skeletal structure for constructing the building. In a similar fashion, the design model acts as a skeletal structure to the design process and enables the development of the intended design of the product. Pugh [8] states that the criteria for a design activity model are

- All must be able to relate to it.
- All must be able to understand it.
- All must be able to practice more effectively and efficiently as a result of using it.
- It must be comprehensive.
- It should preferably have universal application, owing allegiance to neither traditional discipline, industry, nor product.

The activities of the design process largely depend on the magnitude and complexity of the product. As Pugh [8] rightly argues, the design model (or the scaffolding) should be built for each project.

The scaffolding or the design process model is built using the design stage model (described in the following section) as its basis, incorporating various appropriate design methods to complete each stage.

1.6 DESIGN PROCESS AND THE DESIGN MODEL

To design is to create a new product that generates a profit and/or benefits society in some way. As discussed previously, the design process is a sequence of events and a set of guidelines that helps the designer define a clear starting point—from visualizing a product in the designer's imagination to realizing it in real life. It should be done in a systematic manner without hindering the creative process. The *design model* describes the sequence of stages or activities that takes place during the process of converting an abstract set of requirements into the definition of a physically realizable system. The *design stage model* is more stable with a fixed number of stages; the number of activities carried out in each stage varies and depends heavily on the complexity and magnitude of the product.

Considering the two challenges facing the design team (1) in defining the problem and (2) in identifying the solution, the starting point can be only the identified societal need. The design brief outlines the goal or "what the product is," the objectives of the company, and who its customers are. In this process, the first challenge must be met. Once the problem has been well defined, the search for the solution, the second challenge, begins. With these aims in mind, the design stage model is created. At each stage, certain design activities are carried out, and there are several design methods to assist these activities. With these a design model can be devised for any given design project. Figure 1.2 illustrates the design model adopted in this book. It has the following five stages:

1. ***Requirements.*** The *requirements* stage starts with the design brief that comes from the senior management or a client of a design company. It describes what the product is (goal). There are several approaches to establish requirements. One method is described as follows. The design team understands the product as outlined in the design brief and carries out a survey among the customers and records what the customers say. The design team translates the customer's verbatim into customer requirements. The team then identifies measurable parameters or metrics that reflect how the customer requirements are provided in the design. With the metrics and their units, the design team works out the *target specifications*. Target specifications are the ideal specifications, which would satisfy the customers in full since they are drawn by considering the customer requirements only. Engineering and manufacturing considerations are added after an initial round of conceptual design, which often trims the target specifications.

2. ***Product concept:*** The product concept describes the functions the product should perform and the specifications that define the product. It is worth remembering at this stage that a *metric* and a *value* form a single specification. Product specifications are a collection of several individual specifications. Though the specifications are described in a solution-neutral form, they define the product to a great extent, and consequently it is called the *product concept*.

3. ***Solution concept:*** A *solution concept* is a sufficiently developed idea to be converted into structures that can behave as expected and provide the functions that are expected from the product or subsystem. The main task here is the identification of the optimal or near-optimal solution from the design-solution space. It is achieved in several stages, including concept generation, concept evolution, concept prototyping and

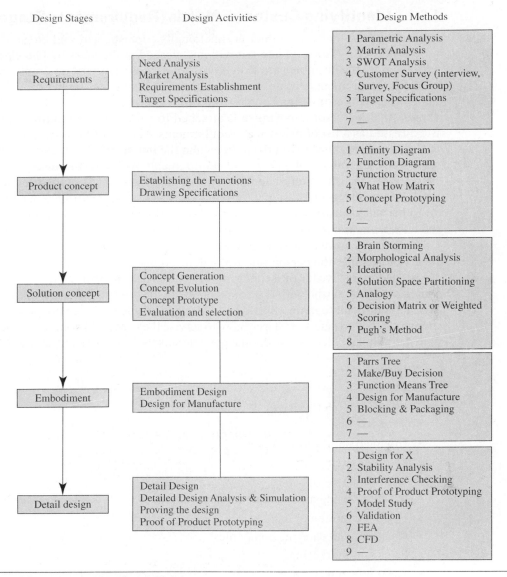

Figure 1.2: Adopted Design Model

proving, and evaluation and selection. Conceptual design is a difficult and important part of the design process, and there are several design methods available to assist.

4. ***Embodiment design:*** Embodiment design defines the hardware or physical form of the product. At this stage the main considerations are manufacturability and design for X. A large number of calculations and cross-checkings are made at this stage. Often the constraints are considered here, and they may condemn a solution as unacceptable even though it is acceptable in all other ways except for a constraint.

5. ***Detail design:*** The detail design comprises two main parts: (1) definition of the geometry, materials, dimensions, and permissible tolerances required for manufacturing and (2) detailed design or the engineering calculations required to ensure correct functionality and safety in operation. The following subsections describe some of the activities and items related to activities in the model proposed.

1.6.1 Identifying Customer Needs (Requirements Stage)

No matter how much a product is functionally sophisticated and elegant in appearance, if it does not meet a societal need, it will not succeed in the market. The elements of the need expressed by the customer are called *customer requirements*, or simply, *requirements*. All of these requirements are not equally important. Some can be fundamentally important or mandatory, while some may come from the list of items "good to have." A weight factor to express the degree of importance is attached to each of these requirements. Thus, the need is expressed as a list of prioritized requirements. The importance ratings are then translated to the specifications derived from them, and the important specifications are used in concept selection. A good set of prioritized requirements will help the design team to distinguish between the principal and secondary function carriers during embodiment design.

The most important step of the design process is identifying the requirements of the customer—the *requirements*. However, before this is done, it is important to establish who the customers are. A vital concept to grasp here is that customers are not the only end users. Customers of a product are all those who will deal with the product at some stage during its lifetime. For example, the person who will sell the product is also a customer. A designer must make the product attractive for the seller to agree to advertise and market it. Another example of a customer is the person who will service and maintain the product during its lifetime in operation. If a product is difficult to maintain and/or service, independent service providers will be keen to recommend other products or charge more to service the item and so on. As a group these people are called the stakeholders. Consider the possible stakeholders of an airplane. The list can include

- Passengers
- Crew
- Pilot
- Airport
- Engineers and service crew
- Fueling companies
- Airlines
- Manufacturing and production departments
- Baggage handlers
- Cleaning and catering companies
- Sales and marketing
- Accounts and finance
- Military/courier/cargo/etc.
- Authorities and official bodies
- Companies involved with the items that will be outsourced

Each of these customers can have entirely different (and sometimes conflicting) needs for the same product. By identifying these customers first, it is possible to identify their requirements as a whole and to arrive at a reasonable compromise according to priority and feasibility. The need for a new design can arise from several sources, including the following.

Client request: In a design company, clients may submit a request for developing an artifact. It is often unlikely that the need will be expressed clearly. The client may know only the type of product that he or she wants; for example, "I need a safe ladder."

Modification of an existing design: Often a client will ask for a modification of an existing artifact to make it simpler and easier to use. In addition, companies may want to provide customers with new, easy-to-use products. For example, a market search will identify many brand names for coffee makers and detail the differences among them, in terms of shape, material used, cost, or special features. Looking at the market will help you to identify gaps and introduce a new product of your own. As another example, Figure 1.3 through 1.5 demonstrate design developments for paper clips. Each of the designs has its own advantages over the other ones. For example, the endless-filament paper clip can be used from either side of the clip. One might argue that the different designs are based on the human evolution of designs and the birth of new ideas. However, one of the major driving forces for the renovation of designs is to keep companies in business. The first patent for a paper clip

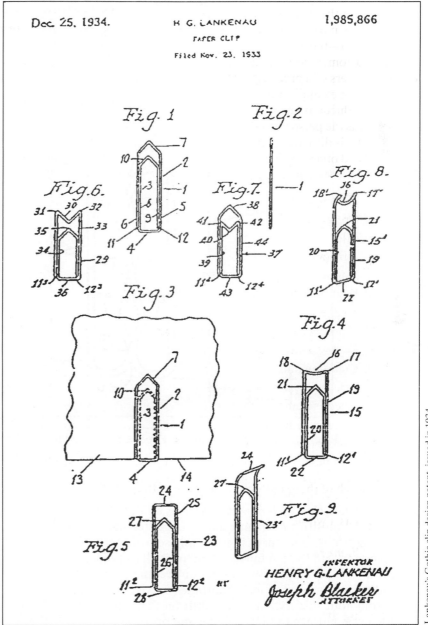

Lankenau's Gothic clip design patent, issued in 1934.

Figure 1.3: Patent of Gothic-Style Paper Clip, Issued in 1934

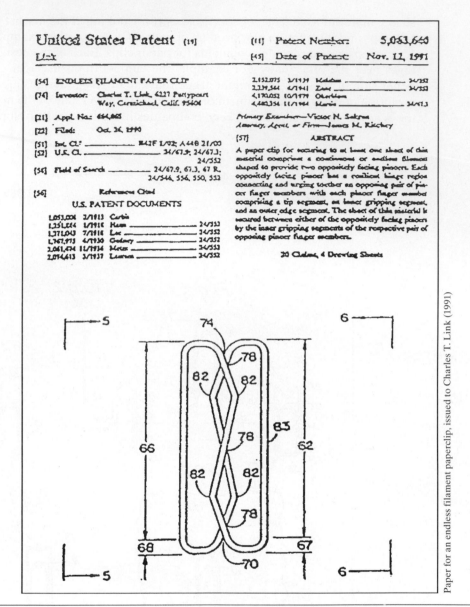

Paper for an endless filament paperclip, issued to Charles T. Link (1991)

Figure 1.4: Patent of an Endless-Filament Paper Clip, Issued to Charles T. Link in 1991

was filed in 1899, and many more have appeared over the years; there have been more than a hundred years of paper clip development and innovation.

Generation of a new product: In all profit-oriented industries, all the attention, talent, and abilities of management, engineering, production, inspection, advertising, marketing, sales, and servicing are focused on getting the product to return profit for the company and in turn for the company's shareholders. Unfortunately, sooner or later every product is preempted by another or becomes unprofitable due to price competition. For an industry to survive, it must continue to grow; it cannot afford to remain static. This growth, throughout history, has been built on new products. New products have a characteristic life cycle pattern in sales volume and profit margins. It will peak out when it has saturated the market and then begin to decline. An industry must seek out and promote a flow of new product ideas. Applying for patents usually protects these new products.

US 20130205565A

(19) **United States**
(12) **Patent Application Publication**   (10) Pub. No.: US 2013/0205565 A1
        Santos   (43) Pub. Date: Aug. 15, 2013

(54) **PAPER CLIP BEARING A PLATE**

(76) Inventor: **Arsenio P. Santos, Aloha, OR (US)**

(21) Appl. No.:   **13/806,882**

(22) PCT Filed: **Jun. 27, 2011**

(86) PCT No.:   **PCT/US11/42059**
      § 371 (c)(1).
      (2). (4) Date: **Dec. 26, 2012**

      Related U.S. Application Data

(60) Provisional application No. 61/359.537.
      filled on Jun. 29, 2010.

Publication Classification

(51) **In. Cl.**
      B42F 1/02   (2006.01)
      F21V 33/00   (2006.01)
(52) **U.S. CL**
      CPC ***B42F 1/02*** (2013.01); ***F21V 33/0048***
                                                    (2013.01)
      USPC **29/428**; 24/67.3; 362/382

(57)   **ABSTRACT**

A paper clip assembly, that includes a wire frame that is
formed into a loop and a jaw that protrudes outwardly from
the loop. Also, a plate insert is shaped to fit into said loop,
and in one embodiment is fit into the loop, to form a paper
clip having a plate that can bear a message.

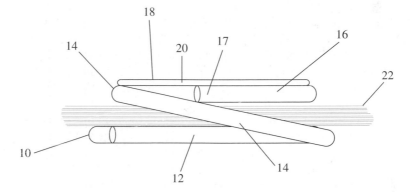

Figure 1.5: Patent Application for a Paper Clip Filed by Santos in 2013

1.6.2 Market Analysis (Requirements Stage)

Market analysis involves gathering and analyzing information that provides a sufficiently
detailed understanding of the market. The information gathered may reveal an available
design solution and the hardware to accomplish the goal. Sometimes, the goal may be altered
to produce a requested product or abandon it if the product already exists. Knowledge of existing products will save the designer and client time and money. Once the designer determines
what is in the market, creativity should be directed toward generating alternatives. Chapter 4
discusses market analysis and the gathering of information in more detail.

Methods to develop customer requirements

Customer requirements are preference details relating to features of a product as expressed by
customers who understand their product well. This is a difficult process for the following reasons:

1. Understanding the product well when the new product or the new version of the cur-
    rent product does not exist is a difficult task for the customers.
2. Getting customer requirements is a resource-consuming process; therefore only a small
    representative sample of customers are consulted during the study. The validity of the

requirements obtained depends primarily on how well the sample represents the population of customers.

3. To maximize the benefits of the information, the customers must be systematically queried about the areas where the information will be useful to the design team as it decides how to move toward the solution.

Methods to develop customer requirements should therefore address these difficulties. This book includes methods to analyze and interpret information, to devise a balanced sample group of customers, and to query the customers in useful areas while recording their requirements and importance ratings.

1.6.3 Defining Goals (Requirements Stage)

In this phase of the design process, the designer defines what must be done to address the need(s). The definition is a general statement about the desired end product. Many of the difficulties encountered in design may be traced to poorly stated goals or goals that were hastily written and resulted in confusion or too much flexibility. The definition of goals spells out what the product will be as opposed to what the product will do. To understand this clearly, consider the design team as a design company. The company takes orders from a client who gives a design brief, which states the goals for the product. The client intends to manufacture the product and sell it to customers. The customers buy the product for what the product will do. These are functions expressed in the verb/noun format. The design team thus has two tasks: (1) to understand the client and establish the goals or objectives and (2) to understand the customers and establish the functional requirements. Figure 1.6 illustrates this schematically.

In almost all cases, the client request comes in a vague verbal statement such as, "I need an aluminum can crusher" or "I need a safe ladder." Designers must recognize that clients' needs are not the same as product specifications. Clients' needs are expressed as "being statements" (e.g., "be inexpensive" or "be in five colors"), and they express what they want the product to be. The designer's function is to clarify the client's design requirements. An *objective tree* may be constructed for clarification. For example, a sample objective tree of a coffee maker is shown in Figure 1.7. The design team assembled the client's objectives in a tree form where related objectives are illustrated as distinct branches. The primary objectives of the coffee maker are (1) safety of the user, (2) quality of coffee, (3) convenience of use, and (4) affordable to the user. They are placed at the first level of the tree. These are further subdivided as sub-objectives, and the tree is formed.

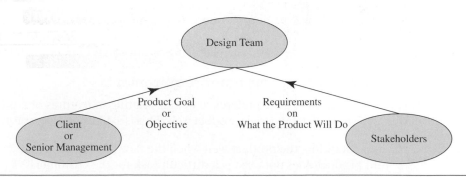

Figure 1.6: Objectives and Functional Requirements

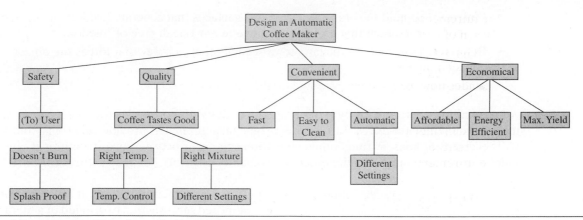

Figure 1.7: Objective Tree for a Coffee Maker

1.6.4   Establishing Functions (Product Concept Stage)

Function is the purpose or intended use of a product and is an abstract formulation of the task, independent of any particular solution. The term *behavior* is defined as "the manner in which something acts under given circumstances." Behavior is viewed as an observable characteristic of the function exhibited or manifested by the designed structure. Function is what the design is used for and behavior is what the design does. It is worthwhile recalling the definition of *engineering design* as "the use of scientific principles, technical information, and imagination in the definition of a mechanical structure, machine, or system to perform pre-specified functions with the maximum economy and efficiency." Consider the bench vice shown in Figure 1.8.

- **Goal**: To hold an object firmly to apply force or moment to shape the object and release it as and when required
- **Action**: To restrict and release the object in all six degrees of freedom as required by tightening and loosening actions by user

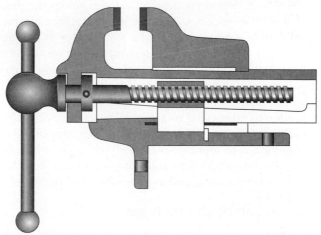

Based on http://practicalengineer.blogspot.ca/2011/02/bench-vice-work-vise.htmlCharles T. Link (1991)

Figure 1.8: A Bench Vice

- **Function**: To hold the object between two surfaces that generate frictional resistance to any force or moment that attempts to move in any one degree of freedom
- **Behavior**: Movement of the jaws toward each other in a way that forces the object between the jaws
- **Structure**: The fixed jaw, sliding jaw, the nut, and the screw

Function is determined by structural connectivity and occurs at the interface of two or more connecting structures. In a product some structures will provide functions, and some will receive functions. A functional connection can occur between two structures if one provides a function required by the other. A product consists of a set of structures (its parts) and a set of functional connections.

A function may be decomposed into sub-functions; when integrated it will provide the intended function. The functions and their decomposed sub-functions arranged in a hierarchical manner provide the function tree of a product. As an example consider a washing machine. The washing machine (1) loosens and dislodges the dirt, (2) rinses away the dislodged dirt and excess soap, and (3) removes the excess water to facilitate drying. The dislodging function is achieved by (a) wetting the clothes, (b) adding and soaking with soap, and (c) agitating the clothes, thereby providing the rubbing function. The functions and sub-functions arranged in the form of a hierarchical tree are shown in Figure 1.9.

A more general way to describe functions is in terms of its inputs and outputs. Function is seen as a black box that has input and output flows. This model can explain complex systems in an easy manner. Figure 1.10 illustrates this concept as a black box showing input and output and as a transparent box in which the sub-functions performed are shown. A product represented as a functional system is characterized by flows of material, energy and information, and the inputs and outputs are these flows as shown in Figure 1.10. A functional system described in this manner is called a *function structure*.

The main purpose of a method to establish functions is to identify the purpose or intended use of a product independent of any particular solution. The technique of establishing functions in solution-neutral form allows for alternatives to be explored that can address the needs and goals rather than unintentionally fixating on a solution that the client or customer provides early on. For example, a client may ask for a traffic light system to be placed at a particular junction, where, in fact, an underpass may be a more viable solution to achieve the real goals of the task—to alleviate traffic congestion. This stage of the design process demonstrates one of the advantages of systematic design in that it guides the designer to a problem-focused design rather than a solution-focused one as in the traditional practice. In another example, a blood bank approached a designer to find a solution for its frequently broken centrifuge. When the centrifuge breaks down, the blood separation unit shuts down. The blood bank has tried replacing the centrifuge, but after a few months it breaks down again. The need statement is to fix the centrifuge in such a way that it will reduce the breakdown time. A designer who recognizes the design process will not jump on fixing the centrifuge; he will ask the following questions: "What is the function we are trying to accomplish with the centrifuge? And is there any better way to accomplish this function?" The answer is to separate blood by enhancing the gravitational pull, and the function is to separate blood cells from the whole blood. This clear statement enables the designer to find alternative solutions other than the centrifuge. Often the functions will be divided into sub-functions, and together they will fulfill the requirements of the artifact.

Methods to establish functions

As shown previously, the functions act on flows of (1) material, (2) energy, and (3) information; in a complex product all three of these flows would be present and a black-box type representation of the flows called the *function structure* would provide a clear understanding.

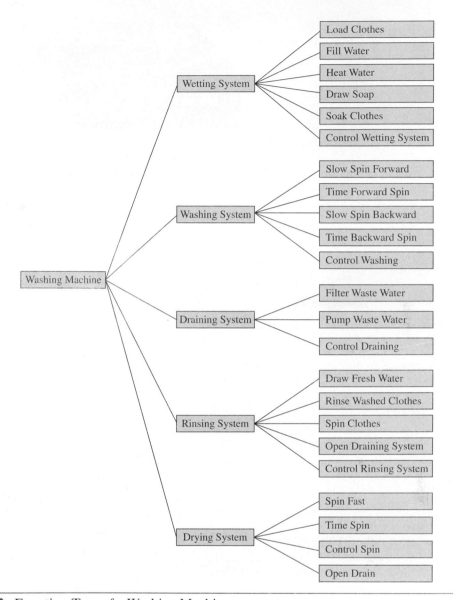

Figure 1.9: Function Tree of a Washing Machine

In a simple passive design a hierarchical representation called the *function tree* would be effective. As a way of differentiating concepts from functions, a *function means tree* can be developed from a *function tree*. This book will describe how to construct (1) a function structure and (2) a function tree.

1.6.5   Target Specifications (Product Concept Stage)

Target specifications are intended to focus the design process by considering the customer requirements. The process requires the designer to list all pertinent data and parameters that will control the design and guide it toward the desired goal; it also sets limits on the acceptable solutions. It should not be defined too narrowly because by so doing the designer

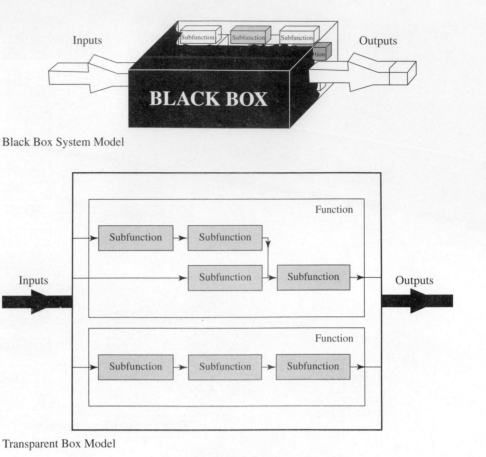

Black Box System Model

Transparent Box Model

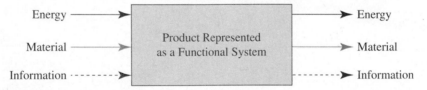

Figure 1.10: Functional System Model

will eliminate some acceptable solutions. However, it cannot be too broad or vague because this will leave the designer with no direction to satisfy the design goal. A typical specification is shown in Table 1.1.

Methods to establish specifications

A specification is an expected measurable performance characteristic (output) of a product with a target value and units. This is a merger and integration of customer requirements and their derived measurable characteristics from the customer side and the function structure from the technical side. Competitive benchmarking results assist it. Value analysis can help to fine-tune the specifications. In Chapter 7 the process steps are described in a systematic way for drawing specifications.

Table 1.1: Specification Table for Automatic Can Crusher

Metric	Value
Dimensions	$20 \times 20 \times 10$ cm
Cans crushed	1/5 original volume
Weight	<10 kg
Sales price	$<\$50$
Number of parts	<100
People able to use	> 5 yrs
Probability of injury	$<0.1\%$
Manufacturing cost	$<\$200$
Steps to operate	1
Maintenance cost	$<\$10$ annually
Efficiency rating	>95 percentile
Internal parts enclosed	100%
Storage of crushed can	60
Loader capacity	>30 cans
Crush cans	≥ 15 cans/min
Crush cans	$\geq 1.2 \times 10^{-2}\,\mathrm{m}^3/\mathrm{min}$
Crush cans	≥ 0.57 kg/min
Noise output	>30 dB
Starts	<10 sec
Runs	>2 hours at a time
Stops	<5 sec
Vibration magnitude	<5 mm
Vibrations	<4 sec
Withstand	... 250 N

1.6.6   Conceptualization (Solution Concept Stage)

If technical elements are arranged in a novel combination forming a system that satisfies a societal need, that combination consists of a conceptual design. The designer's creativity or power of synthesis, a recognized need, and the technical elements available at that point in time are matched to form the conceptual design. The process can be visualized as a technical art of arranging different combinations. The process of generating alternative solutions to the stated goal in the form of concepts requires creative ability. The conceptualization starts with generating new ideas. In this stage, the designer must review the market analysis and the task specifications as he or she engages in the process of creativity and innovation. This activity usually requires free-hand sketches for producing a series of alternative solutions.

As an example consider a shower curtain rail. The need is for a rail to carry a shower curtain and to stand firmly between two walls without being nailed to them. The components are (1) a telescopic aluminum pipe consisting of two pieces, one sliding into the other; (2) a coil spring that just fits into the inner pipe and is locked in position by a crimp; (3) a spring washer, a little bigger than the inner side of the outer pipe, connected to the free end of the helical spring; and (4) two shoes, one for the inner tube and one for the outer tube (see Figure 1.11).

The design consists of a clever use of elastic force and friction. The spring washer is a little larger than the outer pipe; when pulled in, it bends. This bending creates a rebounding force. The generated rebounding force creates a resistive force due to friction when pushed. This

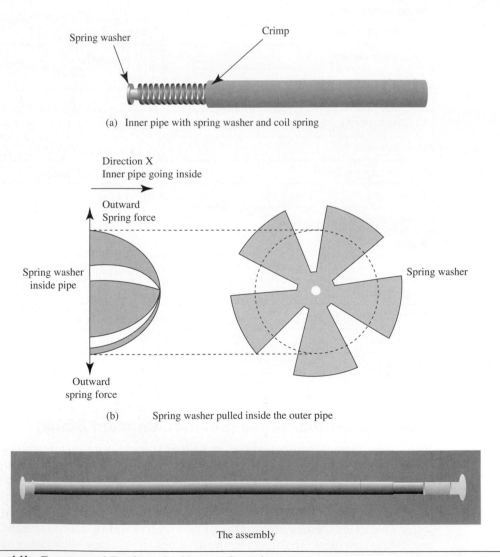

(a) Inner pipe with spring washer and coil spring

(b) Spring washer pulled inside the outer pipe

The assembly

Figure 1.11: Conceptual Design of a Shower Curtain

makes the coil spring compress and forces the pipes against the walls. The friction between the walls and the shoes creates the necessary force to carry the weight of the curtain and the rails.

Tubes forming a telescope, shoes at the free ends, a helical spring, and a radial spring are components known to technical personnel. However, their novel combination creates a solution to a design problem that arose from a societal need. A rubber block can easily replace the radial spring, giving rise to another design solution in the solution space. Chapter 8 covers this stage of the design process in more detail.

Methods to establish conceptual design

Conceptual design thinking can be explained in the following way. Prater [9] describes a machine as "any device that helps to do work." It may help by changing the amount of force or the speed of action to transform energy like a generator, to transfer energy from one place to another like a drive shaft, to multiply force like a press, to multiply speed or to change the direction of a force. There are only six simple machines: (1) the lever, (2) the block, (3) the wheel and axle, (4) the inclined plane, (5) the screw, and (6) the gear. They are explained in Table 1.2.

Table 1.2: Basic Machines as Function Providers

Basic Machine	Description	Function
Lever	A lever is constituted with the fulcrum (F), a force or effort (E), and a resistance (R). Depending on their relative positions they are classified as Class 1 RFE (as shown), class 2 FRE, and class 3 FER.	Levers of the first and second class help in overcoming big resistances with a relatively small effort by reducing the speed of the resistance. Third-class levers speed up the movement of the resistance with the use of a large effort.
The block or Pulley	A pulley is a wheel with a groove along its edge, where a rope or cable can be placed.	A single pulley simply changes the direction of the effort. A combination of pulleys can reduce the effort needed to lift a resistance.

Continued on next page.

Basic Machine	Description	Function
Wheel and Axle	The wheel-and-axle machine consists of a wheel or crank rigidly attached to the axle, which turns with the wheel.	The force on the rim of the wheel causes the axle to rotate. Wheel and axle magnify the effort or speed it up. Variations are possible with separation of axle and the wheel.
Inclined Plane Paul Matthew Photography/ Shutterstock.com	A plank whose one end is resting at a lower level and the other at a higher level	To raise or lower heavy objects by applying a small force over a long distance
Screw	An inclined plane wrapped around a cylinder	To reduce large amounts of circular motion to very small amounts of straight-line motion
Gears	Meshing wheels	Gears can change the direction of motion, increase or decrease the speed of the applied motion, and magnify or reduce the applied force. Gears also produce a positive drive.

Complex machines are merely combinations of two or more simple machines. Consider a wheelbarrow as shown in Figure 1.12. It is a class-2 lever with a variation of the wheel and axle attached at the fulcrum. Functionally the lever helps to lift a larger load with a smaller effort, and the variation of the wheel with the help of friction provides the motion.

The design thinking of the conceptual design of the wheelbarrow can be considered as a scheme to combine the functions of the two basic machines.

Now consider the barrow with a hand mixer. It is a wheelbarrow with a provision to rotate the drum using the wheel and axle where the drum is the axle and the hand crank is the device causing the rim of the wheel to rotate. The design thinking of the conceptual design of the

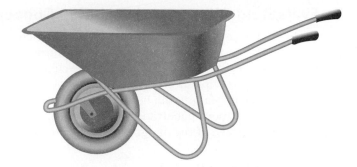

Figure 1.12: Wheelbarrow as the Combination of Two Basic Machines

barrow and hand mixer can be considered as a scheme to combine the functions of the three basic machines.

Now consider a concrete mixer whose drum is rotated by a gear drive. The conceptual design thinking can be considered as a scheme combining the functions of four basic machines—the lever, variation of wheel, wheel and axle, and the gears.

A power-driven gearbox to drive the concrete mixer shown in Figure 1.13 thus will have five basic units, and the scheme for it can be written as follows:

1. The motor will drive the input shaft of the gearbox.
2. The output shaft will drive the axle of the drum.
3. The axle will drive the drum (wheel).
4. The lever will carry the drum.
5. The variation of the axle and wheel will provide the motion.

Conceptual designs are composed by the "harmonious integration of function providers delivering the functions set out in the functional description outlined by the function tree or function structure." Several design methods are available to assist this important process. Chapter 8 describes them in detail.

Kotruro2 / Shutterstock.com

Figure 1.13: A Full Concrete Mixer

1.6.7 Evaluating Alternatives (Solution Concept Stage)

Once a number of concepts have been generated in sufficient detail, a decision must be made about which one or ones will enter the next, most expensive, stages of the design process. The concept that best meets the objectives set out by the client should be chosen. There are several methods and guiding principles for evaluating and choosing a concept. An excellent technique to guide the designer in making the best decision regarding these alternatives is a *decision matrix*, which forces a more penetrating study of each alternative against specified criteria. Chapter 9 covers this stage of the design process in more detail.

1.6.8 Embodiment Design

Once the concept has been finalized, the next stage is known as the *embodiment design*, and this is where the product that is being designed takes shape. Embodiment design is a definite layout that lends to a clear check of function, durability, manufacture, assembling, operation, and cost analysis. It involves a large number of corrective steps in which analyses and syntheses are switched from time to time. It also requires considerable effort to collect a large amount of information about materials, production processes, standard parts, and standards. It works with the broad definition of *design for manufacturing*, which treats it as a philosophy and mindset to use manufacturing input at the earliest stages of design. Embodiment design involves much mathematical and engineering knowledge acquired during high school and university studies. Various authors have given guidelines for effective embodiment design.

This stage does not include any details yet (no firm dimensions or tolerances, etc.), but it will begin to illustrate a clear definition of a part, how it will look, and how it interfaces with the rest of the parts in the product assembly. This stage is separated from both the conceptual design and the detail design in that new technologies can replace old ones to realize the exact same concept. For example, the concept of a traffic light system may remain the same (three lights: red, amber, and green), perform the same functions and specifications, and work conceptually the same way, but as technologies advance, the lights themselves can change from bulb to LED, or the way the lights change can transition from using a timer to cycle through the lights to using a system that is connected to a modern traffic network. Possibly, the future may hold a system whereby the traffic light is able to sense the most efficient light for the junction to alleviate congestion and change the lights accordingly. The concept still remains the same, but the execution and parts or the "embodiment" of the design can change. Chapter 11 discusses embodiment design in more detail.

During *embodiment design* designers determine the overall layout design with all spatial considerations, the shapes of components, their raw material sizes and forms, and the ease of manufacturing them in a progressive fashion. Design for manufacturability and cost are critical at this stage. Packaging all the components within the often-limited product envelope is a critical task of embodiment design. At this stage all the make/buy decisions have to be made, and the definition of specifications for the parts to be bought have to be drawn. It is worthwhile noting that different combinations of components and technologies can realize the same functions. For example, the function required of a portable concrete mixer is "rotate drum that contains the cement, aggregate, and water," and various combinations of technical elements have achieved this as shown in Figure 1.14. Defining the combination and their physical position or packaging is the embodiment design.

Methods to establish embodiment design

Embodiment design is treated as the arrangement of function carriers (the hardware) and is achieved by a stepwise provision of function carriers. The principal function carriers are provided first. The secondary function carriers are provided progressively and are integrated

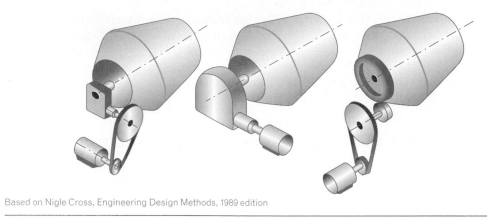

Based on Nigle Cross, Engineering Design Methods, 1989 edition

Figure 1.14: Embodiment of a Concrete Mixture Drive

through interfaces with the already developed embodiment until all the functional units identified by the conceptual design have been provided and integrated.

As an example consider a stepladder, shown in Figure 1.15. The principal function carrier is
the front rail set that provides the height and supports the user at the desired height. The second
most important function provider is the rear rail set, which provides the support and stability to
the front rail set. The next provision is the interface (hinge) between the front rail set and the

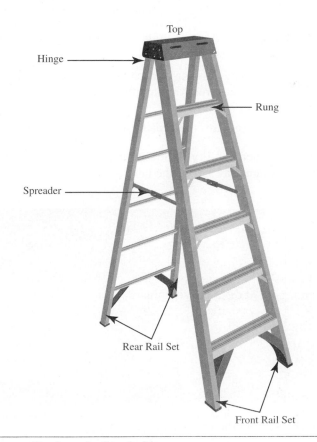

Figure 1.15: Embodiment Design of a Stepladder

rear set. The next item to add after that is the spreader, which permits locking the positions of the front and rear rail sets. The top platform and the rungs can follow this. A lot of calculations about the number of steps, the rough dimensions of features, the load to be carried, and so forth as well as packaging the constituents and interfaces makes it a challenging exercise. Thus the method can be summarized as stepwise addition of providers for functional subsystems and the interfaces. Chapter 11 provides more details on embodiment design.

1.6.9 Detail Design, Analysis, and Optimization

The last step in synthesis is to make the design complete with every detail including the geometric features needed. This is the design that will go into manufacture if analysis proves that the design is safe and operates as intended. In a contemporary design office this will be carried out using CAD software such as CATIA or Solidworks. This amounts to completing the design in a *design and debug* process. Once a possible way to attain the stated goal has been chosen, the synthesis phase of the design has been completed, and the analysis phase begins. This is also known as *detailed design* and is what most of the engineering courses in an undergraduate degree program cover. In essence, the solution must be tested against the laws of physics.

This stage develops the preparation of the technical design, sufficient to coordinate components and elements of the project and information for statutory standards and construction safety. Every part that goes into the product is defined completely at this stage, and production drawings are produced. As an example consider the assembly shown in Figure 1.16. The detail design given by the set of production drawings consists of the following:

1. A labeled, exploded assembly with a list of quantities forms the top drawing, as shown in Figure 1.17.
2. A dimensioned drawing of each labeled part, including all necessary dimensions and information required for manufacturing, is given in separate individual sheets of paper, as shown in Figure 1.18.
3. These drawings are made using a computer-aided design (CAD) system, such as Pro/E.

The manufacturability of the chosen product also must be checked to ensure its usefulness. A product design may satisfy the laws of physics, but if the product cannot be manufactured,

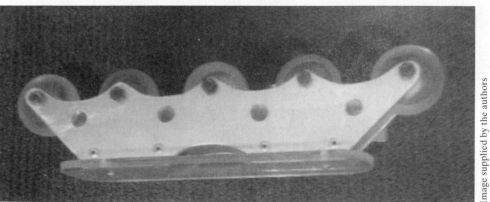

Image supplied by the authors

Figure 1.16: Sample Picture of a Skate Product

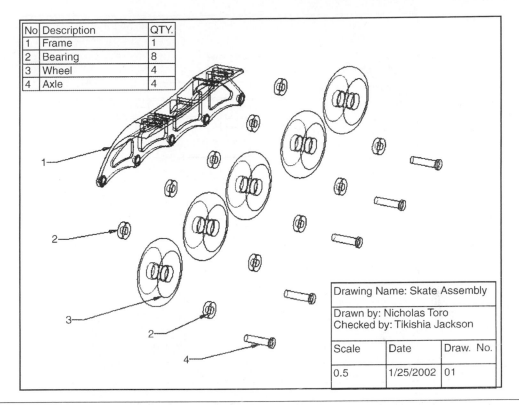

No	Description	QTY.
1	Frame	1
2	Bearing	8
3	Wheel	4
4	Axle	4

Drawing Name: Skate Assembly

Drawn by: Nicholas Toro
Checked by: Tikishia Jackson

Scale	Date	Draw. No.
0.5	1/25/2002	01

Figure 1.17: Exploded Assembly Drawing and Bill of Quantity

the product design is useless. This stage is put into iterative sequencing with the original synthesis phase. Often, analysis requires a concept to be altered or redefined and then reanalyzed so that the design is constantly shifted between analysis and synthesis. Analysis starts with estimation and is followed by order-of-magnitude calculation.

Estimation is an educated guess based on experience. Order-of-magnitude analysis is a rough calculation of the specified problem. The order of magnitude analysis does not provide an exact solution, but it gives the order in which the solution should be expected. Chapter 12 covers some of the needed aspects. However, given that detailed design occupies the greater part of an undergraduate engineering program and the design differs from product to product, details of this stage are beyond the scope of this book.

Engineering systems are made up of many interconnected machine elements such as shafts, pulleys, chains, belts, gears, and bearings. The design of each machine element involves special engineering theories and calculations, considering several operational parameters in both scientific and empirical ways. Companies that specialize in the design and the manufacture of these elements, bearings for example, develop these empirical theories. The system designer has to perform calculations and choose the appropriate one for his application. Then there are situations where the stress levels of structural members or the dynamic behavior of some elements have to be calculated to ensure that the stress levels at the worst-case scenario are well within the limit. The bulk of the engineering curriculum teaches the principles involved and how to perform these calculations. These calculations are fundamental to the safe operation of the system but may not be needed in the

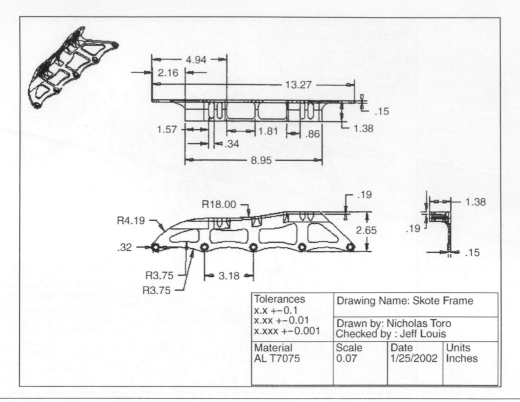

Figure 1.18: Production Drawing for the Skate Frame

detail design of the system. These calculations (sometimes called applied mathematics) are needed for at least some of the machine elements, and they form part of the detailed design. It is the responsibility of the designer to identify the critical elements that must undergo the detailed design and to test them. In short, detail design defines the geometry of the design completely, and *detailed design* carries out the necessary calculations and analyses to confirm the design on both the subsystem level (as a whole) and the overall system.

Methods to perform analysis

In general, all possible analyses are not necessary to carry out in design. The design engineer decides which analyses are needed to ensure safety and performance, and only those analyses are carried out. This is called the context C in the model $D = f(F, B, S, K, C)$ described previously. As Mott [10] rightly asserts, the method to perform analysis depends on both what has already been defined and what can be determined presently.

1. The *detail design* phase would have completed the packaging of all machine elements needed for the system if the geometry was known. Then the analysis should determine the loadings under the worst-case conditions. The designer should determine the kinds of analyses that are needed to ensure performance and safety. The analyst may suggest a suitable material.

2. If the detail design phase has established the dimensions and the manufacturing process selection has defined the material, then the analyses can provide the performance characteristics such as stress and natural frequency to check and ensure that the design is viable.

3. If only a rough idea about the geometry and material has been determined, there is enough room for the analysis to optimize the design.

1.6.10 Experiment

The experiment stage in engineering design requires that a piece of hardware be constructed and tested. It verifies the concept or embodiment (and analysis) of the design as to its workability, durability, and performance characteristics. Here the design on paper is transformed into a physical reality. Three techniques of construction are available to the designer:

1. *Mock-up:* The mock-up is generally constructed to scale from plastics, wood, or cardboard, and the like. The mock-up is often used to check clearance, assembly technique, manufacturing considerations, and appearance. It is the least expensive technique, provides the least amount of information, and is quick and relatively easy to build.

2. *Model:* This is a representation of the physical system through a mathematical similitude. Four types of models are used to predict the behavior of the real system:

   • A true model is an exact geometric reproduction of the real system, built to scale, and satisfying all restrictions imposed in the design parameters.

   • An adequate model is thus constructed to test specific characteristics of the design.

   • A distorted model purposely violates one or more design conditions. This violation is often required when it is difficult to satisfy the specified conditions.

   • Dissimilar models bear no apparent resemblance to the real system, but through appropriate analogies, they give accurate information on behavioral characteristics.

3. *Prototype:* This is the most expensive experimental technique and the one producing the greatest amount of useful information. The prototype embodies the constructed, full-scale working physical system. Here the designer sees his or her idea come to life and learns about such things as appropriate construction techniques, assembly procedures, workability, durability, and performance under actual environmental conditions.

As a general rule, when entering the experimental stage of the design process, one should first deal with the mock-up, then the model, and finally the prototype (after the mock-up and model have proven the real worth of the design), to allow beneficial interaction with concept and analysis. Chapter 10 covers this topic in more detail.

Methods for experimentation

Experimentation has a general objective of establishing a firm proof (validation) to predictions made using approximated mathematical and computational methods. Among these, dimensional-analysis–based *model study* and computer-simulation–based *validation* are two important examples.

1.6.11 Marketing

This stage requires specific information that defines the device, system, or process. Here the designer is required to put his or her thoughts regarding the design on paper for the purpose of communicating with others. Communication is involved in selling the idea to management or the client, directing the shop on how to construct the design, and serving management in the initial stages of commercialization. The description should take the form of one of the following:

- A *report* containing a detailed description of the device—how it satisfies the need and how it works: a detailed assembly drawing; specifications for construction; a list of standard parts; a cost breakdown; and any other information that will ensure that the design will be understood and constructed exactly as the designer intended.
- A *flyer* containing a list of the special features that the design can provide, advertisements, promotional literature, market testing, and so forth.

Although this section is predominantly beyond the scope of this book, it is possible to refer to Chapters 5 and 6, which cover some of the essential skills needed to succeed here.

1.7 MANAGEMENT OF THE DESIGN PROCESS

Following a systematic design process enables designers to convert vague statements to valuable products. However, several factors, beyond the design stages, may influence the successful completion of the design project. These include coordination of the design process among designers as well as people and activities that influence the design process. Chapter 2 presents activities that help coordinate design teams, in particular in academic settings, along with activities to increase effective communication among the design team and management. Chapter 2 also presents tools to help manage the design activities. The successful management of the design process entails management of the design team's activities, output, and stimuli.

1.8 OTHER DESIGN MODELS

Design methodology describes the sequence of activities followed by designers in developing a new product. These are called design models. The sequence, as defined by a researcher on the subject, is called a *prescriptive model*, whereas the description of the sequence of activities followed by experienced, successful designers is called a *descriptive model*. Thus, the sequence of activities or stages followed by the designer in Section 1.5 is a descriptive model. Two prescriptive models are described in the succeeding subsections.

1.8.1 Product Development by Ulrich and Eppinger [11]

This generic product development process consists of six phases. The process begins with a planning phase, which is the link to advanced research and technology development activities. The output of the planning phase is the project's mission statement, which is the input required to begin the concept development phase. It serves as a guide to the development team. The conclusion of the product-development process is the product launch, at which time the product becomes available for purchase in the marketplace. The six phases of the generic development process are

1. **Planning:** The planning activity is often referred to as "phase zero" because it precedes the project approval and the launch of the actual product development process.

This phase begins with corporate strategy and includes assessment of technology developments and market objectives. The output of the planning phase is the project mission statement, which specifies the target market for the product, business goals, key assumptions, and constraints.

2. **Concept development:** In the concept-development phase, the needs of the target market are identified, alternative product concepts are generated and evaluated, and one or more concepts are selected for further development and testing.

3. **System-level design:** The system-level design phase includes the definition of the product architecture and the decomposition of the product into subsystems and components. The final assembly scheme for the production system is usually defined during this phase as well.

4. **Detailed design:** The detailed design phase includes the complete specification of the geometry, materials, and tolerance of all the unique parts in the product and the identification of all the standard parts to be purchased from suppliers. A process plan is established, and tooling is designed for each part to be fabricated within the production system.

5. **Testing and refinement:** The testing and refinement phase involves the construction and evaluation of multiple preproduction versions of the product.

6. **Production ramp-up:** In the production ramp-up phase, the product is made using the intended production system. The purpose of the ramp-up is to train the work force and to work out any remaining problems in the production processes.

1.8.2 Design Model by Ullman [12]

The aim of product development as described by Ullman is to transform the developed concept into products that perform the desired functions. These concepts may be at different levels of refinement and completeness. Refining from a concept to a manufactured product requires work on several elements, and central to this is the function of the product. Surrounding the function, and mutually dependent on each other, are the form of the product, the material used to make the product, and the production techniques used to generate the forms from the materials.

The form of the product is roughly defined by the spatial constraints that provide the envelope in which the product operates. Within this envelope the product is defined as a configuration of connected components. In other words, form development consists of the evolution of components, how they are configured relative to each other and how they are connected to each other. Whatever geometry was developed during the conceptual design process must be questioned and refined. This is usually done on the layout drawing (the drawing in which the product is refined). In this drawing, the configuration of the components and the connections between the components are developed. Therefore, in form generation it is essential to understand several points, which are listed as follows:

- **Understand the spatial constraints:** They are the walls or envelopes for the product. Products must begin with knowledge about the interfaces with other objects. Most products must work in relation to other existing, unchangeable objects. The relationship may define the actual contact between objects in the artifact or define the needed clearances between objects in the artifact. The relationships may be based on the flow of material, energy, or information as well as being physical.

- **Configure components:** Configuration refers to the architecture, structure, or the arrangement of the components and assemblies of components in the product. Developing the architecture or configuration of a product involves decisions that divide the product into individual components and develop the location and orientation of each of them.

- **Develop connections—create and refine interfaces for functions:** This is a key step when embodying a concept because the connections or interfaces between components support their function and determine their relative positions and locations.
- **Develop components:** It has been estimated that less than 20% of the dimensions of most components in a device are critical to performance. This is because most of the material in a component is there to connect the functional interfaces and therefore is not dimensionally critical. Once the functional interfaces between the components have been determined, designing the body of the component is often a sophisticated "connect-the-dots" problem.
- **Refining and patching:** They are major parts of product evolution. Refining is the activity of making an object less abstract. Patching is the activity of changing a design without changing its level of abstraction.

 ## 1.9 STRUCTURE OF THE BOOK AND HOW TO USE THIS BOOK

This book contains fifteen chapters. After this introductory chapter it is divided into seven parts. Other than the general part 2, each part covers a stage in the design process model adopted for the book. Each part has chapters describing specific components of the part. Table 1.3 illustrates the structure of the book.

Chapter 1 gives a complete overview of the design process with appropriate examples. An example (Lab 1), the "Design Model in Action: a tale of developing a sandwich" will be

Table 1.3: Structure of the Book

Part	Description of Contents
	Chapter 1—Introduction
Part 1—General	Chapter 2—Essential Transferable Skills Chapter 3—Ethics and Moral Frameworks
Part 2—Requirements	Chapter 4—Identifying Needs and Gathering Information Chapter 5—Customer Requirements
Part 3—Product Concept	Chapter 6—Establishing Function Structure Chapter 7—Specifications
Part 4—Solution Concept	Chapter 8—Conceptual Design Chapter 9—Concept Evaluation
Part 5—Embodiment Design	Chapter 10—Concept Prototypes Chapter 11—Embodiment Design
Part 6—Detail Design	Chapter 12—Detail Design Chapter 13—Detailed Design or Engineering Analysis
Part 7—Closure	Chapter 14—Case Studies and Closure Chapter 15—Selection of Projects

very helpful in understanding the process. Any beginner in systematic design should read this chapter over and over until he or she is comfortable with the contents and the process being described. After this the reader can start the project. Each part describes a stage in the design process. It will be beneficial to read each part completely and thoroughly before embarking on that stage. This will help to decide on the tools, methods, and approaches to be used at the specific stage of the design project. Labs to support or complement the information presented in each of the design stages are introduced. The problems are designed to help the assimilation of the content and enable use of the knowledge to apply, analyze, evaluate, and create design solutions.

1.10 CHAPTER SUMMARY

1. Various possible solutions to a design problem form the solution space.

2. Acceptable solutions meet or partially meet most of the requirements while optimal and near optimal solutions meet almost all requirements.

3. Definitions for design has been given by several authors. Some are given below:

   - Engineering design is the use of scientific principles, technical information, and imagination in the definition of a mechanical structure, machine, or system to perform pre-specified functions with the maximum economy and efficiency.

   - Design theory refers to systematic statements of principles and relationships, which explain the design process and provide a useful procedural way for design. In the same way the methodology refers to the collection of procedures, tools, and techniques, which the designer may use in applying the design theory to the process of design.

   - Engineering design is the process of devising a system, component, or process to meet desired needs. It is a decision-making process (often iterative) in which the basic sciences, mathematics, and engineering sciences are applied to optimally convert resources to meet a stated objective.

   - An analytical type model for a design can be given in the following form:

     $$D = f(F, B, S, K, C)$$

     Where

     D = Design of the product
     F = Functions intended to be performed by the product
     B = Behavior of the structural elements, which provide the intended functions
     S = Structural attributes
     K = Knowledge used in the design
     C = Context of the product

4. The challenge of design lies in (1) defining the problem and (2) identifying the preferred unknown solution from the unknown solution space.

5. Based on the degree of difficulty, design task is categorized as adaptive design, development design, and new design.

6. Modular design is a technique whereby units or modules that perform distinct functions and easy to assemble are designed and developed so that different combinations of them could be assembled to develop distinctly different products.

7. A product platform embodies the basic design and components that are used in several products of a product family.

8. In conventional design "the problem is assimilated well and work on a solution is started in earnest. No other help is made available other than those provided by the textbooks and handbooks. The solution to the problem lay within himself (the designer). Immersion and long hours of thinking are the working methods. Different solution concepts are considered one after the other. Better insights are achieved through the unsuccessful concepts. The process continues until a working solution to the problem is achieved."

9. The systematic approach to design process divides the process into constituent components and completes the components to achieve the results.

10. The design stage model uses the stage at which the design is, at different points during the product development process, to describe the design process.

11. The design activity model uses the sequence of activities that are performed during the design process to describe the process.

12. The design model adopted in this book has requirement, product concept, solution concept, embodiment and detail design as its stages.

13. The design brief describes what the product is (the goal of the product), and customer requirements describe what the product will do (functions). The design brief and prioritized customer requirements cover the requirements stage.

14. The product concept describes the functions the product would perform and the specifications that define the product.

15. Solution concepts are ideas that are sufficiently developed to convert into structures that can behave as expected and provide the functions that are expected from the product.

16. Embodiment design defines the hardware or physical form of the product.

17. The detail design has two main parts: (1) the geometry, materials, dimensions, and permissible tolerances required for manufacturing and (2) detailed design or the required engineering calculations to ensure correct functionality and safety in operation.

18. Developing an objective tree is a good method to define the goals, and a customer survey is a good method to establish customer requirements.

19. Establishing a function tree and a function structure of flows are useful in establishing the functions provided by the product.

20. Conceptual designs are composed by the "harmonious integration of function providers delivering the functions set out in the functional description outlined by the function tree or function structure."

21. Embodiment design is treated as the arrangement of function carriers and is achieved by the stepwise provision of function carriers. The principal function carriers are provided first. The secondary function carriers are provided progressively and integrated through interfaces with the already developed embodiment until all the functional units identified by the conceptual design are provided and integrated.

22. Detail design describes the entire design in a systematic manner, and detailed design carries out all the necessary analyses and proves that the product in the design is safe for all practical uses.

1.11 PROBLEMS

State whether the following statements are true or false.

Statements	T	F
1. A solution space is the collection of all possible solutions to a design problem.	❏	❏
2. All design problems are well defined.	❏	❏
3. A design problem in general is the characterization of a societal need.	❏	❏
4. Engineering design refers to a product that has been designed or the process that has been adopted to design the product.	❏	❏
5. A design problem characterizing a societal need will have a unique solution.	❏	❏
6. In general the solution space to a design problem is known well.	❏	❏
7. Engineering design consists of the use of scientific principles, technical information, and imagination to define a mechanical structure to perform a prescribed function.	❏	❏
8. Two major challenges in design are (1) defining a problem well and (2) generating the solution space containing the optimal and near-optimal solutions.	❏	❏
9. Development design involves conceptual or embodiment variations of subsystems of an existing design.	❏	❏
10. Adaptive design involves making major modifications to the existing design.	❏	❏
11. The design process is the useful procedural way to design a product.	❏	❏
12. In the analytical model of the design $D = f(F, B, S, K, C)$ F stands for feature.	❏	❏
13. Immersion, working, and evaluating trial solutions and continuous improvements are features of conventional design.	❏	❏
14. Systematic design breaks the design process into constituent components and completes the components to achieve the results.	❏	❏
15. Design using modules that perform distinct functions and are easy to assemble are called modular designs.	❏	❏
16. A platform product is the basic product that can be built as a member of a product family.	❏	❏
17. Modules in modular design, when assembled together in different combinations, form distinctly different products.	❏	❏
18. Immersion and long thinking hours are the design methods of conventional design.	❏	❏
19. Design methods are tools and techniques used during the design process.	❏	❏
20. Morphological analysis is an example of a design method.	❏	❏
21. Like the scaffolding in a building project, a design model provides a skeletal structure to the design process.	❏	❏
22. The design activity model is the description of the sequence of activities carried out during the design process.	❏	❏
23. The design stage model is the description of the sequence of stages the design process passes through.	❏	❏
24. Brainstorming is an example of a design model.	❏	❏

Continued on next page.

Statements	T	F
25. The complexity of a design model will increase as the number of design methods used in each stage increases.	❏	❏
26. For a product to be successful, it need not to meet a societal need	❏	❏
27. The product concept defines the function of a product.	❏	❏
28. Details of the required design activity model will decrease as the complexity of the design project increase.	❏	❏
29. Requirements are elements of a societal need	❏	❏

State whether the following statements are true or false.

Statements	T	F
1. A specification consists of a metric and a value.	❏	❏
2. Requirements define what the product should do.	❏	❏
3. Function is the purpose or intended use of a product.	❏	❏
4. A specification is an expected measurable performance characteristic (output) of a product with a target value and units.	❏	❏
5. The product concept defines the functions of the product.	❏	❏
6. The design brief outlines what the product is.	❏	❏
7. Function occurs at the interface between connecting structures.	❏	❏
8. Behavior is viewed as an observable characteristic of the function exhibited or manifested by the design structure.	❏	❏
9. Function is an abstract formulation of the task independent of any particular solution.	❏	❏
10. In a product some structures will provide and some will receive functions.	❏	❏
11. Behavior is defined as "the manner in which something acts under given circumstances."	❏	❏
12. Function is what the design is used for, and behavior is what the design does.	❏	❏
13. A product consists of structures and a set of functional connections.	❏	❏
14. A function structure shows the flow of material, energy and signal.	❏	❏
15. A functional connection can occur between two structures if one provides a function required by the other.	❏	❏
16. If technical elements are arranged in a novel combination forming a system that satisfies a societal need, that combination is a conceptual design.	❏	❏
17. A function tree is the hierarchical structure showing the functions and their sub-functions.	❏	❏

1. Explain the analytical model $D = f(F, B, S, K, C)$
2. Explain modular design with an example.
3. What is platform design?
4. Give three definitions of "design."
5. What are the challenges in design?

6. Differentiate among adaptive, developmental, and new designs.

7. Explain *design stage model* and *design activity model*.

8. What is the difference between *customer statement* and *problem definition*?

9. Describe conventional design in your own words.

10. Explain systematic design.

11. Why does a good carpenter always cut the wood twice?

12. What is a design method?

13. List three factors that market analysis achieves.

14. Explain the five stages in the design model adopted in this book.

15. Why does function analysis precede the conceptualization?

16. What is *function analysis* and how is it different from the definition of the problem?

17. What is the difference between the specification step and defining the problem step in the design process?

18. Explain the statement "A design model accommodates the two challenges faced by the designer."

GROUP ACTIVITIES

1. The goal or objective for a coffee maker can be stated as "a safe, economical, and easy-to-use coffee maker to produce high quality coffee." Write similar goal or objective statements for (a) a tie clip, (b) a lady's handbag, (c) a cover for an expensive mobile phone, (d) a memorabilia key chain, and (e) a dry travel iron.

2. Draw the objective tree for a high-class cutlery set.

3. Figure 1.19 shows the percent of cost committed and incurred as a function of time during the design of a product. The committed cost is the amount of money allocated

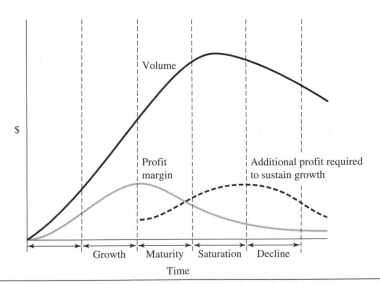

Figure 1.19: Manufacturing cost

for the manufacturing of the product, while the incurred cost is the amount of money spent on the design.

   a. Explain your findings from the figure.

   b. Discuss why it is more important to spend time and money on the early stages of design than on the late stages of design.

4. Identify several commonalities and differences among Figures 1.3 through 1.5.

5. List four factors that may be used to determine quality, and discuss the following statement in light of your listing: "Quality cannot be built into a product unless it is designed into it."

6. Consider the following two statements:

   • What size SAE grade-5 bolt should be used to fasten together two pieces of 1045 sheet steel, each 4 mm thick and 6 cm wide, which are lapped over each other and loaded with 100 N?

   • Design a joint to fasten together two pieces of 1045 sheet steel, each 4 mm thick and 6 cm wide, which are lapped over each other and loaded with 100 N.

   a. Explain the difference between the two statements.

   b. Convert a problem from one of your engineering science or physics classes into a design problem.

7. Explain the engineering design process to a high school student.

8. As we discussed in this chapter, there are several trends and philosophies on design. Perform a technical search and list four different trends. Discuss the differences and similarities among these trends.

LAB 1: Design Model in Action: A Tale of Developing a Sandwich

This example is adapted from an assignment for students in a Product Design and Development course given at the United Arab Emirates University. The students are given the following task:

A sandwich is an item of food consisting of two pieces of bread with a filling between them. A franchise provider wants to develop cheese-salad sandwiches that are sold in his outlets. These sandwiches are made of normal loaves of bread and are sealed as packs of preferably triangular cross-sections. He would like to follow the "Code of Practice and Minimum Standards for Sandwich Bars and Those Making Sandwiches on the Premises" of the BSA (British Sandwich Association). The students working on a team were asked to decide the elements of the design brief needed and to write the design brief and then design and develop the sandwich.

L1.1   Design Brief

The design brief is the written document that the senior management of a given company or client gives to the design team. The items that are included in a design brief depend on the product and its magnitude. It should provide sufficient details for the design team to start the design process. After a team consultation, the elements of the design brief were identified as (1) product description (or *what*), (2) product concept (or *how*), (3) benefits to be delivered (or

Table L1.1: Design Brief of a Cheese-Salad Sandwich

Design Brief Of "*Born To Be Eaten*" Cheese Salad Sandwich	
Drafted by	**Simple Simon Company, Ltd.**
Product Description	A delicious and nutritious cheese-salad sandwich made with commercial sliced bread and a variety of fresh local vegetables and cheese enhanced with different salad dressings. Made in premises that adopt the "Code of Practice and Minimum Standards for Sandwich Bars and Those Making Sandwiches on the Premises" of the British Sandwich Association. Presented in a box with a label showing the ingredients and the date of manufacture and sell-by date.
Product Concept	Made with two slices of bread in various thicknesses and proportion of vegetables, cheese, and dressings and cut into two triangular halves to facilitate eating. The sandwiched vegetables, salad, and cheese would make a nonsoggy and crunchy layer to whet the appetite. The bread would remain nonsoggy and fresh.
Benefits to Be Delivered	The thickness of the bread would vary to cater to the people performing manual labor with more carbohydrates, cater to the office workers who need a medium and balanced mix, and cater to people who are health conscious and consume fewer carbohydrates. Would give 40% of the recommended vegetable consumption per day.
Positioning and Target Price	The sandwich would occupy the middle ground in terms of price and a higher ground near the upper quartile in terms of quality, nutrition, and taste. The selling price should not exceed $3 and, therefore, the manufacturing price should not exceed $1.
Target Market	• Student population in high schools and universities • Office workers
Assumptions and Constraints	Sold through the current outlets A new platform and a new family Fresh ingredients are available and plentiful BSA code will be strictly implemented
Stakeholders	The customers (students and office workers of different age groups) Suppliers of local vegetables and cheeses Wholesalers and retailers Distributors
Possible Features and Attributes	Name of the sandwich is BORN TO BE EATEN Caption reads "Ask not whether you have completed the task; Ask shall we go for a BORN TO BE EATEN"
Possible Area for Innovation	Reflect the tastes of the area where sold Reflect special recipes of the area of manufacture

why), (4) positioning and target price (business goals), (5) target market (or to whom to sell), (6) assumptions and constraints, (7) stakeholders, (8) possible features and attributes, and (9) possible area for innovation. Table L1.1 shows the design brief.

L1.2   Customer Requirements (Verbatim)

The customer requirements recorded in their own words are given Table L1.2.

Table L1.2: Customer Verbatim

Should be enough to satisfy my hunger.
I prefer to have onions in it.
Should have a variety of cheeses.
Should be healthy.
Should have a variety of sandwiches to choose from.
Should be available with different salad dressings.
Should be hygienic.
Should be esthetically pleasing.
Should be able to know the contents easily from outside.
Should know the days on shelf.
Most of my family likes wheat bread.
I exercise, so I need a lot of protein.
Sometimes I like smaller sandwiches.
Should be able to put it in my handbag.
I do not want salad dressing.
I need different types of bread.

L1.3   Customer Needs

Customer needs are derived from the customer verbatim obtained by the design team so that they can bring customers' verbatim into effect in the product. The translation of customer's verbatim into needs is shown in Table L1.3. One verbatim may give rise to more than one need.

Table L1.3: Translating Verbatim to Needs

Verbatim	Needs
Should be enough to fill my hunger.	Provide adequate quantity of bread.
I prefer to have onions in it.	Include onions with the salad.
Should have a variety of cheeses.	Include various varieties of cheese.
Should be healthy.	Should be low-cholesterol and high-vitamin.
Have a variety to choose	Provide a number of varieties.
Should be available with different salad dressings.	Make it available with different salad dressings.
Should be hygienic.	
Should be esthetically pleasing.	Color combination of salad and size and shape of pieces.
Should be able to know the contents easily from outside.	Label should list the ingredients.
Should be able to know the days on shelf.	Needs labels for the manufacture and expiration dates.
Most of my family likes wheat bread.	Provide wheat bread as an ingredient.

I exercise, so I need a lot of protein.	Should have high protein salad mix.
Sometimes I like smaller sandwiches.	Include small size sandwiches.
Should be able to put it in my handbag.	Suitable size and shape for handbag.
I do not want salad dressing.	Provide a no-salad-dressing option.
I like different kinds of bread	Provide different types of bread options.
Should be delicious.	Salad should be spicy. Include sweet and juicy-on-bite fruits. Bread should be dry and soft. Salad should be crunchy.
Premises should meet acceptable standards	Meet BSA Guidelines on premises.

L1.4   Needs and Metrics

A metric is a precise, measurable characteristic of the product that will reflect the degree to which the product satisfies the need from which the metric originated. Metrics form the basis for developing specifications of the product; the underlying principle is that when the specifications are satisfied, the needs from which the metrics and resulting specifications arose are also satisfied. Table L1.4 deploys the metrics for each of the customer's needs.

Table L1.4: Needs and the Deployed Metrics

Needs	Metric	Units
Provide adequate amount of bread.	The slice size (narrow, medium, or thick).	M or T
Include onions with the salad.	Onions present or absent.	Number
Include various varieties of cheese.	Cheese varieties provided.	Number
Should be low-cholesterol and high-vitamin.	Label food values.	mg
Provide a number of varieties.	Number of variety choices.	Number
Available with different salad dressings.	Number of salad dressing choices.	Number
Color combination of salad and size and shape of pieces.	Number of ingredients with different colors.	Number
Size and shape of pieces.	Medium and large triangles.	Q or H
Label should list the ingredients.	Include the list.	Y or N
Label date of manufacture, and date of expiration.	Include both dates.	Y or N
Wheat Bread as an ingredient.	Wheat bread as a variety.	Y or N
High-protein salad mix.	Label the food values.	%
Include small size sandwiches.	Quarter-sized sandwich	Y or N
Suitable size and shape for handbag.	Handbag sizes	Y or N
Provide no-salad-dressing option.	No-salad-dressing as an option.	Y or N
Provide different types of bread options.	Bread variety as an option.	Y or N

Continued on next page

Needs	Metric	Units
Spicy salad.	Spice levels.	L,M, or H
Include sweet and juicy-on-bite fruits.	Type of fruits.	Names
Bread should be dry and soft.	Triangle test.	P or F
Salad should be crunchy.	Triangle test.	P or F
Meet BSA Guidelines on premises.	Audits per year	Number

In a triangle test, a panel of assessors are presented with three products, two of which are identical and the other one different. The assessors are asked to state which product they believe is the odd one out. You should provide your outstanding cheese salad sandwich and provide two other crunchy cheese-salad sandwiches that are identical to each other and that aren't as good the one you're selling. In deciding metrics, often the design team will be presented with qualitative needs, which somehow have to be measured. The task for the design team is to find ingenious ways to measure qualitative needs.

L1.4.1 Importance Ratings

Use the importance ratings of the needs to proportionally match the effort you have to use to deploy specific customer needs. In large projects, go back to the customer and get the importance ratings. For medium-sized and small projects, the design team may work out the importance ratings since they would have met the customers initially to record their verbatim. You also need to determine whether you need to refer back to the customer for *all* the needs for prioritizing their importance. You only have to refer back to the needs where the customers' inputs are necessary for a decision. For example, whether the sandwich should have bread or not is not up for debate because it is a fundamental requirement for a sandwich to have bread, whereas the importance of having onions in the sandwich is an important question where the customers' input would affect the design.

L1.5   Target Specifications

Specifications are an exact statement of the essential characteristics that a product will satisfy and are usually written in a manner that enables the measurement of the degree of conformance. For this reason, a specification contains a metric and a value with units. When specifications are written considering only the customer needs, they represent the ideal situation where the customer is 100% satisfied. You can treat this as the target for that product, so these are called the target specifications Table L1.5 presents the target specifications for the sandwich. In reality these specifications will be moderated when technical, manufacturing, and cost constraints are imposed. When you write the specifications, some of these needs and metrics can be combined.

Table L1.5: Target Specifications Derived from Customers' Specifications

Originating Need	Specifications	Target Value
1	Size of Bread Slices	T or M
	Varieties:	
2	Salad Choice (with or without onion)	2
3	Sizes Available (quarter, half and H. Bag)	3

Originating Need	Specifications	Target Value
4	Dressing Choice (with or without)	2
5	Spice Level Choice (medium or hot)	2
6	Bread Choice (white, brown, or wheat)	3
7	Cheese Choice (mozzarella or cheddar)	2
	Label Details	
8	Date of Manufacture	Y or N
9	Date of Expiration	Y or N
10	List of Ingredients	Y or N
11	Food Values	Y or N
12	Conformity to BSA Standards	Y or N

L1.6   Function Diagram

A function diagram is a representation of the functions performed by the product in a hierarchical manner. It is achieved by constructing an affinity diagram. The process involved in constructing the function diagram is as follows:

1. Prepare a full list of functions the product is to perform in the *verb-noun-phrase* format as shown in Table L1.6.
2. Arrange them as clusters and give them names.
3. Group the clusters into major groups and name them.
4. Arrange them in a tree format as shown in Figure L1.1.

Table L1.6: Functions of a Sandwich

Cover—Salad	Ensure—Dry Salad
Contain—Salad	Contain—Nutrients (vitamins and minerals)
Support—Salad	Contain—Protein
Absorb—Salad Juices	Ensure—Appetizing Colors
Assure—Calorie Requirement	Contain—Spice at Different Levels
Contain—Food Value Nutrients	Allow—Appropriate Sizes
Contain—Bran	Allow—Uniform Mix of Salad
Contain—Grains	Contain—Juicy and Dry Fruits (grapes)
Include—Air Pockets	Contain—Appropriate Cheese
Ensure—Pleasing Appearance	Ensure—Cheese in Right Form
Ensure— Strength to Hold Salad	Protect—Sandwich
Ensure—Handy Size	Lock—Sandwich from Spilling

Continued on next page.

Ensure—Cleanliness According to BSA	Maintain—Visibility of sandwich
Ensure—Appropriate Space Arrangement	Prevent—Contamination
Standardize—Process of Making	Prevent—Drying Up from Sun
Secure—Regular Certification	Prevent—Deterioration
Display—Ingredients	Display—Date of Manufacture
Display—Good Values	Display—Date of Expiration

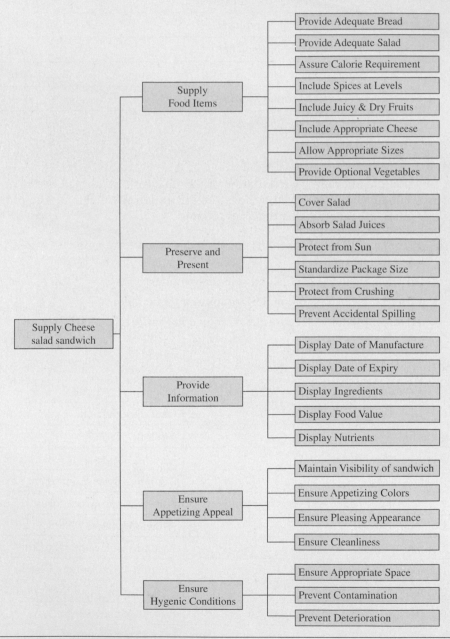

Figure L1.1 Function Tree of a Sandwich

L1.7   Conceptual Design

In the sandwich, partitioning the product into subsystems was not a problem. However proposing conceptual solutions and choosing a concept for further development is tricky. The morphological method is a useful way to identify subsystems. Again, the method of partitioning the product into subsystems is a key for pointing toward the optimal and near-optimal region in quick steps. For example, the subsystems of the sandwich can be (1) bread, (2) salad, (3) packaging, and (4) premises where the sandwiches are manufactured. However, they would not provide useful solutions. Consider the morphological chart as shown in Table L1.7

With this morphological analysis $3 \times 3 \times 2 \times 2 \times 3 \times 2 = 216$ sandwiches are possible. Obviously, making 216 types of sandwiches is not economically viable for small to medium-sized outlets. Yet there are solutions that could be chosen for the small outlets.

L1.8   Concept Selection

Within the 216 designs, some will sell more and some will sell less. According to Pareto's Law 80% of the sales will be from 20% of the designs. The question is which 20%. If you take modular approach to design, all ingredients are kept at the point of sale, and the customer chooses the ingredients for the sandwich. This is like building a computer according to a customer's requirements. The problem with this approach is the requirement of employees to be present at every sale point, which can be costly. An alternative is the platform approach, where you manufacture families of sandwiches of a basic type with variations. Another approach is to choose the 20% with a sample size of few thousand sandwiches on a promotional trial for a month or so. All these approaches aim to partition the design solution space in such a way as to get a smaller portion containing the optimal and near-optimal solution for the given constraints.

To continue the tale, some experimental sale is needed at this point. A group of students made a developmental experiment to obtain ball-park values for parameters. Some variations were kept constant, and 16 types were made and sold at 10 outlets for four weeks. The frozen choices were salad combinations (kept at 1 & 2) size (kept at half), spice level (kept at medium) and bread type (kept as white or brown). The resulting number of varieties is $2 \times 1 \times 2 \times 1 \times 2 \times 2 = 16$. The variations are shown in Table L1.8. The sandwiches were

Table 1.7: Morphological Chart

Feature	Solution 1	Solution 2	Solution 3
Salad Choice	Combination 1	Combination 2	Combination 3
Sizes Available	Quarter	Half	Hand-bag
Dressing Choice	With Dressing	Without Dressing	
Spice-Level Choice	Medium	Hot	
Bread Choice	White	Brown	Wheat
Cheese Choice	Mozzarella	Cheddar	

Table L1.8: The Variants

Type of Cheese	Salad	Dressing	Bread
Cheddar A_1	Lettuce B_1	Present C_1	White D_1
Mozzarella A_2	Lettuce and Tomato B_2	No Dressing C_2	Brown D_2

sold at a price that is three times the cost of manufacture. On the first two days, 800 sandwiches were sent to the chosen outlets, and the numbers sold were as shown in Table L1.9.

The results showed that the quantities sold varied between 30 and 700. To continue the experiment the quantities sent to the outlets were chosen as shown in the far-right column of Table L1.10. The summary results on five working days per week for four weeks are given in Table L1.11.

The results were arranged in descending order and a Pareto analysis was carried out to identify the 20% that yields 80% of the sales.

From Table L1.11 it is clear that the first eight sandwiches count for more than 80% of the sandwiches, and they are the selected ones. Figure L1.2 shows the Pareto graph.

Table L1.9: Types of Sandwiches and the Sales in the First Two Days

Sandwich Type	First Day Sales	Second Day Sales	Average	New Quantity per Day
$A_1 B_1 C_1 D_1$	488	514	501	550
$A_1 B_1 C_1 D_2$	692	712	702	750
$A_1 B_1 C_2 D_1$	250	254	252	300
$A_1 B_1 C_2 D_2$	344	354	349	400
$A_1 B_2 C_1 D_1$	174	168	171	200
$A_1 B_2 C_1 D_2$	234	232	233	250
$A_1 B_2 C_2 D_1$	78	82	80	100
$A_1 B_2 C_2 D_2$	116	124	120	150
$A_2 B_1 C_1 D_1$	185	187	186	200
$A_2 B_1 C_1 D_2$	262	268	265	300
$A_2 B_1 C_2 D_1$	62	62	62	100
$A_2 B_1 C_2 D_1$	62	62	62	100
$A_2 B_1 C_2 D_2$	86	90	88	100
$A_2 B_2 C_1 D_1$	30	30	30	50
$A_2 B_2 C_1 D_2$	36	42	39	50
$A_2 B_2 C_2 D_1$	56	60	58	100
$A_2 B_2 C_2 D_2$	74	74	74	100
Total	3167	3253	3210	3700

Table L1.10: Weekly Sales for Four Weeks

Sandwich Type	First Week	Second Week	Third Week	Fourth Week	Average
$A_1 B_1 C_1 D_1$	2480	2508	2513	2507	2502
$A_1 B_1 C_1 D_2$	3500	3514	3527	3511	3513
$A_1 B_1 C_2 D_1$	1248	1278	1257	1261	1261

$A_1 B_1 C_2 D_2$	1724	1763	1751	1746	1746
$A_1 B_2 C_1 D_1$	852	854	858	856	855
$A_1 B_2 C_1 D_2$	1162	1168	1173	1165	1167
$A_1 B_2 C_2 D_1$	396	403	401	400	400
$A_1 B_2 C_2 D_2$	593	617	602	600	603
$A_2 B_1 C_1 D_1$	921	937	937	929	931
$A_2 B_1 C_1 D_2$	1321	1333	1341	1325	1330
$A_2 B_1 C_2 D_1$	308	312	314	310	311
$A_2 B_1 C_2 D_2$	438	445	437	440	440
$A_2 B_2 C_1 D_1$	144	157	149	150	150
$A_2 B_2 C_1 D_2$	193	198	200	197	197
$A_2 B_2 C_2 D_1$	297	301	297	289	296
$A_2 B_2 C_2 D_2$	369	375	372	368	371

Table L1.11: The Descending Order of the Sandwiches

Sandwich	Rank by Sales (Numbers)	Weekly Average	Cumulative %
$A_1 B_1 C_1 D_2$	1	3513	21.86
$A_1 B_1 C_1 D_1$	2	2502	37.43
$A_1 B_1 C_2 D_2$	3	1746	48.3
$A_2 B_1 C_1 D_2$	4	1330	56.58
$A_1 B_1 C_2 D_1$	5	1261	64.43
$A_1 B_2 C_1 D_2$	6	1167	71.7
$A_2 B_1 C_1 D_1$	7	931	77.5
$A_1 B_2 C_1 D_1$	8	855	82.82
$A_1 B_2 C_2 D_2$	9	603	86.58
$A_2 B_1 C_2 D_2$	10	440	89.32
$A_1 B_2 C_2 D_1$	11	400	91.81
$A_2 B_2 C_2 D_2$	12	371	94.12
$A_2 B_1 C_2 D_1$	13	311	96.06
$A_2 B_2 C_2 D_1$	14	296	97.91
$A_2 B_2 C_1 D_2$	15	197	99.14
$A_2 B_2 C_1 D_1$	16	150	100.08

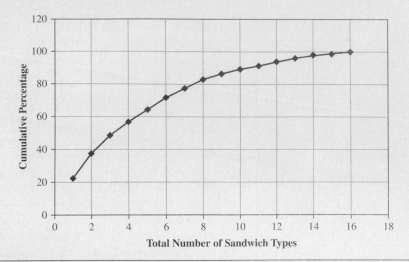

Figure L1.2 Pareto Analysis of Sandwiches

L1.9 Embodiment Design

In embodiment design, the abstract conceptual design is molded into a system with clarity, simplicity, and safety. Here (1) clarity allocates defined roles for each component; (2) simplicity ensures simple shapes that are easy to manufacture, assemble, operate, maintain, and dispose of; and (3) safety provides safety measures. Decisions are justified as much as possible by mathematical and engineering proof. Deliverables of embodiment design include

1. Definite layout, including the parts tree
2. Preliminary form designs
3. Bought-out components and systems and their specifications
4. Production processes

The embodiment-determining requirements are those relating to the supply of food values. This involves the thickness of the salad and cheese (found to be 10 mm) and the thickness of the bread slices (5 mm, 8 mm, and 11 mm). Once the dimensions have been chosen, noncritical items such as spices and types of salad can be added later.

Layout and parts tree: The thickness of a slice in a medium-sized slice of bread is about 8 mm, the thickness of the filling is about 10 mm, and the thickness of one triangle is about 26 mm. Therefore a packing with an overall thickness of about 57 mm with the triangle shape would house the payload. The label would be attached to the seal. The parts tree is given in Figure L1.3.

Preliminary form designs: Only the sandwiches with a half-triangular slice are considered. However sandwiches with thin slices, medium slices, and thick slices were considered.

Bought-out components: The packaging for the given design has to be bought out. A sealing and labeling machine may have to be purchased as well.

Another set of items to be purchased includes the vegetables, cheese, breads, and dressings. Assuming that the franchise has 20 outlets and 3,500 sandwiches are sold daily, the demand can be calculated as 3,500 sandwiches every day. Thus the quantity to be bought out is high, and the quality of these raw materials greatly influences the quality of the finished product; consequently, quality checks must be done periodically to maintain quality. A proper supply chain must be organized.

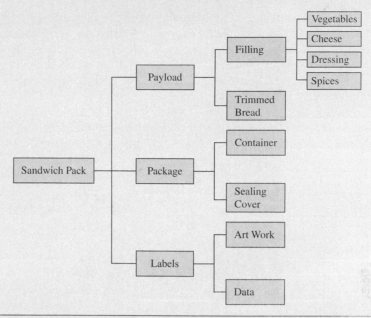

Figure L1.3 Parts Tree for a Sandwich

Production processes: Similar to dividing as (1) component manufacture and (2) assembly in mechanical product manufacture, the sandwich manufacturing process should be classified as (1) raw material processing where the vegetables are washed and cut to size and (2) assembling the sandwiches.

L1.10 Detail Design

Detail design consists of two parts: (1) the definition of all geometries, needed to produce a set of production drawings and (2) all engineering analyses, needed to ensure engineering integrity. In the case of a sandwich two geometries are needed: one to define the shape of the bread piece and one to define the shape of the package. Figure L1.4 shows the sandwich made with medium-sized bread pieces. The thickness of the slice is assumed to be 7 mm.

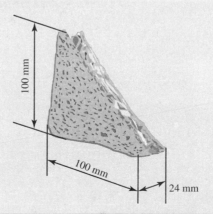

Figure L1.4 Sandwich Made up of Medium-Sized Slices of Bread

Three sizes of sandwiches made out of the thick, medium, and thin bread slices are chosen; that results in three different package sizes. In the detailed design two possible net configurations for the box are given in Figures L1.5 and L1.6. We know from sheet-metal work that near-rectangular shapes will ensure better use of the paper card with minimal off-cuts. Environmental friendliness dictates that the packaging be made of recycled paper card instead of plastic.

In this design the net is near-rectangular, but there are more places that need to be glued. The long card shown in Figure L1.5(a) is indented along the folds, and the triangular cards shown in Figure L1.5(b) are glued to the ears to form the basic box. The sandwiches are

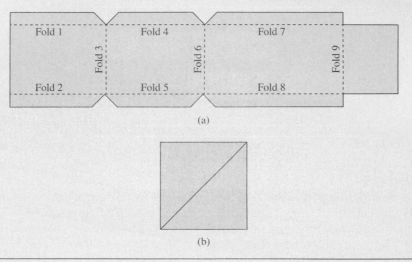

(a)

(b)

Figure L1.5 Near-Rectangular Net Design

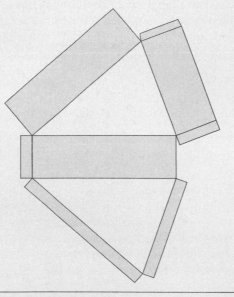

Figure L1.6 Nonrectangular Net Design

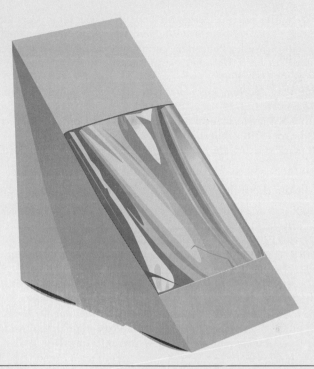

Figure L1.7 A Sandwich Box Similar to the Intended One

inserted inside, and the ears from folds 7, 8, and 9 are glued at the time of packing the sandwich. A triangular block can be used as a jig/fixture to assist the process.

In this design a single card is used to minimize the number of gluing places. The off-cut—and hence waste—will be more. It is a decision that must be made with due consideration to cost. The finished product would be somewhat similar to the one shown in Figure L1.7.

Student Project

Form a team, visit a sandwich or pizza outlet near you, and obtain all variables as demonstrated in this Lab. The ultimate objective is to produce at least three different products with suitable packaging.

References

[1] Feilden, B.R. *Engineering Design: Report of Royal Commission,* 2nd Ed.; HMSO: London, UK, 1963.
[2] Design Theory and Methodology. Task force, U.S. National Science Foundation. http://www.nsf.gov
[3] *Goals and Priorities for Research in Engineering Design, A Report to the Design Research Community,* American Society of Mechanical Engineers, USA 1986. *http://www.abet.org*
[4] Gero J.S. *"Design Prototypes:* "A Knowledge Representation Schema for Design," *AI Magazine,* Vol. 11, No. 4, pp. 26–36, 1990.
[5] Robertson D., and Ulrich K.T. "Planning for Product Platforms," *Sloan Management Review*; Summer 1998.
[6] Gregory S.A., *The Design Method*, London, UK: Butterworths & Co Publishers Ltd, 1966.

[7] Cross N., Engineering Design Methods: Strategies for Product Design. New York: Wiley, 1994.

[8] Pugh S., *Creating Innovative Products Using Total Design*, edited by Don Clausing and Ron Andrade, Addison Wesley Publishing Company, Reading Massachusetts 1996.

[9] Prater E.L., *Basic Machines*, Naval Education and Training Professional Development and Technology Center, USA 1994.

[10] Mott R.L., *Machine Elements in Mechanical Design*, 4th Edition in SI Units, Pearson Education South Asia Pvte Ltd, 2004.

[11] Ulrich K.T. and Eppinger S.D., *Product Design and Development*, 3rd Edition, New Delhi, India: Tata McGraw Hill Publications 2004.

[12] Ullman D.G. *The Mechanical Design Process*, International Edition, New York : McGraw Hill, 2004.

In addition, the following books, articles, and websites were used in preparing this chapter:

AMBROSE, S.A. and AMON, C.H. "Systematic Design of a First-Year Mechanical Engineering Course at Carnegie Mellon University." *Journal of Engineering Education*, pp. 173–181, 1997.

BUCCIARELLI, L.L. *Designing Engineers*. Cambridge, MA: MIT Press, 1996.

BURGER, C.P. "Excellence in Product Development through Innovative Engineering Design." *Engineering Productivity and Valve Technology*, pp. 1–4, 1995.

BURGHARDT, M.D. *Introduction to Engineering Design and Problem Solving*. New York: McGraw-Hill, 1999.

CROSS, N., CHRISTIAN, H., and DORST,K. *Analysing Design Activity*. New York: Wiley, 1996.

DHILLON, B.S. *Engineering Design: A Modern Approach*. Toronto: Irwin, 1995.

DIETER, G. *Engineering Design*. New York: McGraw-Hill, 1983.

DYM, C.L. *Engineering Design: A Synthesis of Views*. Cambridge, UK: Cambridge University Press, 1994.

EEKELS, J., and ROOZNBURG, N.F.M. "A Methodological Comparison of Structures of Scientific Research and Engineering Design: Their Similarities and Differences." *Design Studies,* Vol. 12, No. 4, pp. 197–203, 1991.

FLEDDERMANN, C.B. *Engineering Ethics*. Upper Saddle River, NJ: Prentice Hall, 1999.

HENSEL, E. "A Multi-Faceted Design Process for Multi-Disciplinary Capstone Design Projects." Proceedings of the 2001 American Society for Engineering Education Annual Conference and Exposition, Albuquerque, NM, 2001.

HILL, P.H. *The Science of Engineering Design*. New York: McGraw-Hill, 1983.

HORENSTEIN, M.N. *Design Concepts for Engineers*. Upper Saddle River, NJ: Prentice Hall, 1999.

JOHNSON, R.C. *Mechanical Design Synthesis*. Huntington, NY: Krieger, 1978.

WATSON, S.R. "Civil Engineering History Gives Valuable Lessons." *Civil Engineering*, pp. 48–51, 1975.

JANSSON, D.G., CONDOOR, S.S., and BROCK, H.R. "Cognition in Design: Viewing the Hidden Side of the Design Process." *Environment and Planning B, Planning and Design,* Vol. 19, pp. 257–271, 1993.

KARUPPOOR, S.S., BURGER, C.P., and CHONA, R. *"A Way of Doing Design." Proceedings of the 2001* American Society for Engineering Education Annual Conference and Exposition, Albuquerque, NM, 2001.

KELLEY D.S., NEWCOMER, J.L., and MCKELL, E.K. "The Design Process, Ideation and Computer Aided Design." Proceedings of the 2001 American Society for Engineering Education Annual Conference and Exposition, Albuquerque, NM, 2001.

PAHL, G., and BEITZ, W. *Engineering Design: A Systematic Approach*. New York: Springer-Verlag, 1996.

PUGH, S. *Total Design*. Reading, MA: Addison-Wesley, 1990.

RAY, M.S. *Elements of Engineering Design*. Englewood Cliffs, NJ: Prentice Hall, 1985.

RADCLIFFE, D.F., and LEE, T.Y. "Design Methods Used by Undergraduate Students," *Design Studies*, Vol. 10, No. 4, pp. 199–207, 1989.

ROSS, S.S. *Construction Disasters: Design Failures, Causes and Preventions*. New York: McGraw-Hill, 1984, pp. 303–329.

SICKAFUS, E.N. *Unified Structured Inventive Thinking*. New York: Ntelleck, 1997.

SIDDALL, J.N. *"Mechanical Design."* ASME Transactions: Journal of Mechanical Design, Vol. 101, pp. 674–681, 1979.

SUH, N.P. *The Principles of Design*. Oxford, UK: Oxford University Press, 1990.

VIDOSIC, J.P. *Elements of Engineering Design*. New York: The Ronald Press Co., 1969.

WALTON, J. *Engineering Design: From Art to Practice*. New York: West Publishing Company, 1991.

GENERAL

CHAPTER 2

ESSENTIAL TRANSFERABLE SKILLS

> *"It is possible to fly without motors, but not without knowledge and skill."*
>
> ~Wilbur Wright

Before discussing the design process in detail, it is important to state some necessary prerequisites. Transferable skills are used throughout the entire design process and indeed are needed throughout the entire life of professional engineers no matter what specialization or specific engineering career path they choose. This chapter focuses on the following skills:

- *Working in teams*: The dynamics of a team and knowing how to work within one are vital to the success of any project.

- *Scheduling*: Scheduling is necessary for managing a project and arriving at a successful conclusion in a timely manner. It also forces members of a team to diverge and converge at relevant stages of the project in order to come up with an integrated solution.

- *Research skills*: These skills are important for collecting data, gathering information, and keeping up with the latest developments, both technically and competitively.

- *Technical writing*: Communicating and reporting is a vital component of any design, one that is sometimes dismissed as a minor but necessary burden. However, even with the most innovative design, one that may revolutionize the way people live, if it is not reported to the world, it will remain an unknown entity, never to be adopted or adapted; ultimately it will be a failure.

- *Presentation skills*: These skills are just as important as technical writing skills and are sometimes more powerful. These skills provide the ability to reach that one important person who will agree to support an idea all the way to the mass public who will buy the resulting product. As such, it is an extremely powerful reporting and marketing tool. As presenters directly communicate with potentially interested parties, they have the distinct advantage and opportunity to convince and promote an idea that may otherwise have been overlooked.

(2.1) OBJECTIVES

By the end of this chapter, you should be able to

1. Identify the essential skills that are prerequisite to the design process.
2. Appreciate the importance and dynamics of working within teams.
3. Develop a project schedule utilizing existing tools.
4. Practice and improve your research and communication skills.

(2.2) WORKING IN TEAMS

Working in teams is an inevitable necessity. In a world where time to market has become a competitive component, a team of individuals will undoubtedly complete a project in a much shorter space of time than a single person would take. Furthermore, in many cases the breadth of expertise and knowledge that is required to visualize and realize a project makes it impossible for a single person to achieve it, even if time is not an issue.

Mohrman and Mohrman [1] define a team as a collection of individuals whose work is interdependent and who are collectively responsible for accomplishing a performance outcome. Thus, a collection of individuals in a music consort doesn't necessarily constitute a team, although all of the individuals are in the same place at the same time for the same purpose. The keyword is *collectively*. A group of students assembled at the start of the semester doesn't compose a team unless the students are working collectively toward arriving at an outcome.

2.2.1 Forming a Team

When appropriate individuals are picked to form a successful team, it is important to select individuals who complement each other. Naturally, each person thinks and behaves in ways that are unique to that individual. These dominant thinking styles are the results of the native personality interacting with family, education, work, and social environments. People's approaches to problem solving, creativity, and communicating with others are characterized by their thinking preferences. For example, one person may carefully analyze a situation before making a rational, logical decision based on the available data. Another may see the same situation in a broader context and look for several alternatives. One person will use a very detailed, cautious, step-by-step procedure. Another has a need to talk the problem over with people and will solve the problem intuitively.

Several models have been proposed as to how the human brain works. One of the well-known models is the Herrmann model. Herrmann developed a metaphorical model of the brain that consists of four quadrants. Although all of us are using all four quadrants, some individuals may have more use of certain quadrants of the brain. The following is a listing of these quadrants and their characteristics:

- *Upper left*: The characteristics of this quadrant are analytical, logical, quantitative, and fact based.
- *Lower left*: The characteristics of this quadrant are organized, planned, detailed, and sequential.

- *Upper right*: The characteristics of this quadrant are holistic, intuitive, synthesizing, and integrating.
- *Lower right*: The characteristics of this quadrant are emotional, social, and communicative.

Lab 2 will help students to understand these four thinking styles in more detail as well as allow them to better understand their own thought processes. A team with all four thinking styles will be a better team, and this lab should be used for this purpose before the design teams are created within the classroom.

2.2.2 Dynamics of a Team

Once a team has been formed, the dynamics of the team contribute to its success or failure. Regardless of the performance outcome, McGourty and DeMeuse [2] observed that all teams are characterized by the following features:

1. A dynamic exchange of information and resources among team members
2. Task activities coordinated among individuals in the group
3. A high level of interdependence among team members
4. Ongoing adjustments to both the team and individual task demands
5. A shared authority and mutual accountability for performance.

Many students dread group projects. Past experience with poorly functional teams and a lack of school system training may have instilled this fear in students. Larson and LaFasto [3] studied the characteristics of effective functional teams. They reported that successful teams have the following characteristics:

1. A clear, challenging goal; this goal gives the group members something to shoot for. The goal is understood and accepted by the entire group.
2. A result-driven structure; the roles of each member are clear, a set of accountability measures is defined, and an effective communication system is established.
3. Team members are competent and talented.
4. Commitment; team members put the team goals ahead of individual needs.
5. Positive team culture. This factor consists of four elements:
   a. Honesty
   b. Openness
   c. Respect
   d. Consistency in performance
6. There is a standard of excellence.
7. External support and recognition; effective teams receive the necessary resources and encouragement from outside the group.
8. There is effective leadership.

Design can be considered as a social activity in which a collective effort from a team produces the required output. Keep in mind that successful teams do not occur automatically or overnight. Effort and time must be devoted to nurturing a successful team. In addition, there are stages according to which teams evolve:

- *Forming*: At this stage, the group members still work as individuals; they do not contribute to the group as a whole but look out for themselves.

- *Storming*: At this stage, the group realizes that the task requires collective contribution and not much has been done. This prompts disagreements, blaming, and impatience with the process. Some members try to do it all on their own and avoid collaboration with team members.
- *Norming*: When the team's objectives are worked out collectively, the common problems or goals begin to draw individuals together into a group, although the sense of individual responsibility is still very strong. Conversation among the team members helps direct efforts toward better teamwork and increases members' sense of responsibility for the team objective.
- *Performing*: At this stage, the team members have accepted each others' strengths and weaknesses and have defined workable team roles.

McGourty, and DeMeuse [2] and DeMeuse and Erffmeyer [4] presented four behaviors that are needed for effective team performance.

1. *Communication team behavior*: Team members need to create an environment in which all members feel free to speak and listen attentively. The following behavioral practices for both roles must be followed:
   a. Listen attentively to others without interpreting.
   b. Convey interest in what others are saying.
   c. Provide others with constructive feedback.
   d. Restate what has been said to show understanding.
   e. Clarify what others have said to ensure understanding.
   f. Articulate ideas clearly and concisely.
   g. Use facts to get points across to others.
   h. Persuade others to adopt a particular point of view.
   i. Give compelling reasons for ideas.
   j. Win support from others.

2. *Decision-making team behavior*: Decision making is done *by* the team, not *for* the team. The following behavioral practices must be followed:
   a. Analyze problems from different points of view.
   b. Anticipate problems and develop contingency plans.
   c. Recognize the interrelationships among problems and issues.
   d. Review solutions from opposing perspectives.
   e. Apply logic in solving problems.
   f. Play a devil's advocate role when needed.
   g. Challenge the way things are being done.
   h. Solicit new ideas from others.
   i. Accept change.
   j. Discourage others from rushing to conclusions without facts.
   k. Organize information into meaningful categories.
   l. Bring information from outside sources to help make decisions.

3. *Collaboration team behavior*: Collaboration is the essence of teamwork; it involves working with others in a positive, cooperative, and constructive manner. The following behavioral practices must be followed:
   a. Acknowledge issues that the team needs to confront and resolve.
   b. Encourage ideas and opinions even when they are different from your own.
   c. Work toward solutions and compromises that are acceptable to all.

   d. Help reconcile differences of opinion.
   e. Accept criticism openly and nondefensively.
   f. Share credit for success with others.
   g. Cooperate with others.
   h. Share information with others.
   i. Reinforce the contribution of others.

4. *Self-management team behavior*: The following behavioral practices must be followed:
   a. Monitor progress to ensure that goals are met.
   b. Create action plans and timetables for work session goals.
   c. Define task priorities for work sessions.
   d. Stay focused on the task during meetings.
   e. Use meeting time efficiently.
   f. Review progress throughout work sessions.
   g. Clarify the roles and responsibilities of others.

Each of these behaviors, when followed, will assure an effective team. It is important that distinct team roles be defined. The following roles apply to a team of four members:

1. *Captain*—possesses behaviors and skills described in self-management team behavior.
2. *Chief engineer*—possesses behaviors in decision-making team behavior.
3. *Human resources person*—possesses behaviors described in collaboration team behavior.
4. *Spokesperson*—possess behaviors described in communication team behavior.

2.3 SCHEDULING

Developing a new product according to the design process is always limited by the time available for the entire process. Over the years, several procedures have been proposed to manage and plan for an artifact. Many projects have failed due to lack of detail. Historically, engineering projects have been typically created by one engineer, working alone. The designer, the drafter, and the planner were the same person. In recent years, however, advances in technology require that teams of engineers and technical helpers work beside each other to accomplish a project. In the following subsections, we discuss existing techniques for scheduling.

2.3.1 Gantt Chart

The Gantt chart was introduced by Henry L. Gantt and Frederick Taylor in the early 1900s to facilitate project management. The Gantt chart is in the form of a bar chart and is created as follows:

1. List all events or milestones of the project in an ordered list.
2. Estimate the time required to establish each event.
3. List the starting time and end time for each event.
4. Represent the information in a bar chart.

An example of the Gantt chart is shown in Figure 2.1. Each event is allocated a start time and duration shown by the shaded rectangles. If certain events can begin in parallel, they are

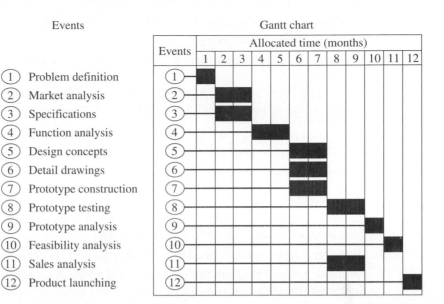

Figure 2.1: Gantt Chart

given the same start date as can be seen with the market analysis and specifications. However, certain events cannot begin until prior events have been completed. An example of this in the figure is prototype testing, which begins in month 8 but obviously only after the prototype construction event has been completed at the end of month 7. Gantt charts can assist a project to arrive at its successful conclusion in an efficient manner by identifying which events are independent of each other and starting them at their earliest possible time. Commercial software is available that can help in developing a Gantt chart for the project (e.g., Microsoft Project Manager™ or Oracle Primavera™). Lab 5 introduces the creation of Gantt charts in Microsoft Project Manager. Even if students do not have this software, they should attempt the Lab, either using an alternative software package or manually, given that the case study included provides good training and practice in creating the charts.

2.3.2 CPM/PERT

The critical path method (CPM) and the program evaluation and review technique (PERT), which were developed during the 1950s and 1960s, are the two most widely used approaches for scheduling projects. PERT was developed under the guidance of the U.S. Navy Special Project Office by a team that included members from the Lockheed Missile Systems Division and the consulting firm Booz, Allen, and Hamilton. The technique was developed to monitor the efforts of 250 main contractors and 9000 subcontractors who were involved with the Polaris missile project. CPM was the result of the efforts of the Integrated Engineering Control Group of E.I. DuPont de Nemours & Co., and it was developed to monitor activities related to design and construction. A survey of manufacturing companies in the United States revealed that CPM/PERT is used over 65% of the time, among all other methods. A CPM/PERT project generally has the following characteristics:

1. There are clearly defined activities or jobs whose accomplishment results in project completion.

2. Once started, the activity or job continues without interruption.

3. The activities or tasks are independent, which means they may be started, stopped, and performed individually in a prescribed sequence.

4. The activities or jobs are ordered, and they follow each other in a specified manner.

Even though the CPM and PERT techniques were developed independently, the basic theory and symbols in both techniques are essentially the same.

2.3.3 CPM/PERT Definitions

Several symbols, terms, and definitions are used in developing the CPM/PERT networks:

1. *Event (node)*: This represents a point in time in the life of a project. An event can be the beginning or the end of an activity. A circle is used to represent an event. Generally, each network event is identified with a number.

2. *Activity*: This is an effort needed to carry out a certain portion of the project.

3. *Network paths*: These are the paths used (or needed) for reaching the project termination point, or the event from the project starting point or event.

4. *Critical path*: This is the longest path with respect to length of time through the PERT/CPM network. In other words, the critical path creates the largest amount of activity time of all individual network paths.

5. *Earliest event time (EET)*: This is the earliest time at which an event occurs, providing that all proceeding activities are accomplished within their estimated times.

6. *Latest event time (LET)*: This is the latest time at which an event could be reached without delaying the predicted project completion date.

7. *Total float*: This is the latest time of an event minus the earliest time of the preceding event and the duration time of the in-between activity.

Figure 2.2 illustrates an example of a CPM or PERT chart. The longest path or the critical path is Task A-B-E-H which takes 14 days. As task D would finish after 12 days (three days for Task A followed by the allocated nine days for Task D), there would be two days remaining until the next event, which in this case is the end of the project. Therefore, there is a two-day float for Task D. Similarly, it can be ascertained that there is a one-day float to complete tasks C, F, and G before Task H can begin.

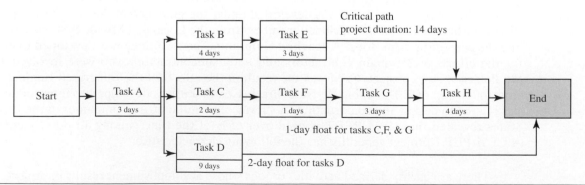

Figure 2.2: CPM/PERT Chart

2.3.4 CPM/PERT Network Development

Steps involved in constructing a CPM network are

1. Break down the design into individual activities and identify each activity.
2. Estimate the time required for each activity.
3. Determine the activity sequence.
4. Construct the CPM network, using the defined symbols.
5. Determine the critical path of the network.

Similarly, the steps involved in developing a PERT network are

1. Break down the design into individual activities and identify each activity.
2. Determine the activity sequence.
3. Construct the PERT network, using the defined symbols.
4. Obtain the expected time to perform each activity, using the following weighted average formula:

$$T_e = \frac{x + 4y + z}{6}$$

where

T_e = the expected time for the activity

x = the optimistic time estimate for the activity

y = the most likely time estimate for the activity

z = the pessimistic time estimate for the activity

5. Determine the network critical path.
6. Compute the variance associated with the estimated expected time of Te of each activity using the formula:

$$s^2 = \left(\frac{z - x}{6}\right)^2$$

7. Obtain the probability of accomplishing the design project on the stated date using the formula:

$$w = \frac{T - T_L}{\sqrt{|\Sigma S_{cr}^2|}}$$

where

S_{cr}^2 = the variance of activities on the critical path

T_L = the last activity's earliest expected completion time, as calculated though the network

T = the design project due date, expressed in time units

Table 2.1 presents the probabilities for selective values of w.

Table 2.1: Probability Table

W	Probability
−3.0	0.0013
−2.5	0.006
−2.0	0.023
−1.5	0.067
−1.0	0.159
0.5	0.309
0.0	0.5
0.5	0.69
1.0	0.84
1.5	0.933
2.0	0.977
2.5	0.994
3.0	0.999

Example 2.1 Based on Robertson and Ulrich

A mechanical design project was broken down into a number of major jobs or activities, as shown in Table 2.2. A CPM network was developed using the data given in the table, and the critical path of the network was determined along with the expected project duration time.

Table 2.2: Design Project

Activity Description	Activity Identification	Immediate Predecessor Activity or Activities	Activity Duration (Week or Weeks)
Needs, goals, and market analysis	A	—	1
Function analysis	B	A	2
Specifications	C	B	1
Alternatives	D	C	4
Evaluations	E	D	3
Prototyping	F	D	4
Analysis	G	E,F	2
	H	G	2
Manufacturing	I	G	4
Marketing	J	H, I, 3	—

Using the defined symbols and the data given in Table 2.2, the CPM network is shown in Figure 2.3. The paths originated at event 1 and terminated at 11 are as follows:

1. A-B-C-D-E-G-I-J time _20
2. A-B-C-D-F-G-I-J time _21
3. A-B-C-D-E-G-H-J time_18
4. A-B-C-D-F-G-H-J time _19

These results indicate that the longest path is (A-B-C-D-F-G-I-J), and the predicted time for the design project is 21 weeks.

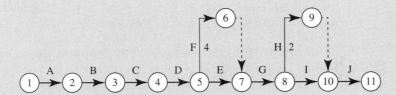

Figure 2.3: CPM Chart

2.4 RESEARCH SKILLS

Research is the ability to collect data, gather the information, and interpret it as knowledge. This is a vital skill for success and lifelong learning. Lifelong learning is the ability to keep developing and learning outside of the classroom. The ability to research allows a person to stay up to date with the latest in technology and the marketplace. Differentiating between data, information, and knowledge is a good start in appreciating what is required to attain and in practicing research skills. Data can be defined as raw, unprocessed material, which can be collected through laboratory testing and through many other activities. Once this data is processed into something meaningful, it becomes information. However, it becomes knowledge only when this information can be applied successfully.

Eisenburg and Berkowitz [5] developed the Big6, which breaks down the research skills into the following six categories:

1. Task definition
2. Information-seeking strategies
3. Location and access
4. Use of information
5. Synthesis
6. Evaluation

The following text gives the interpretation of these points by the authors; however, for more detail you may visit www.big6.com.

- Step 1 defines the task and identifies the data and/or information that must be collected or gathered. Once this step has been completed, the search for information can begin.

- Step 2 provides a means to plan how you will go about seeking or collecting the information required.

- Step 3 identifies the sources (e.g., textbooks, Internet, journals, magazines, etc.), which is where you collect all the data and process it into information relevant to your task. Some material will have been processed already into information at the source, so all that is needed is the ability to be able to access and store it in an organized manner.

- Step 4 extracts the information that is needed.

- Step 5 processes this information into the final body of knowledge. Once knowledge has been attained, evaluation is key to ensuring that the knowledge and the process used to arrive at it is accurate and complete. In many cases, the knowledge may be accurate but incomplete and will require further development.

Within the design process, research is pursued throughout all stages. However, it is most evident immediately after the customers' needs and goals have been identified. This is where the market analysis and information-gathering stage of the project occurs. Chapter 4 discusses this in more detail.

2.5 TECHNICAL WRITING AND PRESENTATION

In the engineering and scientific professions, communication skills are as important as in other fields. Following the development of an innovative design and a cost estimate that predicts large profit, the designer must be able to communicate the findings to other people. An old adage reminds us that a tree falling in a forest doesn't make a sound unless there is someone to listen. Similarly, the best technical design in the world might never be implemented unless the designer can communicate the design to the right people in the right way.

The quality of a report generally provides an image in the reader's mind that, in large measure, determines the reader's impression of the quality of work. Of course, an excellent job of report writing cannot disguise a sloppy investigation, but many excellent design studies have not received proper attention and credit because the work was reported in a careless manner. A formal technical report is usually written at the end of a project. Generally, it is a complete, stand-alone document written for people who have widely diverse backgrounds. Therefore, much more detail is required. The outline of a typical formal report is as follows:

1. *Cover page*:
   a. Title of the report
   b. Name of author(s)
   c. Address

2. *Summary*: This section summarizes the work, and it should also include a short conclusion. This may be the only section some people read (including the executives who may sponsor the project). Therefore, this section needs to stand alone. Writing a technical report is different from writing a novel, where "giving away the ending" at the beginning of the book would not be acceptable. In the case of a technical report, including "the ending" in the summary is not only acceptable; it is actually expected.

3. *Table of contents*

4. *List of figures and list of tables*: This section should include the corresponding page number.

5. *Introduction*: This section provides the background of the work (market analysis) to acquaint the reader with the problems involved with and the purposes of carrying on the work.

6. *Design process*: This section presents the details of the procedure followed in the design process.

7. *Discussion*: This section should contain the comprehensive explanation of the results. The discussion may be divided into several subsections, such as:
   a. Technical analysis (force requirement, speed, etc.)
   b. Details of the build-up of the artifact
   c. Equipment used
   d. Assembly and setup procedures
   e. Experimental setups and results
   f. Details of the final results

8. *Conclusions*: This section states in as concise a form as possible the conclusions that can be drawn from the study.

9. *References*: This section lists all of the documents to which the writers referred. The information on each document must be complete and must follow the same format throughout.

10. *Appendices*: The appendices include material deemed beyond the scope of the main body of the report. The appendix section may be divided into as many subsections as necessary.

2.5.1 Steps in Writing a Report

The five operations involved in the writing of a high-quality report are best remembered with the acronym POWER:

1. **P:** Plan the writing
2. **O:** Outline the report
3. **W:** Write
4. **E:** Edit
5. **R:** Rewrite

Good writers usually produce good reports. Some of the attributes of good writing are presented in the following imperatives:

1. Write as objectively as possible. Do not become emotionally involved or attached to a problem or a solution.
2. Be reasonably methodical.
3. Record whatever is learned, and keep in mind that whatever work is performed must eventually be documented.
4. Always strive for clarity in writing, and keep in mind that the written material should be simple and straightforward.
5. Deliver the written material on time.

There are certain additional qualities associated with good reports, such as the following:

1. Delivered on the due date
2. Effectively answer readers' questions as they arise
3. Give a good first impression
4. Read coherently
5. Contain an effective summary and conclusion
6. Are written clearly and concisely and avoid vague or superfluous phrases
7. Provide pertinent information

2.5.2 Illustration Guidelines

Visual elements (such as figures, charts, and graphs) in technical reports have specific purposes and convey specific information. Visuals are used to explain, illustrate, demonstrate, verify, or support written material. Visuals are valuable only if their presentation is effective. The general guidelines for preparing effective visuals (illustrations) are

1. Reference all illustrations in the text.
2. Reference the data source.
3. Carefully plan the placement of illustrations.
4. Specify all units of measure and the scale used in the drawing.
5. Label each illustration with an identifying caption and title. Include the figure source if the figure was obtained from another document.
6. Spell words out rather than use abbreviations.
7. When a document has five or more figures, include a list of figures at the beginning of the report.
8. Avoid putting too much data in an illustration.

Figures 2.4 and 2.5 demonstrate the same illustration. However, Figure 2.4 shows how not to do it. In this figure, the labeling is missing. The scale does not make efficient use of displaying the curve. The thickness of the line and plot symbols are too big for readers to accurately read the graph. The units and description of what the graph is measuring are also missing. The graph seems to have been poorly scanned and has been displayed at a slight angle.

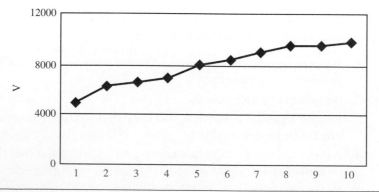

Figure 2.4: Example of Bad Illustration

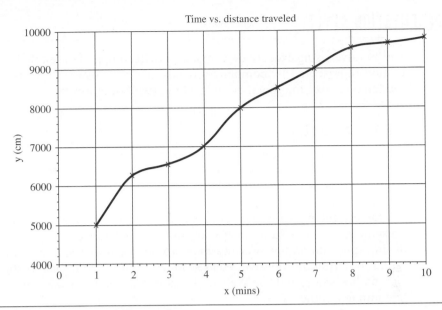

Figure 2.5: Example of Good Illustration

Look at Figure 2.5 for a better example of the same illustration. Try to see what else has been improved.

2.5.3 Mechanics of Writing

The following suggestions will help you avoid some of the most common mistakes in your writing:

- *Paragraph structure*: Each paragraph should begin with a topic sentence that provides the overall purpose of the paragraph. Each paragraph should have a single theme or conclusion, and the topic sentence states that theme or conclusion.

- *Sentence length*: Sentences should be kept as short as possible so that their structure is simple and readable. Long sentences require complex construction, provide an abundance of opportunity for grammatical errors, take considerable writing time, and slow the reader. Long sentences are often the result of putting together two independent thoughts that could be stated better in separate sentences.

- *Pronouns*: There is no room for ambiguity between a pronoun and the noun for which it is used. Novices commonly use *it, this, that*, and so on where it would be better to use one of several nouns. It may be clear to the writer but is often ambiguous to the reader. In general, personal pronouns (*I, you, he, she, my, mine, our, us*) are not used in reports.

- *Spelling and punctuation*: Errors in these basic elements of writing are inexcusable in the final draft of the report.

- *Tense*: Use the following rules when choosing the verb tense:
  a. *Past tense*: Use to describe work done when you were building or designing or in general when referring to past events.
  b. *Present tense*: Use in reference to items and ideas in the report itself.
  c. Future *tense*: Use in making predictions from the data or results that will be applicable in the future.

2.6 PRESENTATION STYLE

Unlike advertising executives, engineers are ill equipped to sell their ideas. Secondhand information represented by company officials may not answer all of the client's questions. This section discusses the specifics and techniques of oral presentations.

2.6.1 Objective

Every presentation should have an objective. The speaker's main objective is to deliver the message (objective) to the audience. The objectives may vary from one presentation to another. To identify the real objective, ask the following question: If everything goes perfectly, what do I intend to achieve?

Determine who your audience is and what their educational level may be. In most cases, presentation time is limited. It is of utmost importance to keep within the scheduled time during question-and-answer sessions. To stay within time restraints, detailed planning is required. Different tools can be used in the presentation, such as slides, models, transparencies, audiovisuals, and the Web. Make sure to account for the time needed to shift from one medium to another.

An easy way to evaluate the effectiveness of a presentation while on a team is to practice through role playing. One person can play the role of the speaker, and the rest of the team can act as the audience and possibly even play devil's advocate. In this way, the team gains valuable experience before attempting the actual presentation.

2.6.2 Oral Presentation Obstacles

To sell your ideas to others, you should first be convinced that your ideas will accomplish the task at hand. Oral presentation requires a high degree of creativity.

People resist change, although they may announce that they embrace it. Humans like familiar methods. Change requires additional effort, which humans generally resist. Typical reactions to change include

- We tried that before.
- It is a too radical change.
- We have never done that before.
- Get back to reality.
- We have always done it this way.
- I don't like the idea.

In addition, keep in mind the following:

- Only 70% of the spoken word is actually received and understood. Complete understanding can come through repetition and redundancy in speech.
- People mostly understand three-dimensional objects. Two-dimensional projections need to be transmitted with added details.
- People usually perceive problems from their own perspectives.
- Convey ideas so that they may be interpreted with the least expenditure of energy.

2.6.3   Oral Presentation Do's and Don'ts

Remember the following in making oral presentations:

1. Know your audience thoroughly.
2. Never read solely from notes, a sheet, or directly from an overhead projector. You can use your notes for reference, but remember to make eye contact with your audience from time to time.
3. Bring the audience up to speed in the first few moments.
4. Stay within the time allotted.
5. Include relevant humorous stories, anecdotes, or jokes, but only if you are good at it.
6. Avoid using specialized technical jargon. Explain the terms you feel the audience may not know.
7. Understand your message clearly. The entire goal is to communicate the message clearly.
8. Practice, practice, practice! You may prefer to memorize the introduction and concluding remarks.
9. The dry run is a dress rehearsal. Use it to iron out problems in delivery, organization, and timing.
10. Avoid mannerisms; speak confidently but not aggressively.
11. Maintain eye contact with audience members, and keep shifting that contact through the talk.
12. Never talk to the board or to empty space.
13. Present the material in a clever fashion, but not in a cheap or sensational fashion. Be genuinely sincere and professional.
14. A logical presentation is much more critical in an oral one than it is in a written one.

2.6.4   Oral Presentation Techniques

The following will help you make your oral presentation as effective as possible:

1. Visual aids (sketches, graphs, drawings, photos, models, slides, transparencies, and the Web) often convey information more efficiently and effectively. Visual aids permit the use of both the hearing and visual senses, and they help the speaker.
2. Limit slides to not more than one per minute.
3. Each slide should contain one idea.
4. The first slide should show the title of your presentation and the names of the collaborators.
5. The second slide should give a brief outline of the presentation.
6. The last slide should summarize the message you just delivered.
7. If you need to show a slide more than once, use a second copy in your slide deck.
8. Avoid leaving a slide on the screen if you have finished discussion on the topic.
9. Never read directly from the slide. Spoken words should complement the slides. Prepare notes for each slide and use them during practice.
10. Use graphs to explain variations. Clearly label the axes, data, and title. Acknowledge the source.

11. Every graph should have a message (idea). Color should enhance the communication, not distract from it.

12. Audiences respond to well-organized information. That means
    a. Efficient presentation
    b. All assumptions clearly stated and justified
    c. Sources of information and facts clearly outlined

13. Begin with the presentation of the problem and conclusion/recommendation (primary goal).

14. Finish ahead of schedule and be prepared for the question-and-answer session.

2.6.5 Question-and-Answer Session

The question-and-answer (Q&A) session is very important. It shows the degree of enthusiasm of the audience and usually reveals interest and attention. In the Q&A session you should

1. Allow the questioner to complete the question before answering it.
2. Avoid being argumentative.
3. Avoid letting the questioner feel that the question is stupid.
4. Adjourn the meeting if the questions slack off.
5. Thank the audience one final time after the Q&A session.

2.7 CHAPTER SUMMARY

1. A team is a collection of individuals whose work is interdependent and who are collectively responsible for accomplishing a task.

2. Forming a team can be done using different models; in this book, we use the brain model to form teams.

3. Features of a team include dynamic exchange of information and resources, coordinated task activities, high level of interdependence among team members, continuous adjustment to both team and task demands, and shared authority and accountability for performance.

4. Successful teams may evolve through different stages including forming, storming, blaming, norming, and performing.

5. Scheduling is essential for successful and timely completion of a project. Various tools are utilized to schedule a project including a Gantt chart and CPM/PERT methods.

6. Research is the ability to collect data, gather information, and interpret it as knowledge. The Big6 breaks down the research skill into six categories.

2.8 PROBLEMS

1. Assume that the optimistic, most likely, and pessimistic activity times are as given in Table 2.3.
   a. Complete the table.
   b. Determine the probability of finishing the job in 32 weeks.

Table 2.3: Problem 1

Pessimistic	Most Likely	Optimistic	Expected Time	Variance
3	2	1		
3	2	1		
3	2	1		
7	5	3		
9	6	4		
3	2	1		
7	5	4		
3	2	1		
3	2	1		
3	2	1		
3	2	1		

2. Table 2.4 breaks down the number of major jobs or activities involved in painting a two-story house.

   a. Develop a CPM network.

   b. Determine the critical path of the network.

   c. Determine the expected project's duration time period.

Table 2.4: Problem 2

Activity	Identification	Predecessor	Duration
Contract signed	A	—	2
Purchase of material	B	A	2
Ladder and staging in site	C	A	2
Preparation of surface	D	C,B	5
Base coat complete	E	D	6
Base coat inspected	F	E	2
Trim coat complete	G	E	5
Trim coat inspected	H	G,F	2
Final inspection	I	H	2
Removal of staging	J	H	2
Final cleanup	K	I,J	2

3. You should work on this activity during lab hours.

   a. Develop a CPM network and determine the critical path for the events defined in the Gantt chart in Figure 2.1.

   b. Develop a Gantt chart for the events defined in Table 2.2.

   c. For the events defined in Table 2.5,

      i. Complete the table.

      ii. Draw the PERT network.

      iii. Determine the probability of finishing the task on time if the design due date is later than 95 days.

Table 2.5: Problem 3

Predecessor	Optimistic	Most Likely	Pessimistic	Expected	Variance
1	3	5	8		
1	4	6	9		
3	3	4	5		
2	2	3	4		
5	3	4	5		
1	8	12	14		
4	14	18	21		
7	5	10	14		
7	5	10	14		
7	5	10	14		
7	5	10	14		
7	5	10	14		
12	4	6	10		
11	4	6	10		
10	4	6	10		
9	4	6	10		
8	4	6	10		
13	10	12	18		
6	16	18	24		
14	7	10	15		
14	10	15	22		
17	5	9	9		
18	4	6	8		
16	6	8	12		
15	3	8	12		
19	3	4	5		

Individual Activities

4. Define the following terms:
   a. CPM
   b. Gantt chart
   c. PERT

5. Assume that, for example, the optimistic, most likely, and pessimistic activity times are as shown in Table 2.6. Calculate each activity's expected time and variance along with the probability of accomplishing the design project in 2.5 weeks. In addition, calculate each event's earliest and latest event times.

Table 2.6: Activity Time Estimates (Weeks)

Pessimistic	Most Likely	Optimistic	Expected Time	Variance
4	2	1	2.16	0.25
3	2	1	2	0.11
4	2	1	2.16	0.25
8	5	3	5.16	0.7
5	4	2	3.83	0.25
7	4	3	4.33	0.44
5	3	2	3.17	0.25
5	3	2	3.17	0.25
9	6	3	6	1
10	4	2	4.17	0.69

6. Assume that an engineering course term project is broken down into a number of major jobs or activities, as shown in Table 2.7.

  a. Draw a Gantt chart.

  b. Develop a CPM network.

  c. Determine the critical path of the network.

  d. Determine the project duration time period.

Table 2.7: Project

Activity Description	Activity Identification	Immediate Predecessor or Activity	Activity Duration (Days)
Literature collection	A	—	7
Literature review	B	A	4
Outline preparation	C	B	1
Analysis	D	B	10
Report writing	E	C	5
Typing	F	E	3
Revision	G	E	4
Final draft	H	G	2

LAB 2: Ice Breaking—Forming Teams

This lab introduces these models and provides a tool with which students can better understand their thought processes. When students understand their own thought processes and those of others, they work better in a group.

The objectives of this lab are as follows:

1. Introduce the model of the brain developed by Ned Herrmann.

2. Present a preliminary self-test to quantify the strength of the four quadrants.

3. Introduce team formation. After the students test which thinking preference they may have, it becomes a tool when the teams form to make a full brain (i.e., one student from group A, another from group B, and so on).

Each person thinks and behaves in ways that are unique to that individual. These dominant thinking styles are the results of native personality interacting with family, education, work, and social environments. People's approaches to problem solving, creativity, and communicating with others are characterized by their thinking preferences. For example, one person may carefully analyze a situation before making a rational, logical decision based on the available data. Another may see the same situation in a broader context and look for several alternatives. Still another person will use a very detailed, cautious, step-by-step procedure.

Some people have a need to talk the problem over with other people and will solve the problem intuitively. Ned Herrmann's metaphorical model divides the brain into left and right halves and into cerebral and limbic hemispheres, resulting in four distinct quadrants. The Hermann profile results have a neutral value. There are no right or wrong answers.

The questionnaire presented in Table L2.1 will help you investigate your own thinking style. Questionnaires like this one are available from specialized companies such as Ned Herrmann's for a nominal charge per student. The idea here is to have this activity and questionnaire as a preliminary step for people who have more interest in finding more and one that is sufficient for people who just want some idea.

The questions in this questionnaire are based on data available in Lumsdaine's book and material distributed through the NSF (National Science Foundation) workshop on introductory engineering design at the Central Michigan University, 1999, presented by Frank Maraviglia.[1]

Procedure

In the rating box to the right, after Questions 1–13, write the numbers 4, 3, 2, or 1, where a 4 indicates most likely and 1 indicates least likely. Use each number *only once* for each question. There is no right or wrong answer; answer to the best of your knowledge and experience.

To find out the totals you need to

1. Multiply the total number in box 14 by 4, and then add the results to the I bracket. Once you have the total sum, call it A.

2. Multiply the total number in box 15 by 4, and then add the results to the II bracket. Once you have the total sum, call it B.

3. Multiply the total number in box 16 by 4, and then add the results to the III bracket. Once you have the total sum, call it C.

4. Multiply the total number in box 17 by 4, and then add the results to the IV bracket. Once you have the total sum, call it D.

In items 14 to 17 you will find four lists. Check the number of items in each list that you enjoy doing. Add up the number of checked items and write the number in the space provided for each item.

---

[1]Based on Lumsdaine, E., and Lumsdaine, M., *Creative Problem Solving*. New York: McGraw-Hill, 1995 and NSF Sponsored Workshop on Introductory Engineering Design, J. Nee (editor). Central Michigan University, 1999.

Table L2.1: Questionnaire

	Question	Options	Rating			
e.g.,	This is an example question	a. Provide a Rating of 1, 2, 3, OR 4 in each box on the right b. 4 indicates most likely and 1 indicates least likely c. Use each number only ONCE for each question d. There is no right or wrong answer. Answer to the best of your knowledge	3	4	1	2
1	In a project setup you prefer to perform the following function:	a. Looking for data and information b. Developing a systematic solution and directions c. Listening to others and sharing ideas and intuitions d. Looking for the big picture and context, not the details				
2	While solving your homework, you	a. Organize the information logically in a framework but not down to the least detail b. Do detailed homework problems neatly and conscientiously c. Motivate yourself by asking why and by looking at personal meaning d. Take the initiative in getting actively involved to make learning more interesting				
3	In studying new material, you prefer	a. Listening to the informational lecture b. Testing theories and procedures to find flaws and shortcomings c. Reading the preface of the book to get clues on the purpose d. Doing simulations and ask what-if questions				
4	Do you like	a. Reading textbooks b. Doing lab work step-by-step c. Talking about, seeing, testing, and listening d. Using visual aids instead of words				
5	In studying new material you like to	a. Analyze example problems and solutions b. Write a sequential report on the results of a lab experiment c. Do hands-on learning by touching and seeing d. Take an open-ended approach and find several solutions				
6	When and after solving a problem you	a. Think through ideas rationally b. Find practical uses of knowledge learned; theory is not enough c. Use group study opportunities and group discussion d. Appreciate the beauty in the problem and the elegance of the solution				
7	Do you prefer to	a. Do research using the scientific method b. Use computers with tutorial software c. Use group study opportunities and group discussion d. Lead brainstorming sessions in which wild ideas are accepted				
8	In project execution, which of the following do you prefer to do?	a. Make up a hypothesis and then test it to find out if it's true b. Plan, schedule, and execute projects according to a set time c. Keep a journal to record what you have seen in the experiment d. Experiment and play with ideas and possibilities				
9	Generally speaking, you absorb new ideas best by	a. Applying them to concrete situations b. Concentrating and conducting careful analysis c. Contrasting them with other ideas d. Relating them to current or future activities				
10	When you read self-help articles, you pay most attention to	a. The ideas that are drawn from the information b. The truth of the finding backed up by information c. Whether or not the recommendations can be accomplished d. The relation of the conclusion to your experience				
11	When you hear people arguing, you favor the side that	a. Presents ideas based on facts and logic b. Expresses the argument most forcefully and concisely c. Reflects your personal opinion d. Projects the future and shows the total picture				
12	When you make a new choice, you rely most on	a. Reality and the present rather than future possibilities b. Detailed and comprehensive studies c. Talking to people d. Intuition				
13	When you approach a problem, you are likely to	a. Try to relate to a broader problem or theory b. Try to find the best procedure for solving it c. Try to imagine how others might solve it d. Look for ways to solve the problem quickly				
Add up the columns vertically in the space provided (not including example question)			I	II	III	IV

14. Total: [ ]
    —Looking for data and information; doing library searches
    —Organizing information logically in a framework
    —Listening to informational lectures
    —Reading textbooks
    —Studying sample problems
    —Thinking through ideas
    —Making up a hypothesis and then testing it to find out if it is true
    —Judging ideas based on facts, criteria, and logical reasoning
    —Doing technical and financial case studies
    —Knowing how much things cost
    —Using computers for math and information processing

15. Total: [ ]
    —Following directions carefully, instead of improvising
    —Testing theories and procedures to find flaws and shortcomings
    —Doing lab work step by step
    —Listening to detailed lectures
    —Taking detailed comprehensive notes
    —Studying according to a fixed schedule in an orderly environment
    —Making up a detailed budget
    —Practicing new skills through frequent repetition
    —Writing a how-to manual about a project
    —Using computers with tutorial software
    —Finding practical uses of knowledge learned

16. Total: [ ]
    —Listening to others and sharing ideas
    —Keeping a journal to record feelings and spiritual values
    —Studying with background music
    —Respecting others' rights and views; people are important, not things
    —Using visual clues to make use of body language
    —Doing hands-on learning by touching and using a tool or object
    —Watching dramatic movies rather than adventure movies
    —Learning through sensory input (i.e., moving, feeling, smelling, testing, listening)
    —Motivating yourself by asking why and by looking for personal meaning
    —Traveling to meet other people and learn about their culture

17. Total: [ ]
    —Looking for the big picture and context, not details, of a new topic
    —Taking the initiative in getting actively involved to make learning more interesting
    —Doing open-ended problems and finding several possible solutions
    —Thinking about trends
    —Thinking about the future and drawing up long-range goals
    —Admiring the elegance of inventions, not the details
    —Synthesizing ideas and information to come up with something new
    —Relying on intuition to find solutions
    —Predicting what future technology may look like
    —Looking for alternative ways of arriving at a solution
    —Using pictures instead of words when learning

Based on Lumsdaine, E., and Lumsdaine, M., Creative Problem Solving. New York: McGraw-Hill, 1995 and NSF Sponsored Workshop on Introductory Engineering Design, J. Nee (editor). Central Michigan University, 1999.

To find out the totals you need to

1. Multiply the total number in box 14 by 4, and then add the results to the I bracket. Once you have the total sum, call it A.

2. Multiply the total number in box 15 by 4, and then add the results to the II bracket. Once you have the total sum, call it B.

3. Multiply the total number in box 16 by 4, and then add the results to the III bracket. Once you have the total sum, call it C.

4. Multiply the total number in box 17 by 4, and then add the results to the IV bracket. Once you have the total sum, call it D.

Sample Result

The following is a sample of one student's questionnaire results:

[1]	[3]	[4]	[2]
[3]	[4]	[1]	[2]
[4]	[3]	[1]	[2]
[2]	[1]	[4]	[3]
[4]	[1]	[3]	[2]
[4]	[2]	[3]	[1]
[4]	[2]	[3]	[1]
[3]	[4]	[1]	[2]
[3]	[4]	[2]	[1]
[4]	[1]	[3]	[2]
[3]	[2]	[4]	[1]
[4]	[1]	[3]	[2]
[3]	[4]	[1]	[2]
I	II	III	IV

- Total [42] [32] [33] [23]
- 14 Total [7]
- 15 Total [5]
- 16 Total [4]
- 17 Total [4]

The sum is expressed as

$A = (4 \times 7) + 42 = 70$

$B = (4 \times 5) + 32 = 52$

$C = (4 \times 4) + 33 = 49$

$D = (4 \times 4) + 23 = 39$

To express the finding graphically, use an Excel spreadsheet. Then use the radar chart within Excel to express the values graphically, as shown in Figure L2.1. People with an A preference are analytical, rational, technical, logical, factual, and quantitative. B-preference people are procedural, scheduled, conservative, organized, sequential, reliable, tactical, and administrative. C-preference students are supportive, interpersonal, expressive, sensitive, symbolic, musical, and reaching out. D-preference thinkers are more strategic, visual, imaginative, conceptual, and simultaneous.

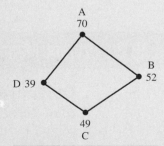

Figure L2-1 Sketch of student's thinking preference.

Discussion

Apparently, the student shown in the example has a strong preference for the A quadrant—analytical. Studies have shown that engineering faculty are predominantly A thinkers. Students change by virtue of practice to A thinkers by the end of their careers. Design students are generally D thinkers.

Everyone is creative and can learn to be more creative. Everyone can learn to use the D-quadrant thinking abilities of the brain more frequently and more effectively. So don't assume that creativity can't be learned and only a few have that ability. Another false assumption is the perception that an intelligent mind is a good thinker. According to Edward de Bono, highly intelligent people who are not properly trained may be poor thinkers for a number of reasons.

1. They can construct a rational, well-argued case for any point of view and thus do not see the need to explore alternatives.

2. Because verbal fluency is often mistaken for good thinking, they learn to substitute one for the other.

3. Their mental quickness leads them to jump to conclusions from only a few data points.

4. They mistake understanding with quick thinking and slowness with being dull witted.

5. The critical use of intelligence is usually more satisfying than the constructive use. Proving someone wrong gives instant superiority but doesn't lead to creative thinking.

Lab 2 Problems

1. Which quadrant would you consider a dominant quadrant for the following well-known figures?

   a. Newton
   b. Galileo
   c. Einstein
   d. Leonardo da Vinci
   e. Malcolm X
   f. Johann Sebastian Bach
   g. Martin Luther King, Jr.
   h. George W. Bush
   i. Bill Clinton
   j. Ben Franklin
   k. Gandhi

2. Would you be able to change your thinking style? Why or why not?

3. If you are organizing a brainstorm session, which thinking styles would you like to have in the session and why?

4. Which role would you like to play in a team? Justify your answer based on your findings in the questionnaire

LAB 3: Project Management (Microsoft Project)[2]

Objectives

In this lab we will introduce some main concepts of project management, and some techniques to control and monitor a project by using Microsoft Project 2016 software. First, some definitions:

- *Management*: The process of getting things done through the effort of other people by planning, organizing, and controlling these processes.

---

[2]Originally developed by Dr. Adnan Bashir, further updated by Ahmad Hayek

- *The project*: A complex effort usually less than three years in duration, made up of interrelated tasks performed by various organizations within well-defined objectives, schedule, and budget.

The objective of the project management is to achieve proper control of the project to assure its completion on schedule and within budget and to achieve the desired quality of the resulting product or services.

To be successful as a project manager requires that you complete your projects on time, finish within budget, and make sure your customers are happy with what you deliver. That sounds simple enough, but how many projects have you heard of (or worked on) that were completed late or cost too much or failed to meet the needs of the customers?

The Project Triangle: Seeing Projects in Terms of Time, Cost, and Scope

This theme has many variations, but the basic idea is that every project has some element of a time constraint, has some type of budget, and requires some amount of work to complete (Figure L3.1). In other words, it has a defined scope. The term *constraint* has a specific meaning in Microsoft Project, but here we're using the more general meaning of a limiting factor. Let's consider these constraints one at a time.

Time

Have you ever worked on a project that had a deadline? (Maybe we should ask, have you ever worked on a project that did not have a deadline?) Limited time is the one constraint of any project with which we are all probably most familiar. If you're working on a project right now, ask your team member what the project deadline is. They might not know the project budget or the scope of work in great detail, but chances are they all know the project deadline.

Cost

You might think of cost as dollar amount, but *project cost* has a broader meaning: *Cost* refers to all the resources required to carry out the project. It includes the people and equipment who do the work, the materials they use, and all the other events and issues that require money or someone's attention within a project.

Scope

You should consider two aspects of scope: *product scope* and *project scope*. Every successful project produces a unique product—a tangible item or a service. You might develop some products for one customer you know by name. You might develop other products for

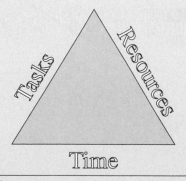

Figure L3.1 The Project Triangle

millions of potential customers waiting to buy them (you hope). Customers usually have some expectations about the features and functions of products they consider purchasing. Product scope describes the intended quality, features, and functions of the product—often in minute detail. Documents that outline this information are sometimes called product specifications. A service or an event usually has some expected features as well. We all have expectations about what we'll do or see at a party, a concert, or the Olympic Games.

Project scope, on the other hand, describes the work required to deliver a product or a service with the intended product scope. Although product scope focuses on the customer or the user of the product, project scope is mainly the measure expressed in terms of *tasks* and *phases*.

A Typical Project Life Cycle

All projects can be described using a four-phase life cycle, as shown in Figure L3.2.

In the first phase, a need is identified by the client, customer, or another stakeholder. This results in a process of describing and defining the needs and requirements, sometimes soliciting information or proposals from vendors, contractors, or consultants. We can call this phase *initiation*.

The second phase is characterized by the development of proposed solutions. This can be initiated by a structured bid from which requests are specific items of information related to project costs, staffing, timescales, description of the activities, compliance with technical standards, and key deliverables.

The third phase occurs when the project is actually executed covering detailed planning and implementation.

The final phase involves terminating the project—or *closure*. In some cases this is marked by formal acceptance by the customer or the client with signed documentation.

Figure L3.3 illustrates how the concept of project phases is incorporated into a new product development methodology.

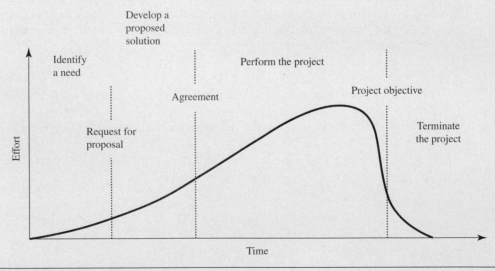

Figure L3.2 Four-Phase Project Cycle

Figure L3.3 Illustration of project concept maturation to product

Reasons for Project Planning

- To eliminate or reduce the uncertainty
- To improve the efficiency of the operation
- To obtain a better understanding of the project's objectives
- To provide a basis for monitoring and controlling work
- To establish directions for the project team
- To support the objectives of the parent organization
- To make allowance for risk
- To put controls on the planned work

If the task is not well understood, then during the actual task execution more knowledge is learned, which leads to change in resource allocation schedules and properties. While if the task is well-understood prior to being performed, then work can be preplanned.

Project Scheduling

Project scheduling involves sequencing and allotting time to all project activities. Managers decide how long each activity will take and compute the resources (manpower, equipment, and materials) needed for each activity.

One popular project scheduling approach is the Gantt chart. **Gantt charts** are planning charts used to schedule resources and allocate time. Gantt charts are an example of a widely used, nonmathematical technique that is very popular with managers because it is so simple and visual. The major discrepancy with Gantt charts is the inability to show the interdependencies between events and activities.

The purposes of project scheduling are the following.

- Shows the relationship of each activity to others and to the whole project.
- Identifies the precedence relationships among activities.
- Encourages the setting of realistic time and cost estimates for each activity.
- Helps make better use of people, money, and material resources by identifying critical bottlenecks in the project.

Interdependencies are shown through the construction of networks. Network analysis can provide valuable information for planning, scheduling, and resource management. Program Evaluation and Review Technique (PERT) and Critical Path Method (CPM) are very popular and widely used network techniques for large and complex projects.

PERT: A technique to enable managers to schedule, monitor, and control large and complex projects by employing three time estimates for each activity.

CPM: A network technique using only one time factor per activity that enables managers to schedule, monitor, and control large and complex projects.

PERT and CPM were developed in 1958 and 1959, respectively, and spread rapidly throughout all industries. Mainly, CPM was concentrated in the construction and process industries.

Framework of PERT and CPM

- Define the project and prepare a work break structure (WBS).
- Develop the relationship between the activities.
- Draw the network connecting all the activities.
- Assign time/cost estimate to each activity.
- Compute the longest path through the network (**critical path**).
- Use the network to help plan, schedule, monitor, and control the project.

Example

Consider the following list of activities with its duration and predecessors. Find the critical path, and then draw the network, as in Figure L3.4.

Computerized PERT/CPM reports and charts are widely available today on personal computers. Some popular software are Primavera™ (by Primavera Systems, Inc.), MS Project™ (by Microsoft Corp.), MacProject™ (by Apple Computer Corp.), Pertmaster™ (by Westminster Software, Inc.), VisiSchedule™ (by Paladin Software Corp.), and Time Line™ (by Symantec Corp.). In this chapter we will use MS Project 2016™.

Activity	Predecessor	Duration
a	—	5 days
b	—	4
c	a	3
d	a	4
e	a	6
f	b, c	4
g	d	5
h	d, e	6
i	f	6
j	g, h	4

Using MS Project 2016[3]

The first step to start the project is to build the project calendar that it will be used in the project. Figure L3.5 shows the standard calendar, which is the default in MS Project 2016.

---

[3]The MS 2007 example was developed by Adnan Bashir. Ahmad Hayek updated the example to MS 2016

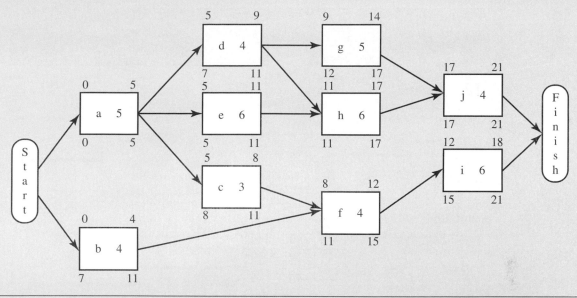

Figure L3.4 The Critical Path and Time for the Project Is a-e-h-j with Completion Time of 21 Days

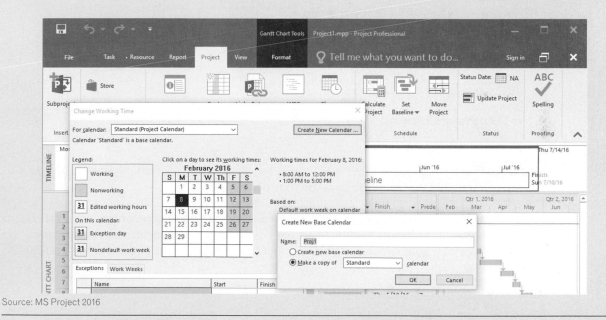

Source: MS Project 2016

Figure L3.5 Standard Project Calendar

You can create a new calendar to suit our project working days, vacations, number of shifts, etc. To create a new calendar, follow the following steps.

STEP 1 Go to the **Project Tab.**

STEP 2 Go to **Change Working Time.**

STEP 3 Select **Create New Calendar.**

STEP 4 Insert the new name of the calendar (proj1).

STEP 5 Choose **Copy from Standard**

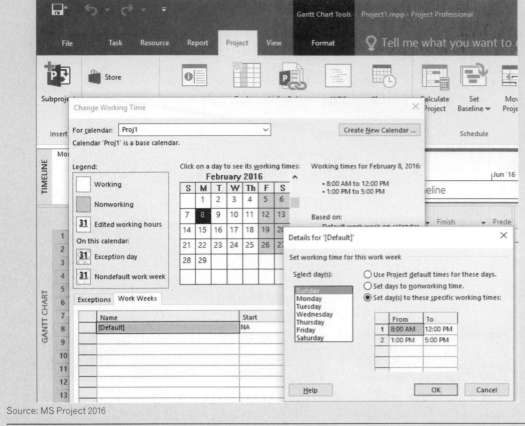

Source: MS Project 2016

Figure L3.6 Proj.1. Calendar

Select the **Work weeks** tab, click on [default], select details, and accordingly change the default working timing and non-working days for your project.

You will get a new calendar called proj1, as shown in Figure L3.6. You can easily change the nonworking days to Thursday and Friday by assigning them to nonworking time and assign the Saturday and Sunday to default (working days). Also, you can increase or decrease the number of working hours in a day by using nondefault working time and fill in for appropriate working hours.

Entering Tasks and Subtasks

When you have identified the majority of the tasks in the project, preferably in a work break structure (WBS), you should enter them into Microsoft Project. Do not worry about putting the tasks in the exact correct order in the beginning. It is easy to reorganize them later on.

You can enter a task using one of the following methods.

1. Click the correct row in the **Task Name** column and enter the name of the Task.
2. Write the name of the Task in the **Entry Bar.** Click the correct row in the **Task Name** column and press the green check box next to the **Entry Bar.**

After you have entered the major tasks of the project it is time to add the details in terms of subtasks. When adding subtasks the upper-level task becomes a summary task. This is a good way of structuring the project and displaying information easily. You are able to have up to nine levels of tasks. You add a subtask (Figure L3.7) using the following steps:

1. Click the row below the task that you would like to become the summary task in the task name column.
2. Write the name of the task and create a new task.
3. Click on the ID number of the row to mark the entire row.
4. Right click on the row and choose indent, or simply click (Shift, Alt, Right) to indent your task.

Once you set up the calendar to your project, you easily can now enter the tasks and the duration of your project, as shown in Figure L3.8

Once the task name and duration have been entered, you can insert the relation between the tasks by inserting a precedence column; then enter the task ID to identify

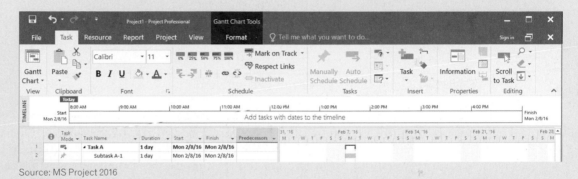

Source: MS Project 2016

Figure L3.7 The summary task is in bold and the subtask is in indent. The summary task is displayed as a black bar on the Gantt chart.

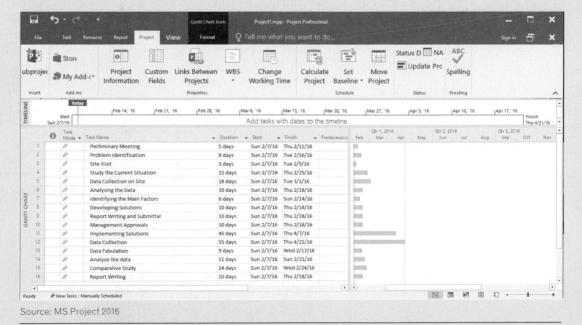

Source: MS Project 2016

Figure L3.8 Task Name and Duration

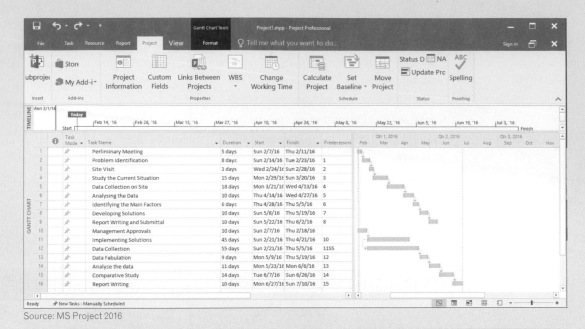

Source: MS Project 2016

Figure L3.9 Gantt chart: Task name and duration

the task. Figure L3.9 shows the task, duration, and the relationship between the tasks. The first activity shows a task summary with the duration covering all the subtasks. The arrows between the bars represent the relation between the tasks—in this case the relationship is Finish to Start (FS); it means that the activity can not start until the preceding activity has been completed.

Figure L3.9 shows the Gantt chart (scaled time bar chart for each task), task name, task duration, start, finish, and precedence. If you did not specify the start date of the project, MS Project 2016 will consider the current date as the project start date (as default). This Gantt chart is commonly used for most of the projects because it is simple, easy to understand, and easy to use to extract a summary information for the project. You can add any information to appear in the Gantt chart, such as Resources, Float (Slack), Percentage completion, Actual start, Actual finish, Late start, Late finish, Actual cost, Baseline cost, Milestones, Remaining duration, Remaining cost, etc.

All of these can be added by

1. Going to the **task** tab
2. Under the **properties** group clicking **Information** (or simply double-clicking the task name to open it directly).

If you double-click on any task, a new window will appear. This window contains the following sub-windows:

- *General*: Gives the name, duration, the percentage completion, and Start and Finish dates of that task.
- *Predecessors*: All the predecessors of this task will appear under this window showing its ID, Name, Type, and the Lag ID available.

Resources: You can assign the resources required to complete the task by entering the resource name and how many units are needed.

- *Advanced*: This will give us the ability to insert time constraints, the task type, and the calendar. It shows also the work break structure (WBS) of the task.

Figure L3.10 shows the task information window for the third task. This task information window consists of the following buttons, each button contains full details of that button:

- General
- Predecessors
- Resources
- Advanced
- Notes
- Custom fields

To enter the resources of the project the following should be followed: Go to the *resource* tab and under *view* group click on the **Team planner** and select **Resource sheet** from the drop-down menu as shown in Figure L3.11.

Start to fill the sheet as shown in Figure L3.12.

1. Go to **Resources**; assign resources.
2. Pick the resource and units for that task.

There is another way to assign the resources to tasks. Once you define the resources, go to the Gantt chart. Insert a new column "Resources." For each single task enter the resources

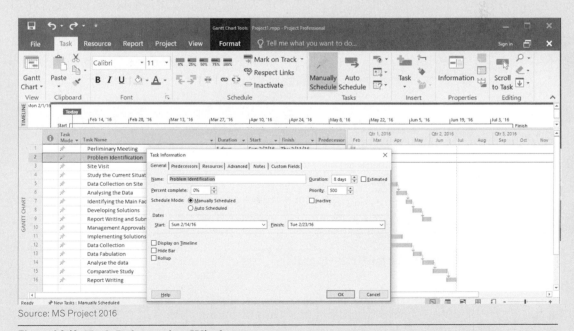

Source: MS Project 2016

Figure L3.10 Task Information Window

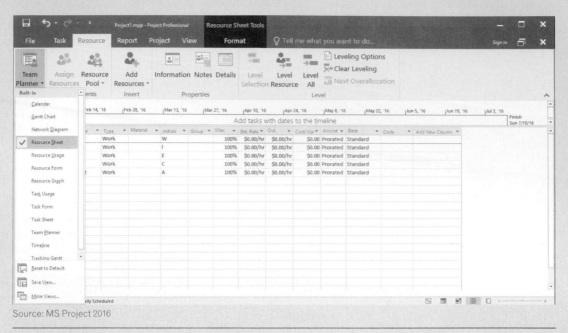

Source: MS Project 2016

Figure L3.11 Resource Sheet

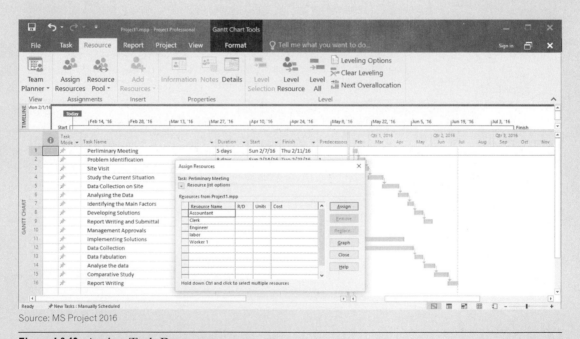

Source: MS Project 2016

Figure L3.12 Assign Task Resources

needed to complete that task You are also able to see the planned workload for the different resources by clicking the **Graphs** button. You will then see the following dialog box (Figure L3.13).

The gray bars in Figure L3.13 indicate the selected task, and the blue bar to the far right shows another task. This graph is useful if you are uncertain whether a specific resource is available for the task or not.

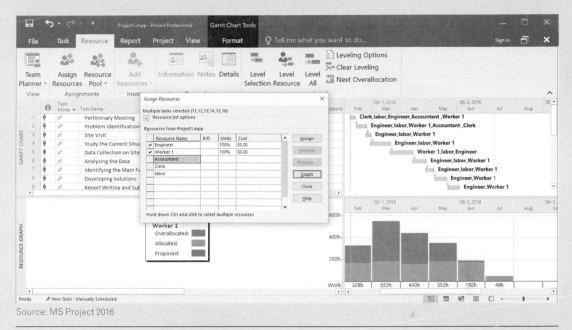

Source: MS Project 2016

Figure L3.13 Graph Dialog Box

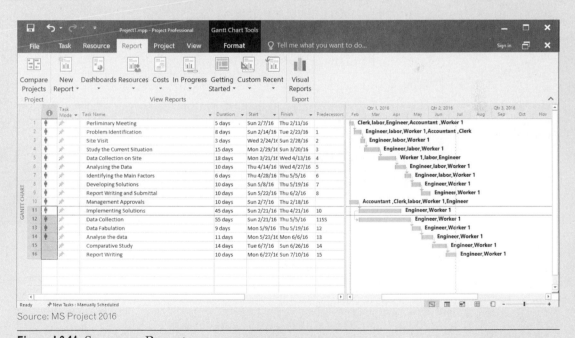

Source: MS Project 2016

Figure L3.14 Summary Report

Reports in MS Project 2016

To obtain the output and the reports, MS Project 2016 provides several predefined reports, as shown in Figure L3.14.

In addition, there is an option to customize any report you choose; for example: click on the *report* tab and all reports can be found under the **View reports** group.

Example

Consider the following list of activities for a project of building a small house.

Activity	Description	Predecessor	Duration (days)
A	Clear site	—	1
B	Bring utilities to site	—	2
C	Excavate	A	1
D	Pour foundation	C	2
E	Outside plumbing	B,C	6
F	Frame house	D	10
G	Electric wiring	F	3
H	Lay floor	G	1
I	Lay roof	F	1
J	Inside plumbing	E,H	5
K	Shining	I	2
L	Outside sheathing insulation	F,J	1
M	Install windows / outside doors	F	2
N	Brick work	L, M	4
O	Install walls and ceiling	G,J	2
P	Cover walls and ceiling	O	2
Q	Install roof	I,P	1
R	Finish interior	P	7
S	Finish exterior	I,N	7
T	Landscape	S	3

- Determine the ES, LS, EF, and LF.
- Determine the Total Float.
- Determine the critical path and the total project duration.

Solution

- For this example, you will keep the default calendar (Standard).
- The project will start on 2/8/16. To enter this information
  a. Go to the **Project** tab.
  b. Select project information under the **Properties Group** and put Start date of the project.

- Enter the task name, duration, and the predecessor for each task in MS Project software.
- Put the grid line for the major and minor columns.
  a. Click the right button of the mouse while the cursor is in the bar chart area.
  b. Select the gridlines option.
  c. A window will pop up allowing you to choose any gridline and edit to your own liking.
- In order to get a full view of the Gantt chart:
  a. Go to **view** tab.
  b. Under the **Zoom group** click on **Entire project**.

The outputs of these steps are shown in Figure L3.15.

- To find the ES, EF, LS, LF, and the Total Float:
  a. Click the Gantt Chart until a **format** tab appears on the top of the screen.
  b. Open the **format** tab and under the **columns** group click **insert column** and choose **Early Start.**
  c. Repeat for the rest of the columns.
- To show the critical path and the critical activities:
  a. Click on the **format** tab.
  b. Under bar styles group check the **Critical tasks** check box.
- You can change the color by clicking on **format** under the same group, a window should pop up, search for **Critical** in the window and change the color to red. Press OK.

Note that the critical tasks should have total slack = 0. The outputs are shown in Figure L3.16.

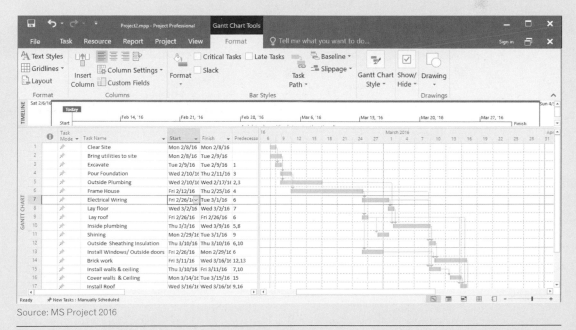

Source: MS Project 2016

Figure L3.15 Gantt Chart for Lab 3 Example

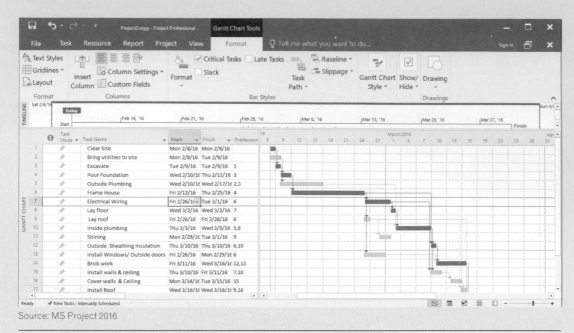

Source: MS Project 2016

Figure L3.16 Gantt Chart with ES, EF, LS, LF, and Total Slack

- The total project duration can be obtained as follows:
  a. Go to the **Project** tab.
  b. Under the **Properties Group** select **Project Information.**
  c. Select **Statistics** on the window that popped up; there will be much more information about the project, as shown in Figure L3.17.

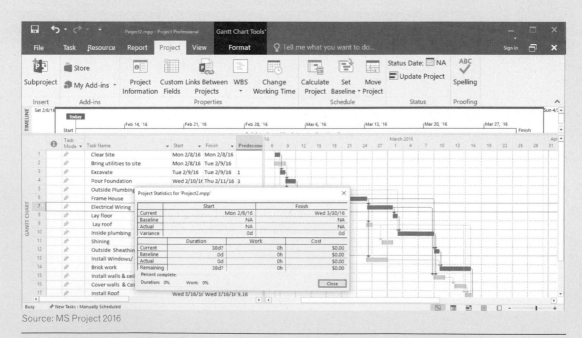

Source: MS Project 2016

Figure L3.17 Project Information Statistics

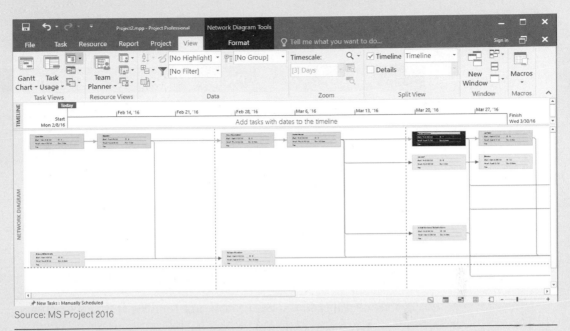

Source: MS Project 2016

Figure L3.18 Part of the Network Diagram for the Lab 3 Example

- To show the network diagram:
  a. Go to the **view** tab.
  b. Under the **Task Views** group select **Network Diagram.**
  c. Use zoom in and zoom out to bring up the desired view.

Figure L3.18 shows the network diagram for the previous example, as well as the critical path on the network with the red color. Each node (box) contains: task name, duration, start date, finish date, and resources. Each node is connected to another node according to the relationship between the tasks.

Lab 3 Problems

1. What are the differences between Gantt, PERT, and CPM?
2. What is the dummy activity, and why do you use it?
3. Why is planning important in projects?
4. What are the main reasons for project scheduling?
5. Construct a project network consisting of 12 activities (from 10 to 120) with the following relationships:
    a. 10, 20, and 30 are the first activities and are begun simultaneously.
    b. 10 and 20 must precede 40.
    c. 20 precedes 50 and 60.
    d. 30 and 60 must precede 70.
    e. 50 must precede 80 and 100.
    f. 40 and 80 must precede 90.
    g. 70 must precede 110 and 120.
    h. 90 must precede 120.
    i. 100, 110, and 120 are terminal activities.

6. Draw the CPM network associated with the following activities for a certain project. How long will it take to complete the project? What are the critical activities?

Activity	Predecessor(s)	Time (min.)
A	—	2
B	A	4
C	A	6
D	B	6
E	B	4
F	C	4
G	D	6
H	E, F	4

7. For Problem 2, find the ES, EF, LS, LF, and Total Float for each activity.

8. ABC Company has a small project with the listed activities below. The manager of the project has been very concerned about the amount of time it takes to complete the project. Some of his workers are very unreliable. A list of activities, their optimistic completion time, the most likely completion time, and the pessimistic completion time (all in days) are given in the following table. Determine the expected completion time and the variance for each activity. Draw the PERT network diagram.

Activity	a	m	b	Predecessor(s)
A	3	6	8	—
B	2	4	4	—
C	1	2	3	—
D	6	7	8	C
E	2	4	6	B, D
F	6	10	14	A, E
G	1	2	4	A, E
H	3	6	9	F
I	10	11	12	G
J	14	16	20	C
K	2	8	10	H, I

9. In Problem 6, the project manager would like to determine the total project completion time and the critical path for the project. In addition, determine the ES, EF, LS, LF, and Total Slack for each activity.

10. In Problem 6, by using MS Project, produce a report showing the list of tasks that have an expected duration of more than five days.

11. Redo Problem 6 using the following calendar:
    a. Working hours per day is 10 hrs.
    b. Working days per week is six days, (only Sunday is off).

References

[1] Mohrman, S. A., and Mohrman, A. M. *Designing and Leading Team-Based Organization: A Workbook for Organizational Self Design.* San Francisco: Jossey-Bass, 1997.

[2] McGoutry J., DeMeuse K. P., *The Team Developer: An Assessment and Skill Building Program Student Guidebook*, Hoboken, NJ: Wiley, 2001.

[3] Larson C. E. and LaFasto F. M. J. *Teamwork: What Must Go Right/What Can Go Wrong*, Thousand Oaks, CA: SAGE Publications, 1989.

[4] DeMeuse, K. P. and Erffmeyer, R. C. "The Relative Importance of Verbal and Nonverbal Communication in a Sales Situation: An Explanatory Study." *Journal of Marketing Management*, 4, 41230, 1994.

[5] Eisenberg M. B. and Berkowitz R. E. *Information Problem Solving: The Big Six Approach to Library and Information Skills*, New York: Ablex Publishing Corp, 1990.

In addition the following books, articles, and websites were used in preparing this Chapter.

BEARD, P. D., and TALBOT, T. F. "What Determines If a Design Is Safe?" *Proceedings of ASME Winter Annual Meeting*, pp. pp. 90–WA/DE-20, New York, 1990.

BLAKE, A. *Practical Stress Analysis in Engineering Design.* New York: Marcel Dekker, 1982.

BURGESS, J. H. *Designing for Humans: The Human Factor in Engineering.* Princeton, NJ: Petrocelli Books, 1986.

BURR, A. C. *Mechanical Analysis and Design.* New York: Elsevier Science, 1962.

COLLINS, J. A. *Failure of Materials in Mechanical Design.* New York: Wiley, 1981.

COOK, N. H. *Mechanics and Materials for Design.* New York: McGraw-Hill, 1984.

CULLUM, R. D. *Handbook of Engineering Design.* London: Butterworth, 1988.

DEUTSCHMAN, A. D. *Machine Design: Theory and Practice.* New York: Macmillan, 1975.

DHILLON, B. S. *Engineering Design: A Modern Approach.* Toronto: Irwin, 1996.

DIESCH, K. H. *Analytical Methods in Project Management.* Ames, Iowa: Iowa State University, 1987.

DIETER, G. E. *Engineering Design: A Material and Processing Approach.* New York: McGraw-Hill, 1983.

DOYLE, L. E. *Manufacturing Processes and Material for Engineers.* Englewood Cliffs, NJ: Prentice Hall, 1985.

DREYFUSS, H. *The Measure of Man: Human Factors in Design.* New York: Whitney Library of Design, 1967.

EISENBERG, M. B. and Berkowitz, R. E. Big6™. http://www.big6.com.

FREDERICK, S. W. "Human Energy in Manual Lifting." *Modern Materials Handling*, Vol. 14, pp. 74–76, 1959.

FURMAN, T. T. *Approximate Methods in Engineering Design.* New York: Academic Press, 1980.

GLEGG, G. L. *The Development of Design.* Cambridge, UK: Cambridge University Press, 1981.

GRANDJEAN, E. *Fitting the Task of the Man: An Ergonomic Approach*. London: Taylor and Francis, 1980.

GREENWOOD, D. C. *Engineering Data for Product Design*. New York: McGraw-Hill, 1961.

HAJEK, V. *Management of Engineering Projects*. New York: McGraw-Hill, 1984.

HILTON, J. R. *Design Engineering Project Management: A Reference*. Lancaster, PA: Technomic Publishing Co., 1985.

HINDHEDE, A. *Machine Design Fundamentals*. New York: Wiley, 1983.

HUBKA, V., and EDER, W. E. *Principles of Engineering Design*. London: Butterworth Scientific, 1982.

LUPTON, T. *Human Factors: Man, Machine and New Technology*. New York: SpringerVerlag, 1986.

MCCORMICK, E. J. and SANDERS, M. S. *Human Factors in Engineering and Design*. New York: McGraw-Hill, 1982.

MEREDITH, D.D. *Design and Planning of Engineering Systems*. Englewood Cliffs, NJ: Prentice Hall, 1985.

MOHRMAN S. A. and MOHRMAN A. M. *Designing and Leading Team-Based Organization: A Workbook for Organizational Self Design*. San Francisco: Jossey-Bass 1977.

OBORNE, D. J. *Ergonomics at Work*. New York: Wiley, 1982.

PAPANEK, V. *Design for Human Scale*. New York: Van Nostrand Reinhold, 1983.

PHEASANT, S. T. *Bodyspace: Anthropometry, Ergonomics and Design*. London: Taylor and Francis, 1986.

SCHMIDTKE, H. *Ergonomic Data for Equipment Design*. New York: Plenum Press, 1985.

SHIGLEY, J. E., AND MISCHKE, C. R. *Mechanical Engineering Design*. New York: McGraw-Hill, 1983.

WOODSON, W. E. *Human Factors Design Handbook*. New York: McGraw-Hill, 1981.

CHAPTER 3

ETHICS AND MORAL FRAMEWORK

"There is not...a more dangerous trait than the deification of mere smartness unaccompanied by any sense of moral responsibility."

~Theodore Roosevelt

Before you delve into the rest of this book, it is important to understand how you, as a student, will develop into a professional engineer. This will influence the way you deal with professional issues. Understanding the concept of being a professional engineer may sound easy, but adopting it and living by it involves far more depth and effort.

3.1 OBJECTIVES

By the end of this chapter, you should be able to

1. Identify codes of ethics.
2. Define professionalism.
3. Apply ethical and moral frameworks to ethical dilemmas.

3.2 PROFESSIONALISM

Professionalism is a way of life. A professional person is one who engages in an activity that requires a specialized and comprehensive education and is motivated by a strong desire to serve humanity. The work of engineers generally affects the day-to-day life of all humans. Developing a professional frame of mind begins with your engineering education. Being a professional should imply that in addition to providing the specialized work, the professional engineer should provide such services with honesty, integrity, and morality. In this spirit, many professional communities have tried to be more explicit about the codes of ethics and have developed rules and guidelines to adhere to. Providing a set of rules to be followed in all circumstances is not as straightforward as it might seem, as some of the rules will be problem-, profession-, and situation-dependent. Inevitably, other authors have discussed the ethical implications of the profession as a set of moral values that are associated with culture and religion. With a subject like this, debates will continue, and there will possibly never be a definitive set of rules that the entire world can agree on. However, several engineering societies have developed a code of ethics that must be followed by its member engineers. This serves as an acceptable compromise, and from time to time these codes are reviewed and updated when necessary.

In such a small introduction, it is immediately apparent that ethics is a complex and still emerging subject. Because of its importance, we have designed a Lab that deals with ethics (**Lab 4: Ethics**) and presents several case studies for examination, discussion, and debate. Students should do this Lab immediately after covering this chapter. For additional reading, several online resources deal with ethics, including

- http://onlineethics.org

3.3 NSPE CODE OF ETHICS

As mentioned in the previous section, many engineering societies now include their version of a code of ethics to which their members must adhere. This book refers to the code of ethics promulgated by the National Society of Professional Engineers (NSPE).[1] It is updated from time to time, and this section refers to the January 2006 *Code of Ethics*. Other engineering societies have very similar codes of ethics, which vary in style or wording, but all invariably require that engineers uphold and advance the integrity, honor, and dignity of the engineering profession by

---

[1] Reprinted by Permission of the National Society of Professional Engineers (NSPE) www.nspe.org.

- Using their knowledge and skill for the enhancement of human welfare.
- Being honest and impartial, and serving with fidelity the public, their employers, and clients.
- Striving to increase the competence and prestige of the engineering profession.

The NSPE Code of Ethics is divided into three main sections:

1. *The Fundamental Canons*: These are the main issues that govern a professional engineer from an ethical and professional standing.
2. *Rules of Practice*: This section discusses the first five points of the fundamental canons in more detail.
3. *Professional Obligations*: This section discusses the final point of the fundamental canons in more detail and is focused on professional conduct from a legal, ethical, and societal viewpoint.

The remaining part of this chapter reprints excerpts from the NSPE *Code of Ethics*. Students should read and discuss the points raised and try to think of examples whereby they could be applied in the workplace.

3.3.1   The Fundamental Canons

While fulfilling their professional duties, engineers shall

1. Hold paramount the safety, health, and welfare of the public.
2. Perform services only in areas of their competence.
3. Issue public statements only in an objective and truthful manner.
4. Act for each employer or client as faithful agents or trustees.
5. Avoid deceptive acts.
6. Conduct themselves honorably, responsibly, ethically, and lawfully so as to enhance the honor, reputation, and usefulness of the profession.

3.3.2   Rules of Practice

1. Engineers shall hold paramount the safety, health, and welfare of the public.
   a. If engineers' judgment is overruled under circumstances that endanger life or property, they shall notify their employer or client and such other authority as may be appropriate.
   b. Engineers shall approve only those engineering documents that are in conformity with applicable standards.
   c. Engineers shall not reveal facts, data, or information without the prior consent of the client or employer except as authorized or required by law or this Code.
   d. Engineers shall not permit the use of their names or associates in business ventures with any person or firm that they believe is engaged in fraudulent or dishonest enterprise.
   e. Engineers shall not aid or abet the unlawful practice of engineering by a person or firm.
   f. Engineers having knowledge of any alleged violation of this Code shall report thereon to appropriate professional bodies and, when relevant, also to public authorities, and cooperate with the proper authorities in furnishing such information or assistance as may be required.

2. Engineers shall perform services only in the areas of their competence.
   a. Engineers shall undertake assignments only when qualified by education or experience in the specific technical fields involved.
   b. Engineers shall not affix their signatures to any plans or documents dealing with subject matter in which they lack competence, nor to any plan or document not prepared under their direction and control.
   c. Engineers may accept assignments and assume responsibility for coordination of an entire project and sign and seal the engineering documents for the entire project, provided that each technical segment is signed and sealed only by the qualified engineers who prepared the segment.

3. Engineers shall issue public statements only in an objective and truthful manner.
   a. Engineers shall be objective and truthful in professional reports, statements, or testimony. They shall include all relevant and pertinent information in such reports, statements, or testimony, which should bear the date indicating when it was current.
   b. Engineers may express publicly technical opinions that are founded upon knowledge of the facts and competence in the subject matter.
   c. Engineers shall issue no statements, criticisms, or arguments on technical matters that are inspired or paid for by interested parties, unless they have prefaced their comments by explicitly identifying the interested parties on whose behalf they are speaking, and by revealing the existence of any interest the engineers may have in the matters.

4. Engineers shall act for each employer or client as faithful agents or trustees.
   a. Engineers shall disclose all known or potential conflicts of interest that could influence or appear to influence their judgment or the quality of their services.
   b. Engineers shall not accept compensation, financial or otherwise, from more than one party for services on the same project, or for services pertaining to the same project, unless the circumstances are fully disclosed and agreed to by all interested parties.
   c. Engineers shall not solicit or accept financial or other valuable consideration, directly or indirectly, from outside agents in connection with the work for which they are responsible.
   d. Engineers in public service as members, advisors, or employees of a governmental or quasi-governmental body or department shall not participate in decisions with respect to services solicited or provided by them or their organizations in private or public engineering practice.
   e. Engineers shall not solicit or accept a contract from a governmental body on which a principal or officer of their organization serves as a member.

5. Engineers shall avoid deceptive acts.
   a. Engineers shall not falsify their qualifications or permit misrepresentation of their or their associates' qualifications. They shall not misrepresent or exaggerate their responsibility in or for the subject matter of prior assignments. Brochures or other presentations incident to the solicitation of employment shall not misrepresent pertinent facts concerning employers, employees, associates, joint ventures, or past accomplishments.
   b. Engineers shall not offer, give, solicit, or receive, either directly or indirectly, any contribution to influence the award of a contract by public authority, or which may be reasonably construed by the public as having the effect or intent of influencing the

awarding of a contract. They shall not offer any gift or other valuable consideration in order to secure work. They shall not pay a commission, percentage, or brokerage fee in order to secure work, except to a bona fide employee or bona fide established commercial or marketing agencies retained by them.

3.3.3 Professional Obligations

1. Engineers shall be guided in all their relations by the highest standards of honesty and integrity.
   a. Engineers shall acknowledge their errors and shall not distort or alter the facts.
   b. Engineers shall advise their clients or employers when they believe a project will not be successful.
   c. Engineers shall not accept outside employment to the detriment of their regular work or interest. Before accepting any outside engineering employment, they will notify their employers.
   d. Engineers shall not attempt to attract an engineer from another employer by false or misleading pretenses.
   e. Engineers shall not promote their own interest at the expense of the dignity and integrity of the profession.
2. Engineers shall at all times strive to serve the public interest.
   a. Engineers shall seek opportunities to participate in civic affairs; career guidance for youths and work for the advancement of the safety, health, and well-being of their community.
   b. Engineers shall not complete, sign, or seal plans and/or specifications that are not in conformity with applicable engineering standards. If the client or employer insists on such unprofessional conduct, they shall notify the proper authorities and withdraw from further service on the project.
   c. Engineers shall endeavor to extend public knowledge and appreciation of engineering and its achievements.
   d. Engineers shall strive to adhere to the principles of sustainable development in order to protect the environment for future generations.
3. Engineers shall avoid all conduct or practice that deceives the public.
   a. Engineers shall avoid the use of statements containing a material misrepresentation of fact or omitting a material fact.
   b. Consistent with the foregoing, engineers may advertise for recruitment of personnel.
   c. Consistent with the foregoing, engineers may prepare articles for the lay or technical press, but such articles shall not imply credit to the author for work performed by others.
4. Engineers shall not disclose, without consent, confidential information concerning the business affairs or technical processes of any present or former client, employer, or public body on which they serve.
   a. Engineers shall not, without the consent of all interested parties, promote or arrange for new employment or practice in connection with a specific project for which the engineer has gained particular and specialized knowledge.
   b. Engineers shall not, without the consent of all interested parties, participate in or represent an adversary interest in connection with a specific project or proceeding in which the engineer has gained particular specialized knowledge on behalf of a former client or employer.

5. Engineers shall not be influenced in their professional duties by conflicting interests.
   a. Engineers shall not accept financial or other considerations, including free engineering designs, from material or equipment suppliers for specifying their product.
   b. Engineers shall not accept commissions or allowances, directly or indirectly, from contractors or other parties dealing with clients or employers of the engineer in connection with work for which the engineer is responsible.

6. Engineers shall not attempt to obtain employment or advancement or professional engagements by untruthfully criticizing other engineers, or by other improper or questionable methods.
   a. Engineers shall not request, propose, or accept a commission on a contingent basis under circumstances in which their judgment may be compromised.
   b. Engineers in salaried positions shall accept part-time engineering work only to the extent consistent with policies of the employer and in accordance with ethical considerations.
   c. Engineers shall not, without consent, use equipment, supplies, laboratory, or office facilities of an employer to carry on outside private practice.

7. Engineers shall not attempt to injure, maliciously or falsely, directly or indirectly, the professional reputation, prospects, practice, or employment of other engineers. Engineers who believe others are guilty of unethical or illegal practice shall present such information to the proper authority for action.
   a. Engineers in private practice shall not review the work of another engineer for the same client, except with the knowledge of such engineer or unless the connection of such engineer with the work has been terminated.
   b. Engineers in governmental, industrial, or educational employ are entitled to review and evaluate the work of other engineers when so required by their employment duties.
   c. Engineers in sales or industrial employ are entitled to make engineering comparisons of represented products with products of other suppliers.

8. Engineers shall accept personal responsibility for their professional activities, provided, however, that engineers may seek indemnification for services arising out of their practice for other than gross negligence, where the engineer's interests cannot otherwise be protected.
   a. Engineers shall conform with state registration laws in the practice of engineering.
   b. Engineers shall not use association with a non-engineer, a corporation, or partnership as a "cloak" for unethical acts.

9. Engineers shall give credit for engineering work to those to whom credit is due and will recognize the proprietary interests of others.
   a. Engineers shall, whenever possible, name the person or persons who may be individually responsible for designs, inventions, writings, or other accomplishments.
   b. Engineers using designs supplied by a client recognize that the designs remain the property of the client and may not be duplicated by the engineer for others without express permission.
   c. Engineers—before undertaking work for others in connection with which the engineer may make improvements, plans, designs, inventions, or other records that may justify copyrights or patents—should enter into a positive agreement regarding ownership.

d. Engineers' designs, data, records, and notes referring exclusively to an employer's work are the employer's property. The employer should indemnify the engineer for use of the information for any purpose other than the original purpose.

e. Engineers shall continue their professional development throughout their careers and should keep current in their specialty fields by engaging in professional practice, participating in continuing education courses, reading in the technical literature, and attending professional meetings and seminars.

3.4 THEORY—CODE OF ETHICS AND MORAL FRAMEWORKS

Although strongly related to each other, there is a distinct difference between the terms "ethics" and "morals." In most cases, both are needed together to make a well-balanced judgment. Ethics relates to the philosophy behind a moral outcome and determines the working of a social system. These are usually presented as a set of rules that dictate right or wrong behavior. The NSPE Code of Ethics is one such example. The Code of Ethics alone, though, does not cover everything, and at times individual canons can conflict with each other. For example, fundamental canon 4 in the NSPE code of ethics states: "Act for each employer or client as faithful agents or trustees." This dictates that you should remain faithful to your employer. However, in some cases your employer may be deceiving the public, selling a product they claim performs a certain function whereas it actually does not. This will be in direct conflict with fundamental canon 5: "Avoid Deceptive Acts." In this case, what do you do?

Morals, on the other hand, define our character and usually address "appropriate" and "expected" behavior. Morals deal with adopted codes of conduct or frameworks within a given environment, conception, and/or time. Such codes can deal with controversial behavior, prohibitions, standards of belief systems, and social conformity. Moral frameworks can be abstract, and the same outcome can be deemed "appropriate" or "inappropriate" depending on the situation. For example, people can argue that "killing a person is immoral," but during war and on the battlefield maintain that "killing a person is acceptable."

In some cases, moral frameworks are too abstract to point to a conclusive ethical resolution. Parallel to this, ethics can be regarded as an application of morality. This is why both are usually needed together to form a balanced opinion. We will briefly introduce five common moral frameworks or approaches.

3.4.1 The Utilitarian Approach (Utilitarianism)

Some ethicists emphasize that ethical action is the one that provides the most good, does the least harm, or (put another way) produces the greatest balance of good over harm. The ethical corporate action then is the one that produces the greatest good and does the least harm for all who are affected—customers, employees, shareholders, the community, and the environment. The utilitarian approach deals with consequences; it tries both to increase the good done and to reduce the harm done. A typical example usually discussed with this approach is given here.

You are hiking alone in a forest, and you come up to a village where there is a terrorist holding 20 people hostage. The terrorist is about to kill all 20 people, but somehow you convince him not to kill anyone. He agrees as long as you take the gun and kill one of them to prove his point (whatever that may be). If you choose not to kill one person, then he will proceed to kill all 20. There are no other options in this situation. What would you do? Discuss this within your group.

3.4.2 The Rights Approach

This approach bases decision-making on actions that best protect and respect the moral rights of those affected. It begins with the belief that humans have a dignity based on their human nature as well as the ability to choose freely what they do with their lives. On the basis of such dignity, they have a right to be treated as ends and not merely as means to other ends. The list of moral rights, which includes the rights to make one's own choices about what kind of life to lead, to be told the truth, not to be injured, to have a degree of privacy, and so on, is widely debated. Recently, moral development has progressed toward providing rights for nonhuman beings as well.

3.4.3 The Fairness or Justice Approach

This approach is based on contributions from Aristotle and other Greek philosophers and revolves around the idea that all equals should be treated equally. This also suggests that it may be fair to treat all that are unequal unequally. We pay people more based on their harder work or the greater amount that they contribute to an organization and say that is fair. But what about CEO salaries that are hundreds of times larger than the pay of others? Many ask whether the huge disparity is based on a defensible standard or whether it is the result of an imbalance of power and hence is unfair.

3.4.4 The Common Good Approach

Greek philosophers also have contributed the notion that life in a community is good in itself, and our actions should contribute to that life. This approach suggests that the interlocking relationships of society are the basis of ethical reasoning and that respect and compassion for all others—especially the vulnerable—are requirements of such reasoning. This approach also calls attention to the common conditions that are important to the welfare of everyone. This may be a system of laws, effective police and fire departments, health care, a public educational system, or even public recreational areas.

3.4.5 The Virtue Approach

A very ancient approach to ethics is that ethical actions ought to be consistent with certain ideal virtues that provide for the full development of our humanity. These virtues are dispositions and habits that enable us to act according to the highest potential of our character and on behalf of values like truth and beauty. Honesty, courage, compassion, generosity, tolerance, love, fidelity, integrity, fairness, self-control, and prudence are all examples of virtues. Virtues ask of any action, "What kind of person will I become if I do this?" or "Is this action consistent with my acting my best?"

3.4.6 Putting the Approaches Together

Each of these approaches helps us determine which standards of behavior can be considered ethical. Although seemingly straightforward, it unfortunately is not so simple. The first problem is that we may not agree on the content of some of these specific approaches. We may not all agree to the same set of human and civil rights. We may not agree on what constitutes the common good. We may not even agree on what is good and what is harmful.

The second problem is that the different approaches may not all answer the question "What is ethical?" in the same way. Nonetheless, each approach gives us important information with which to determine what is ethical in a particular circumstance. More often than not, the different approaches do lead to similar answers.

3.5 MORAL REASONING AND APPROACHING ETHICAL DILEMMAS

Now that you are aware of both the Code of Ethics and moral frameworks, you can try to tackle some ethical problems. To do that, you will need to follow a process such as the one presented here. The more you practice, the more natural this process will become; ultimately, this should become an integral and habitual aspect of your profession.

STEP 1. Identify the Ethical Dilemma.

    a. Will any of your options be damaging to someone, some group, or some particular thing? Do the potential decisions involve a choice between a good and bad alternative, or between two "goods" or even two "bads?"

    b. What are the relevant facts of the case? What facts are not known? Try to establish a timeline, if relevant. Can you learn more about the situation? Do you know enough to make a decision?

    c. Who are the involved parties? What individuals and groups have an important stake in the outcome? Are some concerns more important? Why?

STEP 2. Identify the Relevant Codes of Ethics.

    a. List all of the relevant codes of ethics that affect both the problems identified in Step 1 as well as the codes relevant to any potential decisions you may make.

    b. List the possible solutions that you may take based on the codes of ethics alone.

STEP 3. Establish any Possible Potential Conflicts within the Identified Relevant Codes of Ethics.

    a. Do any of the identified codes conflict with each other? Group them together.

    b. Is there one that is more important to address over the other? Why?

STEP 4. Evaluate the Moral Frameworks to Resolve Any Conflicting Codes of Ethics.

    a. Which option will produce the most good and do the least harm? (The utilitarian approach)

    b. Which option best respects the rights of all who have a stake? (the rights approach)

    c. Which option treats people equally or proportionately? (the justice approach)

    d. Which option best serves the community as a whole, not just some members? (the common-good approach)

    e. Which option leads you to act as the sort of person you want to be? (the virtue approach)

STEP 5. Make a Recommendation/Decision

    a. Considering the Code of Ethics and the given moral frameworks, which option best addresses the situation? You should structure your recommendation as a well-thought-out argument based on a model such as the one provided by Toulmin (explained next).

    b. Ask yourself: If you told someone you respect or announced to the public the recommendation you have chosen, what would they say?

    c. How can your decision be implemented with the greatest care and attention to the concerns of all stakeholders?

3.5.1 Toulmin's Model for Argumentation and Moral Reasoning

Delving further into these problem-solving steps, you need to be able to reason and argue (both to yourself and to others) the decision you will make and to justify your choices. Stephen Toulmin broke down the structure of an argument into a logical model that includes all of the elements necessary to complete an argument from start to finish. These points should be addressed each time you make a "claim" (as described next) at any stage of the five steps described previously. The model also should be used to present your final argument and decision.

Essential Elements of an Argument:

1. *Claim.* This is your claim or statement that has no merit yet. Most arguments will begin with a claim. (For example, You should buy an aluminum can crusher for your home.)

2. *Data.* These are the facts that one uses as a foundation to later establish validity to the claim. (For example, environmentally, recycling aluminum cans produces 95% less air pollution.)

3. *Warrant.* This is the link made from the data to the claim in order to authorize or validate it. (For example, buying an aluminum can crusher for your home will be better for the environment by causing less air pollution.)

Optional Elements of An Argument:

4. *Backing.* These are facts that can be used to give credit to the warrant. This can be done by providing evidence to the statement made or by making another statement that adds credibility to the warrant. (For example, you told me last week that you wanted to be more responsible to the environment after reading about the new environmental legislation.)

5. *Rebuttal.* These are statements that recognize restrictions or limitations to the claim. (For example, unless of course you have changed your mind about being more responsible to the environment.)

6. *Qualifier* These are words that qualify how certain you are about your claim. Words such as "certain," "probable," and "presumably," and so forth are typical examples. (For example, I am certain that the can crusher will be more friendly to the environment, which is what I presume you want.)

3.6 CHAPTER SUMMARY

1. An engineer is a professional person who engages in activities to serve humanity.

2. Professional organizations have developed code of ethics to guide their members. The professional code of ethics contains fundamental canons, rules of practice, and professional obligation sections.

3. Ethics relates to the philosophy behind a moral outcome and determines the workings of a social system.

4. The five main moral frameworks are the utilitarian approach, the rights approach, the fairness approach, the common-good approach, and virtue approach.

5. The process to follow analyzing ethical dilemmas includes
   a. Identify the dilemma.
   b. Identify the relevant code of ethics.
   c. Establish possible conflicts.

    d. Evaluate the moral framework.

    e. Make a recommendation.

6. The essential elements of an argument are

    a. Claim

    b. Data

    c. Warrant

7. Optional elements of an argument are

    a. Backing

    b. Rebuttal

    c. Qualifier

3.7 PROBLEMS

1. Read through the ASME code of ethics and list three findings.

2. Define professionalism.

3. Discuss the following scenario, which is divided into two parts. Follow the steps described previously for approaching ethical dilemmas, moral reasoning, and presenting an argument. Refer to the NSPE Code of Ethics as well as the appropriate moral frameworks. Justify your reasoning and be prepared to debate your viewpoint with those who disagree with you in the class.

    a. High Concept Manufacturing (HCM) has an engineering plant in a small town that employs 12.4% of the community. It provides approximately $10 million dollars of salaries to its community workers and pays $2 million in taxes to the local government. As a consequence of some of its manufacturing procedures, the HCM plant releases bad-smelling fumes. These fumes annoy HCM's residential neighbors, hurt the local tourism trade, and have been linked (although not conclusively) to a rise in asthma in the area. The financial impact of this to the town is estimated to be around $3 million as a result of a decrease in tourism and lower house prices. The town is considering issuing an ultimatum (final warning) to HCM; "Clean up your plant, or we will fine you $1 million." HCM had previously made it known that the business will close down and go somewhere else if it is fined by the town. What should the town do?

    b. There will be a town meeting that all concerned parties have agreed to attend and discuss the matters at hand. You are an engineer and a respected member of the town who has been recently offered an excellent job opportunity at HCM. You have signed a contract with HCM, and you are officially one of their new employees. However, this is as of yet not public knowledge. HCM asks you to try to convince the town to drop the case and to say that the town is better off with HCM's presence. What are you going to do?

4. The following ethical scenario was obtained from http://www.cwru.edu/wwwethics.

    You are an engineer charged with performing safety testing and obtaining appropriate regulatory agency or outside testing laboratory approvals of your company's product. The Gee Whiz Mark2 (GWM2) has been tested and found compliant with both voluntary and mandatory safety standards in North America and Europe. Because of a purchase-order error and subsequent oversights in manufacture, 25,000 units of GWM2 were built that are not compliant with any of the North American or European safety standards. A user would be more vulnerable to electric shock from one of these units than from a compliant unit. Under some plausible combinations of events, the user of the bad unit could be electrocuted. Retrofitting these products to make them compliant is not feasible because the rework costs would exceed the profit margin by far. The company agrees that because of this defect the agency safety labels will not be attached to the bad units, as per the requirements of the several agencies. Only two options exist: (a) Scrap the units and take the loss, or (b) sell the units. An employee of the company notes that many countries have no safety standards of any kind for this type of product. It is suggested that the bad units be marketed in these countries. It is pointed out that many of these nations have no electrical wiring code; or if codes exist, they are not enforced. The argument is thus advanced that the bad GWM2 units are no worse than the modus operandi of the electrical practice of these countries and their cultural values. Assume that no treaties or export regulations would be violated.

   a. What would be your recommendation?

   b. Suppose one of the countries under consideration was the country of origin for you or your recent ancestors. Would this affect your recommendations?

   c. Now suppose you are not asked for a recommendation, only an opinion. What is your response?

   d. Suppose it is suggested that the bad units be sold to a third party, who would very likely sell the units to these countries. Your comments?

   e. You are offered gratis one of the bad units for your use at home, provided that you sign a release indicating your awareness of the condition of the unit and that it is given to you as a test unit. Assume you cannot retrofit it and that the product could be very useful to you. Would you accept the offer?

   f. Suppose it is suggested that the offer stated in part (e) be made to all employees of the company. Your comments?

5. Discuss the following statements or claims in class. Some are *true* statements; some are *false*, while others may be *either*. What is your opinion and why? Try to remember Toulmin's model when putting your argument forward.

   a. Ethics is not the same as feelings.

   b. Ethics is following the law.

   c. Ethics is not a science.

   d. Ethics is following culturally accepted norms.

   e. In some cases, it may be ethical to make and act on an unethical decision.

   f. Designing and/or selling a gun is unethical.

LAB 4: Ethics and Moral Frameworks

Ethics is a part of all professional careers but plays an extremely important role in engineering. In this Lab you will use the process presented in Chapter 3 to approach and resolve ethical dilemmas that can and will present themselves many times during a design project and your professional career as an engineer. It is important to note that, although there are certain scenarios that are clear cut "ethical" or "unethical," there are other times when this line is not so clear. It is beyond the scope of this book to delve into too much depth, and there are other books that cover this vast area sufficiently. The aim of this Lab is to give you a foundation and an opportunity to identify, discuss, and provide an ethical resolution for ethical issues.

Before you start this Lab, you will need to visit the following Web pages and familiarize yourself with their content:

1. http://www.onlineethics.org/Resources/instructessays/probcase.aspx by Caroline Whitbeck.
2. http://www.onlineethics.org/

Procedure

You will need to submit two reports in the course of this lab.

1. *Report I assignment:* In this report you should (as a group)
   a. List the major points that were discussed in the Caroline Whitbeck paper.
   b. Discuss the paper and point out where you agree or disagree.
   c. Discuss your scenario and report your findings and arguments (the instructor will assign an ethical scenario to your team).
2. *Report II assignment*
   a. Identify experts from your university and local community.
      Interview at least three experts and report your interview discussion.
   b. Analyze the three interviews.
   c. Make a recommendation/final argument based on your opinion and the interviews.

The following are the different scenarios that were obtained from outside sources. The discussion at the end of each scenario will help you get started; do not limit your scope to those questions and discussion points when you report.

Scenario I[2]

You are an engineer charged with performing safety testing and obtaining appropriate regulatory agency or outside testing laboratory ("agency") approvals of your company's product. The Gee-Whiz Mark 2 (GWM2) has been tested and found compliant with both voluntary and mandatory safety standards in North America and Europe. Because of a purchase-order error and subsequent oversights in manufacture, 25,000 units of GWM2 ("bad units") were built that are not compliant with any of the North American or European safety standards. A user would be much more vulnerable to electric shock from a bad unit than from a compliant unit. Under some plausible combinations of events, users of the bad unit could be electrocuted.

Retrofitting these products to make them compliant is not feasible because the rework costs would exceed the profit margin by far. All agree that, because of this defect, the agency safety labels will not be attached to the bad units, as per the requirements of the several agencies. Only two options exist:

1. Scrap the units and take the loss.
2. Sell the units.

An employee of the company notes that many countries have no safety standards of any kind for this type of product. It is suggested that the bad units be marketed in these countries. It is pointed out that many of these nations have no electrical wiring codes; if codes exist, they are not enforced. The argument is thus advanced that the bad GWM2 units are no worse than the modus operandi of the electrical practice of these countries. Assume that no treaties or export regulations would be violated in marketing the bad units to these countries.

---

[2]Scenarios in Business and Engineering Settings by Joseph H. Wujek and Deborah G. Johnson, from http://www.onlineethics.org/cms/7335.aspx. Used by permission.

Discussion

1. What is your recommendation?

2. Suppose one of the countries under consideration was the country of origin for you or your recent ancestors. Would this affect your recommendation?

3. Now suppose you are not asked for a recommendation, only an opinion. What is your response?

4. Suppose it is suggested that the bad units be sold to a third party, who would very likely sell the units to these countries. What is your comment?

5. You are offered gratis one of the bad units for your use at home, provided that you sign a release indicating your awareness of the condition of the unit and that it is given to you as a test unit. (Assume that you can't retrofit it, and that the product could be very useful to you.) Would you accept the offer?

6. Suppose it is suggested that the offer in item 5 be made to all employees of the company. Your comment?

Scenario II[3]

The U.S. Federal Communications Commission (FCC) Rule Part 15J applies to virtually every digital device (with a few exceptions) manufactured in the United States. The manufacturer must test and certify that the equipment does not exceed FCC-mandated limits for the generation of communications interference caused by conducted and radiated emissions. The certification consists of a report sent to the FCC for review. It is largely an honor system, because the FCC has only a small staff to review an enormous number of applications. The FCC then issues a label ID to be attached to each unit that authorizes marketing of the product. Prior to receiving the label the manufacturer cannot offer for sale or advertise the product. An EMC (electromagnetic compatibility) consultant operating a test site installs a new antenna system and finds that it results in E-field measurements consistently higher than those obtained with the old antennas. Both track within the site-calibration limits, and both antenna vendors claim National Institute of Standards and Technology (formerly National Bureau of Standards) traceability. Which system is the better in absolute calibration is thus unknown. There is not enough time to resolve this discrepancy before a client's new product must be tested for FCC Rules Part 15J compliance.

Discussion

1. Which antenna system should be used to test the product?

2. Is averaging the results ethical, assuming that engineering judgment indicates that this procedure is valid?

3. Suppose the site was never properly (scientifically or statistically) calibrated. Should this fact be made known voluntarily to the FCC?

Scenario III[4]

You are an engineer working in a manufacturing facility that uses toxic chemicals in processing. Your job has nothing to do with the use and control of these materials.

The chemical Mega X is used at the site. Recent stories in the news have reported alleged immediate and long-term human genetic hazards from inhalation or other contact with the

---

[3]Scenarios in Business and Engineering Settings by Joseph H. Wujek and Deborah G. Johnson, from http://www.onlineethics.org/cms/7335.aspx. Used by permission.

[4]Scenarios in Business and Engineering Settings by Joseph H. Wujek and Deborah G. Johnson, from http://www.onlineethics.org/cms/7335.aspx. Used by permission.

chemical. The news items are based on findings from laboratory experiments, done on mice, by a graduate student at a well-respected university physiology department. Other scientists have neither confirmed nor refuted the experimental findings. Federal and local governments have not made official pronouncements on the subject. Several employee friends have approached you on the subject and asked you to do something to eliminate the use of MegaX at your factory. You mention this concern to your manager, who tells you, "Don't worry; we have an industrial safety specialist who handles that." Two months pass, and MegaX is still used in the factory. The controversy in the press continues, but there is no further scientific evidence (pro or con) in the matter. The use of the chemical in your plant has increased, and now more workers are exposed daily to the substance than was the case two months ago.

Discussion

1. What, if anything, do you do?
2. Suppose you again mention the matter to your manager and are told, "Forget it; it's not your job." What should you do now?
3. Your sister works with the chemical. What is your advice to her?
4. Your pregnant sister works with the chemical. What is your advice to her?
5. The company announces a voluntary phasing out of the chemical over the next two years. What is your reaction to this?
6. A person representing a local political activist group approaches you and asks you to make available to them company information regarding the amounts of MegaX in use at the factory and the conditions of use. Do you comply? Why or why not?

Scenario IV[5]

The Zilch Materials Corporation employs you as a test engineer. The company recently introduced a new two-component composition-resin casting material, Megazilch, which is believed to have been well tested by the company and a few selected potential customers. All test results prior to committing to production indicated that the material meets all published specifications and is superior in performance and lower in estimated cost than competitors' materials used in the same kinds of applications.

Potential and committed applications for Megazilch include such diverse products as infants' toys, office equipment parts, interior furnishings of commercial aircraft, and the case material for many electronic products. Marketing estimates predict a 25% increase in the corporation's revenues in the first year after the product was shipped in production quantities.

The product is already in production, and many shipments have been made when you discover, to your horror, that under some conditions of storage temperature and other (as yet) unknown factors, the shelf life of the product is seriously degraded. In particular, it will no longer meet specifications for flame retardation if stored for more than 60 days before mixing, instead of the 24 months stated in the published specifications. Its tensile and compressive strengths are reduced significantly as well.

Substantial quantities have been shipped, and the age and temperature history of the lots shipped are not traceable. To recall these would involve great financial loss and embarrassment to the company, and at this point, it is not clear that the shelf life can be improved. Only you and a subordinate, a competent test technician, know of the problem.

Assume that no quick fixes by chemical or physical means are possible and that the problem is real. That is, there are no mistakes in the scientific findings.

---

[5]Scenarios in Business and Engineering Settings by Joseph H. Wujek and Deborah G. Johnson, from http://www.onlineethics.org/cms/7335.aspx. Used by permission.

Discussion

1. What is the first action you would take relevant to this matter?

2. Suppose you express concern to your immediate supervisor, who tells you, "Forget it! It's no big deal, and we can correct it later. Let me handle this."

3. Suppose further that you detect no action after several weeks have passed since you told your supervisor. What now?

4. In item 3, assume you speak to your supervisor, who then tells you, "I spoke to the executive staff about it, and they concur. We'll keep shipping product and work hard to fix it. We've already pulled out all the stops, people are working very hard to correct the problem." What, if anything, do you do?

5. It is now three months since you told your supervisor, and in a test of product sampled from current shipments, you see that no fix has been incorporated. What now?

Scenario V[6]

Marsha is employed as the city engineer by the city of Oz, which has requested bids for equipment to be installed in a public facility. Oz is bound by law to purchase the lowest bid that meets the procurement specifications except "for cause." The low bidder, by a very narrow margin, is Diogenes Industries, a local company. The Diogenes proposal meets the specifications. Marsha recommends purchase of the equipment from Diogenes.

After the equipment has been installed, it is discovered that John, the chief engineer for Diogenes, is the spouse of Marsha. John was the engineer who was in charge of the proposal to Oz, including the final authority on setting the price. As a result of this, Marsha is requested to resign her position for breach of the public trust.

Discussion

1. Was the city justified in seeking Marsha's termination of employment?

2. Suppose Marsha had never been asked to sign a conflict of interest statement. Would this affect your response to question 1?

3. Given the conditions of question 2, suppose Marsha had mentioned, before going to bid, in casual conversation with other persons involved in the procurement, that she was married to the chief engineer at Diogenes. Would this affect your response?

4. Suppose Marsha and John were not married but shared a household. Would this affect your response?

5. Now suppose Marsha had made known officially her relation to John and the potential for conflict of interest before soliciting bids. Then suppose Marsha rejects the Diogenes bid because she is concerned about the appearance of conflict of interest. She then recommends purchase of the next lowest bid, which meets the specifications. Comment on Marsha's action.

References

The following articles and websites were used in preparing this chapter:

Toulmin, S., *The Uses of Argument. Cambridge*, UK: Cambridge University Press, 1958.

http://www.scu.edu/ethics/practicing/decision/framework.html

http://www.ethicsandbusiness.org/pdf/strategy.pdf

http://ocw.usu.edu/English/intermediate-writing/english-2010/2010/toulmins-sohema

http://www.cwru.edu/wwwethics

---

[6]Scenarios in Business and Engineering Settings by Joseph H. Wujek and Deborah G. Johnson, from http://www.onlineethics.org/cms/7335.aspx. Used by permission.

PART 2

REQUIREMENTS

CHAPTER 4

IDENTIFYING NEEDS AND GATHERING INFORMATION

"*It is a very sad thing that nowadays there is so little useless information.***"**

~Oscar Wilde

Many statements can be classified as statements of fact—for example, "many people have itching in their backs which they cannot scratch." Design problems or *needs* can be derived from such statements of fact to form imperatives, such as "identify and develop a chemical that will eliminate itching" or "design and develop a back scratcher that would enable scratching the back." The *need* thus is a description (what it is) of a solution to a problem derived from a statement of fact. A *design brief* best describes a need for a design solution. However, the need has to be considered and chosen among the various possibilities (such as the chemical and mechanical solution for itch alleviation) as well as alternative solutions offered by others, cost, and the like. The solution has to meet the requirements of various national and international standards. It has to provide information relating to the users. This demands gathering and analyzing information before establishing a need described by a design brief. This chapter describes four main things: (1) the design brief and its elements; (2) standards and standardization; (3) human factors; and (4) the methods employed to gather and use information for identifying needs.

4.1 OBJECTIVES

At the end of this chapter you will be able to

1. Identify and understand statements of facts.
2. Derive a problem statement or need from a statement of facts.
3. Compile and establish the elements of a design brief.
4. Conduct a search to identify and obtain both the relevant standards and the information required to clearly define needs.
5. Conduct a search to identify needs related to human factors.

4.2 PROBLEM DEFINITION: THE DESIGN BRIEF OR NEED STATEMENT

Dictionary meanings of need include inadequacy, insufficiency, and shortage. A societal need is an inadequacy experienced by a society or group of people. Designers try to rectify these inadequacies with products, and the ones that appeal most to society are hailed as successful products. A successful product may lose its appeal after some time when products are introduced that better meet the inadequacy as a result of better insight or technology. Thus societal need is the genesis of all products. Lincoln Steffens [1] wrote, "The world is yours, nothing is done, nothing is known. The greatest poem isn't written, the best railroad isn't built yet, the greatest state hasn't been thought of. Everything remains to be done right, everything." To understand Steffens, consider the following: society needed a calculation aid. The abacus was the first solution. It lasted for a long time before logarithmic tables were used as calculation aids. The logarithmic tables were succeeded by slide rules, which in turn were overtaken by the modern electronic calculators. Originally products were developed to meet a societal need, and improved generations of the product succeeded them. The societal need expanded as additional types of calculations became available. With the rapid emergence of modern technologies and growing global markets, new societal needs are appearing in abundance. An engineer is a person who applies scientific knowledge and experience in manufacturing (technologies) to satisfy humankind's needs.

One serious difficulty engineers must overcome involves the form in which problems are often presented to them. Even if some goals are given, often they are not specifically stated. As an example, consider the situation where the designer is presented with the waste of irrigation water in public parks. "The park keepers forget to turn off the water." This is a statement of fact. It points to a problem that needs a solution. A general formulation of the problem is "what can we do to minimize the possibility of workers forgetting to turn off the water before the end of their shifts?" An engineer could ask the following questions: Why do workers continue to forget to turn off the water? What is the sequence of events that workers do during their daily activities? Does the water have to be turned on and off manually?" A more precise form of the problem statement is "How do we prevent waste of irrigation water in public parks?" Even this formulation lacks details, but it can give rise to several solutions. An administrator may see this as an opportunity to employ an inspector who checks whether the water has been turned off after specified times. An engineer may come up with a solution that uses automatic switching on and off controlled by a timer. The senior management of a company looks at the statement of the problem and its possible solutions and chooses one of them; this becomes a *need*. A design brief is the description of this need. It is the first document produced that helps the engineers design something to solve the problem.

4.2.1 Elements of a Design Brief

A design brief is a written document that the senior management of a company or a client gives to the design team. The items that are included in a design brief depend on the product and its size. It should give sufficient details for the design team to start the design process. A typical design brief will include (1) a product description (what); (2) a product concept (how); (3) the benefits to be delivered (goals—these can be used as criteria for concept selection); (4) positioning and target price (business goals); (5) the target market (to whom to sell); (6) assumptions, constraints, and standards; (7) stakeholders; (8) possible features and attributes; and (9) possible areas for innovation. The design brief should give sufficient details to identify the paths the design team can follow in the design process. As an example, consider the design brief for a single-stage scissor lift shown in Table 4.1.

The design brief identifies the stakeholders, the selling market, the intended position in the market, and the benefits to be delivered by the special features of the product. This enables the design team to work on identifying the requirements or what the customers expect the product to do. The person who originates the design brief (usually the senior management, or a client if you are working in a design company), must identify the relevant standards. He has to gather and sift through a lot of information. The following sections describe standards and standardization.

Table 4.1: Design Brief

Design Brief of "Professional Painter's Pal" a Single-Stage Scissor Lift	
Drafted by	Simple Simon Company, Ltd.
Product Description	A portable, compact, power-driven, and firm-access platform that enables both professional and amateur painters to reach the ceiling of rooms in domestic houses to carry out painting.
Product Concept	A single scissor-type power-driven mechanism carrying a platform of size about 0.64 m^2.
Benefits to Be Delivered	Access to paint heights up to the ceiling of the rooms of domestic houses with accessories Ability to be located at different intermediate positions Easily movable from place to place Compact for transportation in cars
Positioning and Target Price	Middle of the range 400 kg payload and price of $1000
Target Market	DIY enthusiasts; professional painters
Assumptions, Constraints, and Standards	Sold through the current outlets EN280 will be followed wherever appropriate.
Stakeholders	End users Wholesalers and retailers Distributors
Possible Features and Attributes	Provision of hand rails Caption reads "BORN TO CARRY"
Possible Area for Innovation	Compact packaging Aesthetically pleasing

4.3 STANDARDS AND STANDARDIZATION

An important item for consideration during the identification of a need is the standards. Imagine the development of a small mobile crane or access platform that helps people to reach heights to do maintenance work inside houses. The access platform has to reach the required height. But to serve the purpose, it should be able to go inside rooms through the doors. This is where the standard comes into the scene. If there are no standard dimensions for the doors, then the designer has to carry out a survey measuring the dimensions of the doors and decide the dimensions of the wheelbase for the platform. These dimensions are governed by national standards. If there are international standards for door dimensions, a standard door made in the Philippines can fit in a house in Europe or the United States, and access platforms designed for the European or American standards would pass through it. This is the case with almost all consumer goods such as electric bulbs, plug point sockets, plugs, computer parts, and so forth. Frequently the country of origin is far and remote from the country of consumption. This has been made possible only by the existence of standards. Standards facilitate the global trade of goods and services. Standards play an important role in daily life by establishing size or shape or capacity of a product, process, or system and specifying performance. They define terms so that there is no misunderstanding among those using the standard. A standard is a technical expression of how to make a product safe, efficient, and compatible. Conformity assessment determines whether the relevant requirements are fulfilled. Conformity assessment provides assurance to consumers, and it is for this reason that products often display statements saying that they conform to this or that standard.

The British Standards Institution (BSI) defines "standard" as "a document, established by consensus and approved by a recognized body, that provides, for common and repeated use, rules, guidelines or characteristics for activities or their results, aimed at the achievement of the optimum degree of order in a given context." There is an accompanying note stating "Standards should be based on the consolidated results of science, technology and experience, and aimed at the promotion of optimum community benefit." BSI defines "standardization" as an activity of establishing, with regard to actual or potential problems, provisions for common and repeated use, aimed at the achievement of the optimum degree of order in a given context.

Government regulations and private-sector standards affect a majority of world trade. Countries have been working together to establish international standards in almost every field. As a result, workers in all sectors need to have an understanding of standards. Engineering and technology students must have a knowledge and understanding of engineering standards. They also must learn how to apply them in designing, developing, testing, and servicing products, processes, and systems.

4.3.1 Institutions Providing Standards

A number of national and international organizations provide guidance for developing and implementing standards. Some of them are

- American National Standards Institute (ANSI)
- American Society of Mechanical Engineers (ASME)
- American Society for Testing and Materials (ASTM)
- British Standards Institution (BSI)
- Association of Electrical, Electronic and Information Technology, Germany (VDE)
- Canadian Standards Association (CSA)

- European Commission of the European Union (CE Standards called EN)
- Federal Communications Commission (FCC)
- Institute of Electrical and Electronics Engineers (IEEE)
- National Transportation Safety Board (NTSB)
- U.S. Food and Drug Administration (FDA)

4.3.2 ANSI Standards

The American National Standards Institute (ANSI) coordinates the U.S. voluntary consensus standards system and acts as the watchdog for standards development and conformity assessment programs and processes. ANSI itself does not develop standards, but it establishes the consensus procedures for the development of American national standards. It accredits qualified organizations whose standards development process meets all of ANSI's requirements to develop American national standards. Presently, there are more than 220 ANSI-accredited standards developers; there are about 10,000 ANSI standards covering various fields and topics. The standards are called American National Standards (ANSs). They provide dimensions, ratings, terminology and symbols, test methods, and performance and safety requirements for personnel, products, systems, and services needed for industries.

4.3.3 Locating Relevant Standards

In order to locate the relevant standards if there are any, try the NSSN search engine at http://www.nssn.org/search/IntelSearch.aspx. A prompt as shown in Figure 4.1 will appear.

Choose the FIND TITLE ABSTRACT OR KEYWORD button and type in the keyword. The relevant standards will appear. Choose the one required. For example, type in "MEWP," standing for Mobile Elevated Work Platforms. The results will be shown as they appear in Figure 4.2. If you are involved in a project where the objective is to design a mobile elevated work platform, you will need to access the guidelines given in the standard EN280—2009 and incorporate them into the need statement so that specifications can be written to meet the standard. Passing the test is mandatory to get the Conformité Européenne (the European Conformity) or the (CE) marking, the official approval needed to sell the machinery in the European Union countries.

It is very important to read the standard thoroughly as there can be additional constraints. These constraints can make an otherwise feasible solution an infeasible solution.

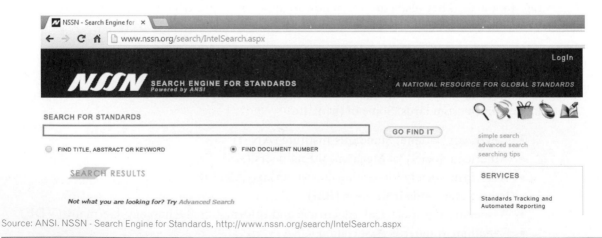

Source: ANSI. NSSN - Search Engine for Standards, http://www.nssn.org/search/IntelSearch.aspx

Figure 4.1: Search Engine for Global Standards

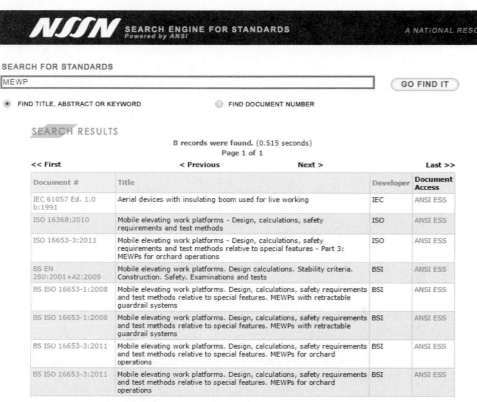

Source: ANSI. NSSN - Search Engine for Standards, http://www.nssn.org/search/IntelSearch.aspx

Figure 4.2: Result from the Search Engine for Standards

4.4 HUMAN FACTORS

Human engineering, human performance, and *ergonomics* are all terms used to describe the science that considers humans and their reactions to both familiar and strange environments. Adapting a human to fit a given environment is almost impossible, so it is better to adapt the environment to suit the human. Human behavior is critical to the success of an engineering system. Therefore, it is important to consider typical human behaviors during the design phase. Examples of typical human behaviors include the following:

1. People are often reluctant to admit errors.
2. People usually perform tasks while thinking about other things.
3. People frequently misread or overlook instructions and labels.
4. People often respond irrationally in emergency situations.
5. A significant percentage of people become complacent after successfully handling dangerous items over a long period of time.
6. Most people fail to recheck outlined procedures for errors.
7. People are generally poor estimators of speed, clearance, or distance. They frequently overestimate short distances and underestimate large distances.
8. People are generally too impatient to take the time needed to observe.
9. People are often reluctant to admit that they cannot see objects well, due either to poor eyesight or to inadequate illumination. They generally use their hands for examining or testing.

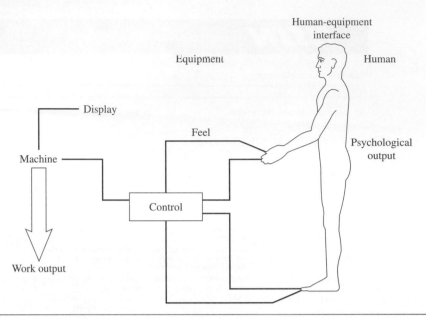

Figure 4.3: Human–Machine System

10. Every time people dial telephones, drive cars, study oscilloscopes, use computers, or form parts on a lathe, they have joined their sensing, decision-making, and muscular powers into an engineering system. Figure 4.3 shows the coupling of components that is necessary for considering a human as an integral part of an efficient human-machine system.

4.4.1 Human Sensory Capabilities

Human sensory capabilities include the following:

1. *Sight:* Human eyes see differently from different angles or positions. For example, looking straight ahead, the eyes can perceive all colors. However, with an increase in the viewing angle, human perception begins to decrease. The limits of color vision are given in Table 4.2. Furthermore, at night or in the dark, small-sized orange, blue, green, and yellow lights are impossible to distinguish from a distance.

2. *Noise:* The performance quality of a task requiring intense concentration can be affected by noise. It is an established fact that noise contributes to people's feelings, such as irritability, boredom, and well-being. A noise level of 90 dB can be tolerated for up to two hours of exposure, whereas a level above 100 dB can cause damage after 15 minutes of exposure. Levels in excess of 130dB could cause immediate nerve damage.

3. *Touch:* Touch adds to, or may even replace, the information transmitted to the brain by eyes and ears. For example, it is possible to distinguish knob shapes by touch alone. This ability can be valuable if, for instance, the power goes out and there is no light.

4. *Vibration and motion:* When equipment operators perform physical and mental tasks poorly, it can be partially or fully attributed to vibrations. For example, eye strain, headaches, and motion sickness can result from low-frequency, large-amplitude vibrations.

Table 4.2: Human Color Vision

Situation	Green	Blue	Yellow	White	Green-Red	Red
Vertical	40°	80°	95°	130°	—	45°
Horizontal	—	100°	120°	180°	60°	—

5. Useful guidelines for the effects of vibration and motion are as follows.

   a. Eliminate vibrations with an amplitude greater than 0.08 mm.

   b. Use devices such as shock absorbers and springs whenever possible.

   c. Use damping materials or cushioned seats to reduce vibration wherever possible.

   d. Vertical vibrations most affect people sitting, so use this information to reduce vertical vibrations.

   e. The resonant frequency of the human vertical torso in the seated position is between three and four cycles per second. Therefore, avoid any seating that would result in or would transmit vibrations of three to four cycles per second.

4.4.2 Anthropometric Data

Anthropometry is the science that deals with the dimensions of the human body. Anthropometric data are divided into statistical groups known as percentiles. If 100 people were lined up from smallest to largest in any given respect, they would be classified from the 1st percentile to the 100th percentile. The 2.5 percentile means that designs based on this series of dimensions would include up to 2.5% of the population in the system, and the remaining 97.5% would be excluded. Figure 4.4 shows the dimensions of a standing adult male.

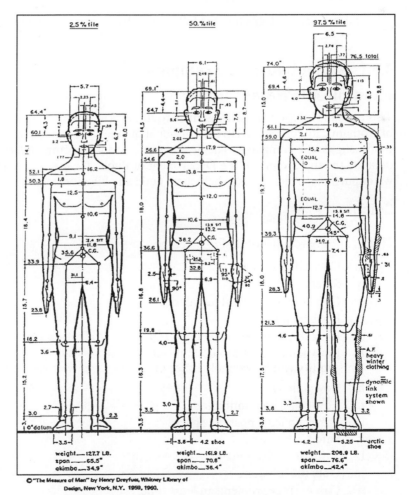

Figure 4.4: Standing Adult Male

4.5 ORGANIZING THE GOAL—OBJECTIVE TREE

One of the fundamental requirements in drawing the design brief is a clear objective of the design. You can achieve this by developing an *objective tree*. The objective tree method procedure was summarized by Cross [2].

1. *Prepare a list of design objectives*. These are prepared from the market survey and analysis reports, questions to the client, and from discussion with the design team. Remember to ask as many questions as you possibly can to enable you to better understand what exactly is the requirement in the market. Remember, a vague statement is equal to a vague understanding of the need, which may lead you to develop a product that does not match the market needs. This is the most important step in developing the objective tree. Remember, there is no limitation for what you can put in the product at this time.

2. *Organize the list into sets of higher-level and lower-level objectives*. Group the expanded list of objectives and subobjectives into roughly hierarchical levels.

3. *Draw a diagrammatic tree of objectives*. This shows hierarchical relationships and interconnections. The branches in the tree represent relationships, which suggest means of achieving objectives.

Two examples are given in the following subsections to explain this important tool.

4.5.1 Objective Tree Examples

Example 4.1: Water Purifier

This example is based on an assignment to students at Florida State University. After the residents of Gotham City complained to their mayor about the city's water quality, he ordered the health unit to investigate the complaint. The health unit based its recommendation on the chemical analysis done by the top chemist at the Department of Public Health. City engineers now have to design water purifiers. Your task is to help them build an objective tree.

Solution

STEP 1. Prepare a list of design objectives.

a. Cost effectiveness
b. Safety
c. Can detect chemical imbalance
d. Fewer repairs
e. Easy to repair when needed
f. Long lasting
g. Affordable
h. Low damage
i. Low or no contamination
j. Takes up least possible space
k. Safe for humans
l. Safe for environment
m. Gets the job done
n. Can correct problems in least time

o. Low maintenance
p. Cleans high volume of water
q. Efficient
r. . . . . (complete per project objectives)
s. . . . .

STEP 2. Order the list into sets.

Safety	Cost Effectiveness	Efficiency
Safe for humans	Few repairs	Can detect chemicals
Safe for environment	Easy to repair	Long lasting
	Affordable	Low damage
	Takes least possible space	Gets job done
	Low maintenance	Corrects problems in minimal time

STEP 3. Draw an objective tree (Figure 4.5)

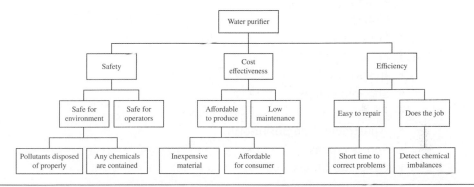

Figure 4.5: Objective Tree for the Water Purifier

The objective tree will help senior management to clearly state (1) the product description, (2) the product concept, and (3) the benefits to be offered.

Example 4.2: Automatic Aluminum Can Crusher

This example is based on an assignment to students at Florida State University; the details are shown subsequently in Figures 4.6-4.9.

The Need Statement

Design and build a device or machine that will crush aluminum cans. The device must be fully automatic—that is, all the operator needs to do is load cans into the device; the device should switch on automatically. The device should automatically crush the can, eject the crushed can, and switch off unless more cans are loaded. The following guidelines should be adhered to:

- The device must have a continuous can-feeding mechanism.
- Cans should be in good condition when supplied to the device (not dented, pressed, or slightly twisted).

- The can must be crushed to one-fifth of its original volume.
- The maximum dimensions of the device should not exceed $20 \times 20 \times 10$ cm.
- Performance will be based on the number of cans crushed in one minute.
- Elementary-aged school children (age 5 and up) must be able to operate the device safely.
- The device must be a stand-alone unit.
- The total cost of the device should not exceed the given budget ($200).

Market Research

Although the need statement is relatively clear, the design team did several interviews with the client, asked questions, and carefully listened to the client's responses in order to determine the goal of the intended device. In parallel, the design team conducted a full market survey to assess similar products as well as to consider all the potential "stakeholders" or customers.

Here is a short summary of the market analysis.

Potential Customers

- Schools
- Colleges
- Hospitals
- Hotels
- Resorts
- Shopping malls
- Playgrounds and recreational areas
- Apartments and dormitories
- Sports arenas
- Office buildings
- Residential homes

Companies That Have Similar Devices (Selection)

- Edlund Company, Inc. (159 Industrial Parkway, Burlington, Vermont), Mr. R. M. Olson (President)
- Prodeva, Inc. (http://prodeva.com)
- Enviro-Care Kruncher Corporation (685 Rupert St., Waterloo, Ontario, N2V1N7, Canada)
- Recycling Equipment Manufacturer (6512 Napa, Spokane, Washington, 99207)
- Kelly Duplex (415 Sliger St., P.O. Box 1266, Springfield, Ohio, 45501)
- Waring Commercial (283 Main St., New Hartford, Connecticut, 06057)
- DLS Enterprises (P.O. Box 1382, Alta Loma, California, 91701)

Standard Industrial Classification (SIC) Code

- Food service industry 3556
- Recycling 3599

Trends

The aluminum industry produces approximately 100 billion cans a year in the United States. This number has been flat for the past 13 years. In 2007, 54 billion of the cans were returned for recycling. According to the Aluminum Association, at a recycling rate of 53.8 percent the aluminum can is by far the most recycled beverage container in the United States. Although this recycling figure has been rising steadily for the past six years, it actually represents a drop in recycling rates from the previous decade (66.8% in 1997 and 62.1% in 2000), even though the same quantities were produced per year.

Market Information

Growth opportunities for aluminum beverage cans exist throughout the world. Global aluminum beverage can shipments represents more than 70% of the metal can shipment in the world. In North America alone, more than 100 billion beverage cans are consumed each year. In China, the consumption exceeds 10 billion cans per year. Seventy-four percent of U.S. beverage sales occurred in convenience stores, drugstores, clubs, mass-merchandise stores, vending machines, and grocery markets. For aluminum beverage cans, the initial market is schools.

Patents

A Google search on "can crusher" showed more than 20,000 items in 2015. Fifteen years ago (coincides with the first edition of this book) only 56 patents are listed; they are shown in Section 4.6.1. Here is an example of an old design from 1981.

US Patent No: 4,436,026, Empty Can Crusher:[1]

*An empty **can crusher** for crushing and flattening empty cans, comprising an inlet, a chute, a stopper device, a pressing device and a forked chute. Empty cans supplied in the crusher are crushed and flattened by the pressing device and are sorted into **aluminum** cans and steel cans by means of a magnet embedded in the pressing device, which fall down into respective receptacles through the forked chute.*

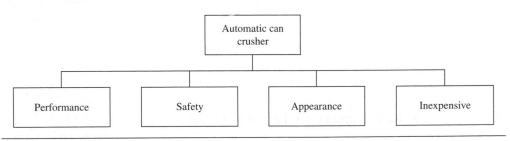

Figure 4.6: Main Objectives

---

[1]U.S. Patent Application Publication, Pub. No.: US4436026 A, Yoshinobu Imamura, Shigeki Kamei, Tetuo Yamagata, Hiroshi Fujii, Publication date Mar 13, 1984.

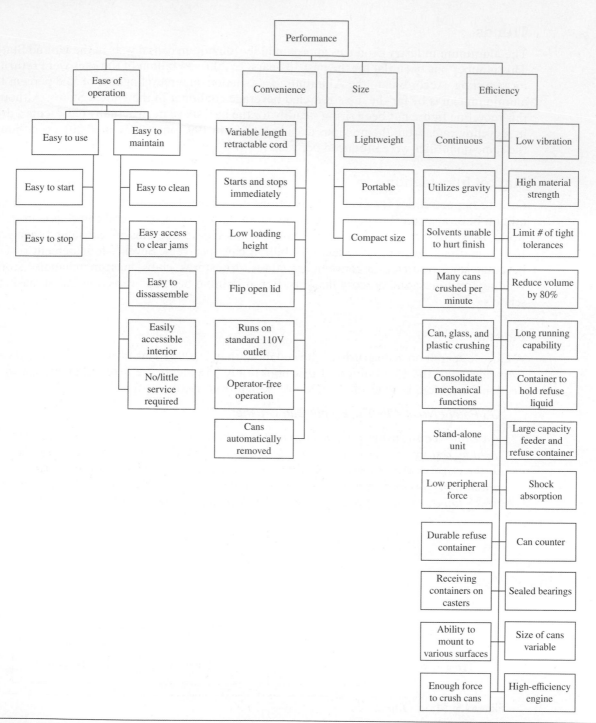

Figure 4.7: Expansion of the Performance Branch

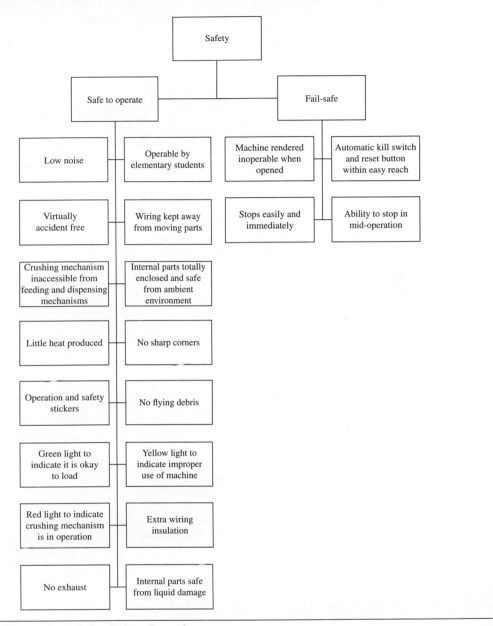

Figure 4.8 Expansion of the Safety Branch

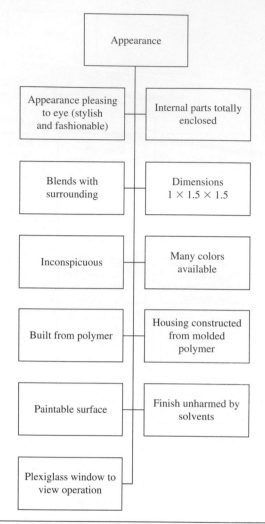

Figure 4.9: Expansion of the Appearance Branch

4.6 GATHERING INFORMATION: CLARIFYING THE NEED

Needs can be identified in several ways. Either the client approaches the designer directly with a specific problem, or the company employing the designer finds an opportunity in the market by identifying a need for a new or improved product. In the design process, it is important to determine whether such a business opportunity exists for a product before time and money are invested to develop the product. Most likely, products will fail when the market analysis only skims the surface. The analysis requires a thorough study of the total market. This includes trends, competition, volume, profit, opportunities, consumer needs, and some indication of customer feeling about the product. Market analysis must be conducted at the start. It serves as a mechanism to define the design brief and provides an opportunity for designers to review other attempts at solving the problem at hand. Since the majority of designs are *development designs*, knowing what others have presented as a solution is very important before attempting to offer one's own solution.

King [3] defines a market as "a group of potential customers who have something in common." Two techniques are employed in market analysis.

1. *Direct search:* This involves obtaining information directly from the consumer, manufacturers, salespeople, and so forth. The information is collected by interviews and surveys.
2. *Indirect search:* Information is collected from public sources, such as patents, journal reports, government analysis, and newspapers.

The market search should be done in a systematic and objective manner, considering all information relevant to the product. An unbiased outlook is necessary when analyzing the data to formulate a market-analysis report. Any biases will be reflected in the report's view of the market, and the results could be disastrous.

Thousands of resources all present some variation of the same information; the trouble is that it is almost never presented in the form needed. Before beginning any project, it is important to plan the process for information collection and organization. Information collection is usually one of the first steps, and modern engineers have realized that it is an essential element of the design process if they wish to remain competitive in a global market. For the engineering student, product and market research often requires the use of unfamiliar information sources and often produces huge quantities of information. Apply engineering skills when planning and dealing with this deluge. These skills include

- Critical thinking
- Strategy
- Analysis
- Time management (It always takes longer than the original estimate.)

In the product-development process, conduct market research initially to assess market potential, market segments, and product opportunities. Market research provides production cost estimates and information on product cost, sales potential, industry trends, and customer needs and expectations. The search has the following steps:

1. Define the problem
2. Develop a strategy
3. Organize and check the information gathered

4.6.1 Define the Problem

Knowing what to look for is very important before starting to gather information. Answer questions similar to the following:

1. Am I developing a new product or solving a problem in an existing product? It is worth remembering that a solution may not be needed and the task is to redefine a problem.
2. Who are the customers, and why would they want or need to buy the product—for example, for its efficiency, utility, or unique value?
3. What are the main needs of these customers?
4. In one sentence and in your own words (abstraction of the need statement), define the problem at hand.
5. How are you going to go about getting the product to customers? Things to consider are development costs, time, manufacturing, and production investment.

Establishing who the customers are is one of the most important initial steps that a designer needs to take. One of the vital concepts to grasp is that customers are not just the end users. Everyone who will deal with the product at some stage during its lifetime are the customers of a product, including the person who will manufacture the product, the person who will sell the product, the person who will service the product, and the person who will maintain the product during its lifetime in operation. All these people are indicated as stakeholders in the design brief.

4.6.2 Develop a Strategy

Set a plan for the search process. Looking through every reference book in a library hoping to find the right piece of information will inevitably lead to an inefficient waste of time. First identify what pieces of information you need, then select where to begin the search.

- **Identify Keywords** Have some relevant terms with which to begin the search.
- **Write a Plan** The search process is not linear. For example, while searching the business literature for information on industry trends, you may come across the text of an interview that directly identifies some customer needs for the product under investigation. This information can be very useful if properly contextualized. It is important to know the framework within which you are working.

4.6.3 Organize and Check the Information Gathered

The following list [4] can be used either as a planning tool for information collection or as a checklist for locating relevant information.

a. Products
    i. Product names
    ii. Patents
    iii. Pricing
    iv. Parts breakdown
    v. Product features
    vi. Development time
b. Companies
    i. Major players
    ii. Company financials for major players
        - *Annual reports.* Yearly record of a publicly held company's financial condition (Information such as the company's balance sheet, income, and cash flow statements are included.)
        - *10K reports.* This is a more detailed version of the annual report and is the official annual business and financial report filed by public companies with the Securities and Exchange Commission. The report contains detailed financial information, a business summary, list of properties, subsidiaries, and legal proceedings.
c. Industry
    i. Trends
    ii. Labor costs
    iii. *Market-size industry facts.* Pieces of information gained from various sources that help to clarify anything about the industry
d. Market information
    i. Market reports
    ii. Market share of major companies in industry

iii. Target markets of major competitors
iv. Demographics
- Age
- Geographic location
- Gender
- Political, social, and cultural factors

e. Consumer trends

4.7 RELEVANT INFORMATION RESOURCES

There are literally hundreds of resources where you can find information, but several stand out because of their superiority of scope, quality, and overall usefulness. Resources can be divided into four categories.

1. Product information
2. Industry information
3. Company information
4. Market information

Most journal databases will provide information relevant to all of the areas discussed in the previous section. Keep this in mind when searching and analyzing. This will help eliminate the repetition of resources.

4.7.1 Patent Information

Patents grant the patent holder exclusive rights to a new idea or invention. In Chapter 1, Figures 1.3 through 1.5 show three different patents of paperclip (1934, 1991, and 2013, respectively). There are two main types of patents: *utility patents* and *design patents*. Utility patents deal with how the idea works for a specific function. Design patents cover only the look or form of the idea. Hence, utility patents are very useful, since they cover how the device works, not how it looks. Patents can serve as an excellent source of ideas for products and as a place to protect your own ideas if you come up with a novel one. Although a patent prevents others from making and selling your novel idea, a patent does not give you the immediate rights to make and sell it, since it still has to go through the regular legal and regulatory channels.

It is difficult to identify the specific patent that contains an idea you are looking for. Since there are more than eight million utility patents, use some search strategies to hone in on the one you may be able to use. Use a web search, such as that provided by the Patent Office. Keyword searches as well as patent numbers, inventors, classes, or subclasses are available. Good websites for patent searches include

- http://www.wipo.int/patentscope/en/
- http://www.uspto.gov/patft/index.html

4.7.2 Industry Information

It is essential to obtain information about the industry you will be designing for. Who are the major players? What are the current trends? How large (in dollar amounts) are the relevant industries? What materials are used by the relevant industries? The following resources can provide access to information that will answer these questions.

Standard Industrial Classification (SIC) Code

SIC codes classify a company's type of business. Many business information sources are organized by the SIC codes. The following publications are guides to these codes:

1. *Standard Industrial Classification Manual:* The official U.S. government manual that provides SIC codes at the two-digit and four-digit levels.
2. *Web version of the SIC code:* This provides access to all classifications.

Trade Associations

The *Encyclopedia of Associations* gives address, telephone, and contact information as well as a brief description of trade and industry associations and their publications. Descriptions usually include dates and locations of conferences and trade shows.

Industry Overview

1. *Manufacturing Worldwide: Industry Analyses, Statistics, Products and Leading Companies*: This annual publication provides statistical data on leading manufacturing industries around the world.
2. *Forbes*: Provides an annual report on U.S. industries in the January issue of each year.
3. *Standard and Poor's Industry Surveys*: Provide a textual analysis for 22 broad industry categories, which are broken down into more specific subsets.
4. *Encyclopedia of American Industries*: This publication is organized by SIC code and has a list of related publications for each described industry.
5. *Moody's Industry Review*: This statistical source contains key financial information and operating data on about 3500 companies. The information is arranged by industry in 137 industry groups. The companies within an industry group may be compared with one another, and they may also be measured against certain averages for that industry. The statistics are updated twice each year.
6. *Predicasts Basebook*: Compiled annually since 1970, but with statistics that go back to 1967, this annual statistical source contains about 29,000 time series arranged by a proprietary seven-digit SIC-code–based system. The data includes economic indicators and industry statistics. The industry statistics usually include production, consumption, exports and imports, wholesale prices, plant and equipment expenditures, and wage rates.
7. *Predicasts Forecasts*: This quarterly statistical source provides short and long range forecasts for economic indicators as well as industries and products. The data are arranged by a proprietary seven-digit SIC-code–based system. Each forecast includes the date and page reference of the source from which the data are taken. The front of each annual accumulation includes composite forecasts, which present historical data for over 500 key series.

Statistics

1. *Annual Survey of Manufacturers*: This survey supplements the Census of Manufacturers in the United States.
2. *American Statistics Index (ASI)*: This contains indexes and abstracts statistics published by the U.S. government.
3. *Statistical Reference Index (SRI)*: This contains indexes and abstracts statistics and is published by trade and industry organizations and state governments.

4.7.3 Company Information

Company sources provide information on the players within an industry. They provide information on the company's financials, products (primary and secondary, such as Pepsi® and Mountain Dew®), brands and trade names, along with other "nuggets" that (in context) can help with overall market assessment and analysis.

When searching for company information, it is important to determine whether the company is public or private. Public companies are required to register and file reports with the Securities and Exchange Commission (SEC), whereas private companies are not. Generally, it is much easier to find information on public companies.

Some companies publish their 10K information as required by the U.S. Securities and Exchange Commission and provide comprehensive summary of their company's financial performance on the Web. Examples include

1. Allied Products Corporation
2. American United Global, Inc.

Directories

1. *Hoovers Corporate Directory.* This provides the company's name, street address, phone number, and fax number; the names of the Chief Executive Officer, Chief Financial Officer, and Human Resources director; the most recently available annual sales figures; the percentage change in sales from the previous year; the number of employees; a description of what the company does; and the company's status (private or public).

2. *Directory of Corporate Affiliations.* This lists more than 4000 parent companies with divisions, subsidiaries, and affiliates. It also provides geographic and SIC indexes.

3. *Duns Million Dollar Disc:* This lists more than 160,000 businesses in the United States with net worth greater than $500,000. It can be used to identify companies by state, industry, or geographic location. It also provides thumbnail bios of top executives for public and private companies and basic company information.

4. *Standard & Poor's Register of Corporations, Directors, and Executives, Volume 1; Corporations:* This gives corporate addresses, telephone numbers, names and titles of officers, SIC codes, products, annual sales, and number of employees for 55,000 public and private companies. Volume 3 provides an index by SIC code.

5. *Corporate Technology Directory.* This is a directory for high-tech companies with very specific product indexes. It also provides basic company information.

6. *Thomas Register of American Manufacturers and Thomas Register Catalog File:* Volumes 1 through 8 list companies arranged by products and services. Volumes 9 and 10 provide an alphabetic list of manufacturers with addresses, subsidiaries, products, and asset estimates. Volume 10 provides an index of brand names. Selected company catalogs are reproduced in Volumes 11 through 16.

7. *Verizon Yellow Pages:* This is a free service that allows users to look up names, addresses, and phone numbers of more than 2.1 million businesses.

8. *Ward's Business Directory:* Ward's is a good source for private companies. It provides information about 142,000 companies, 90% of which are private. Volumes 1 through 4 provide individual company information. Volume 5 ranks companies by SIC code.

Corporate Reports

Moody's manuals also provide detailed information on company history, subsidiaries, business and products, comparative balance sheets, stock and bond descriptions, and more. Geographic indexes classify companies by industry and product. The data in Moody's manuals are obtained from annual and 10K reports.

4.7.4 Market Information

Market information may be found in many places. Use keywords related to the industry, company, and/or product in your search. This is true when using either a paper source or an electronic source. Examples of information to look for include market share, target market, demographics, and market potential.

Market Research Reports

These are ready-made reports that have been commissioned by companies or industrial groups. They are compiled by one of several market research companies. These reports are provided by consulting and advisory firms or by companies and different sectors of industry. The information can be found from firms such as

1. Findex
2. Frost & Sullivan Reports

Market Share and Other Information

1. *Market Share Reporter.* This is an annual publication providing market share data for companies, products, and industries.
2. *World Market Share Reporter.* This is an annual publication providing market share information.
3. *Statistical Abstract of the United States.* Statistics on the industrial, social, political, and economic aspects of the United States are found here. It is a great source for all types of useful statistical information.
4. *Industrial Statistics Yearbook:* This contains industrial statistics published by the Department of Economic and Social Affairs, which is a statistical office of the United Nations.

Demographic Information

1. *American Demographics.* This journal is searchable on ABI/Inform on First-Search.
2. *State and Metropolitan Area Data Book.* Social and economic data are provided for SMSAs, cities, states, and census regions. It is prepared by the U.S. Bureau of Census.
3. *Demographics USA, County Edition:* This contains information on purchasing power, consumption rates, economic conditions, and population statistics at http://www.tradedimensions.com

4.8 WEB TOOLS

At this time, the Web does not provide access to all of the resources needed to conduct a product/market research. However, there are some useful sites that point you to the resources that are available:

1. Business Researcher's Jumpstation at http://www.brint.com/sites.htm
2. Marketing Resource Directory at http://www.ama.org

Web search engines are increasing in sophistication, as are the websites of companies, distributors, industrial association, and retailers. Therefore, it is valuable to spend some time sifting through the information available on the Web.

4.9 DESIGN METHODS FOR INFORMATION ANALYSIS

Design methods can be useful for information analysis at this stage. They provide the following advantages:

- Useful for getting insight into the area of concern
- Useful for getting insight into the place for the product in the market
- Useful for getting insight into the constituent parameters

4.9.1   Matrix Analysis – The Design Method

Matrix analysis is a mechanism to enlist the features or attributes of a particular product provided by companies in the market. This will identify the features provided by all companies, features that are specific to certain companies, and so on.

In the 1980s nearly all big companies outsourced manufacturing. One of the major automobile manufacturers found that it had several hundreds of parts and assemblies to be outsourced. It opened up a contracting department to draft and administer contracts. At the beginning the contract department drafted contracts individually for each company. One engineer noticed that several of the contract clauses were common to many companies. He took 400 such companies and their contracts and produced a matrix analysis. Figure 4.10 shows part of his matrix. The analysis showed the common mandatory clauses and the company-specific special clauses straight away. Using the analysis, he developed a contract-drafting system.

	Acceptance criteria	Allowance for expansion	Autonomy	CE marking	Comp damage	Comp supply	Components	Cycle time	Delivery (*)	Description	Deviations	Documentation	Electrical standards	Environmental conditions	Equipment usage	Ergonomics	Guarding and finishing	Interface and services	Intro	Mechanical standards	On-site activity	Operator attendance	Operator controls	Payment terms (*)	Pneumatics; hydraulics standards	Points of contact	Price of goods (*)	Process FMEA	Project review points	Proprietary equipment	Quality	Risk and property	Scope of supply	Software standards	Spares (included)
Company 1	○			○	○	○	○	○	○	○	○	○	○	○		○	○	○	○			○		○	○	○	○		○	○	○	○	○	○	○
Company 2	○			○	○	○	○		○		○	○	○	○		○	○	○	○			○		○		○	○		○						
Company 3	○	○						○		○	○	○	○	○	○	○	○	○	○			○		○		○			○						
Company 4	○				○	○	○	○	○	○	○		○		○	○	○					○			○	○	○		○	○			○		
Company 5	○		○			○	○		○		○	○	○	○		○			○	○			○		○		○				○	○	○
Company 6	○		○			○	○		○		○	○	○	○		○			○				○				○	○	○
Company 7	○				○	○	○	○	○	○	○	○	○	○		○	○						○			○			
Company 8	○				○	○	○	○		○		○	○	○										○	○				
Company 9	○	○		○	○	○	○	○	○	○	○	○	○	○	○	○	○	○	○	○	○	○	○	○	○	○	○	○	○	○	○	○	○	○	○
Company 10	○		○			○	○	○		○			○									○			○			○		
Company 11		○	○			○	○						○			○				○	○								
Company 12	○		○			○	○	○		○		○						○				○			○			

Figure 4.10: Sample of Matrix Analysis

4.10 MARKET ANALYSIS REPORT

A market analysis report provides information needed to make decisions about a new platform, variant, or module of a product. It should contain the following information:

1. A purpose statement, which is used to explore the launch of a new product family, product variant, or new product module. It is always good to start with an infinitive phrase.
2. The methodology of the research process, which explains whether it is a passive search using the published data (outlined in Sections 4.4 onwards) or an active search through customer interviews, survey etc.
3. A product description
4. The product unit price
5. The market potential. In this part you express the size of the market, current players or competitors, the intended share of the market to be acquired, and the level of difficulty in doing so. It should provide supporting data to help inform decisions.
6. The most-liked features. If a competitor sells the product, what are the features users like.
7. The most-wanted features or challenges. These are the features users want but that competitors do not provide.
8. The most-disliked features or opportunities. These are the features competitors provide but users do not like.
9. The standards that must be complied with. Different countries have different standards that have to be complied with to sell the product in those countries. The U.S. Food and Drug Administration (FDA) is a typical example.
10. Market trend. State whether it is a growing market, a saturated market, or a market that requires the next generation of the product.
11. Technological advantages and alternatives

4.11 CHAPTER SUMMARY

1. A design brief describes a need for a design solution. Typically it will include
   a. A product description
   b. A product concept
   c. The benefits to be delivered
   d. The positioning and target price
   e. The target market
   f. Assumptions and standards
   g. Stakeholders
   h. Attributes and possible features
   i. Possible area of innovation
2. A standard is a technical expression of how to make a product safe, efficient, and compatible with others.
3. Human behavior is critical to the success of an engineering system. Human equipment interface and fitting the environment to assist humans in performing tasks must be taken into consideration in design.
4. An objective tree is a method used to clarify the design objectives and to assist in drawing up a clear design brief.

5. Detailed market analysis is essential for successful products. Market search can be direct, by obtaining the information from the consumers, or indirect, by searching public sources.

6. Market search follows three major steps: (1) define the problem; (2) develop a strategy; and (3) organize and check the gathered information.

7. Matrix analysis is a useful method for comparing companies and product attributes using a matrix format.

8. The market analysis report should be detailed and should include all pertinent information gathered in the search.

4.12 PROBLEMS

1. Write down the elements of the design brief for a family dinner table used for hosting an important visitor.

2. Often you can develop a statement of fact into more than one design problem. Describe two possible design problems that could be derived from the following statements of fact.

   a. Eating more than the recommended portion is bad for the health.

   b. A river runs between two resorts A and B. A mechanism is needed to move people from A to B and B to A.

3. List at least six clarifying questions for the following statements:

   a. The door won't stay open while I bring my shopping bags into the house.

   b. I have an itch in my back, and I can't scratch it.

   c. I need to know where is that smoke is coming from.

   d. Empty boxes take up a lot of space.

   e. I want to do some gardening next summer.

   f. We need a dining table.

4. Write design briefs for the following

   a. A trolley to lift engines from cars for repair

   b. A jack to lift cars weighing up to 1 ton to change wheels

   c. A spoon tree to hang washed spoons in a kitchen

   d. A reading assistant to hold and present multiple open books for cross referencing

   e. A single-stage scissor lift for decorators to use

5. Consider the following problem statement and state whether it is adequate or not, giving reasons. Also state the next steps in the design process.

   Design a table for the CEO of a large company. It should have a drawer for pencils, pens, erasers, and other stationery items, a drawer for high-value items l (e.g., the keys of the safe), and a drawer for arranging important documents currently under review. There should be ample space on the tabletop to keep three or four open documents as well as hand and leg spaces for two persons sitting opposite to the CEO. There should be provision for and space for a desk lamp. The table should be attractive and act as a status symbol for the company.

6. As technology advances and new materials are synthesized, there are certain products that receive only minimal, if any, modernization attempts. Among them are the shopping carts used in grocery stores. There is a tendency to save parking spaces by not designating a return cart area. Leaving carts in the parking lots may lead to serious accidents and car

damage. Furthermore, although many customers do not fill their carts that much when shopping, they do not like to carry baskets. Other customers like to sort products as they shop. Develop an objective tree to clarify the need statement and prioritize the objectives laid out in the problem statement. You may add features that may not be listed clearly in the problem statement but will give an additional advantage to the proposed design.

7. Draw an objective tree for the following: If you drive in the state of Florida, you may notice some traffic congestion when crews trim trees on the state roads and freeways. Assume that the Florida Department of Transportation is the client. Usually, branches 6 inches or less in diameter are trimmed. The allowable horizontal distance from the edge of the road to the branches is 6 feet. The material removed from the trees must be collected and removed from the roadside. To reduce the cost of trimming, a maximum of only two workers can be assigned for each machine. The overall cost, which includes equipment and labor must be reduced by at least 25% from present cost. The state claims that the demand for your machines will follow the price reduction—that is, if you are able to reduce the cost by 40%, the demand will increase by 40%. Allowable working hours depend on the daylight and weather conditions.

8. Draw an objective tree for the following statement: We have a mountain-sized pile of wood chips that we want to process into fire logs for home use. We are located in Tallahassee, Florida. We can have a continuous supply of wood chips throughout the year. We should be able to produce 50 fire logs per minute. The shareholders require us to have a large profit margin, and our prices should be lower than our competition for the same size logs. The logs should produce enough heat to keep our customers satisfied, and they should have environmentally clean exhaust.

9. Table 4.3, which follows, shows anthropometric measured data in mm of U.S. adults (19 to 60 years of age).

   Many sites report such data for example http://www.cdc.gov/nchs/data/nhsr/nhsr010.pdf

Table 4.3: Anthropometric Data

	Men				Women			
Dimension	5th %ile	Mean	95th %ile	SD	5th %ile	Mean	95th %ile	SD
1. Stature	1647	1756	1867	67	1528	1629	1737	64
2. Eye height, standing	1528	1634	1743	66	1415	1516	1621	63
3. Shoulder height (acromion), standing	1342	1443	1546	62	1241	1334	1432	58
4. Elbow height, standing	995	1073	1153	48	926	998	1074	45
5. Hip height (trochanter)	853	928	1009	48	789	862	938	45
6. Knuckle height, standing	Na	Na	Na	Na	Na	Na	Na	Na
7. Finger height, standing	591	653	716	40	551	610	670	36
8. Sitting height	855	914	972	36	795	852	910	35
9. Sitting eye height	735	792	848	34	685	739	794	33
10. Sitting shoulder height (acromion)	549	598	646	30	509	556	604	29
11. Sitting elbow height	184	231	274	27	176	221	264	27
12. Sitting thigh height (clearance)	149	168	190	13	140	160	180	12

13. Sitting knee height [73]	514	559	606	28	474	515	560	26
14. Sitting popliteal height	395	434	476	25	351	389	429	24
15. Shoulder-elbow length	340	369	399	18	308	336	365	17
16. Elbow-fingertip length	448	484	524	23	406	443	483	23
17. Overhead grip reach, sitting	1221	1310	1401	55	1127	1212	1296	51
18. Overhead grip reach, standing	1958	2107	2260	92	1808	1947	2094	87
19. Forward grip reach	693	751	813	37	632	686	744	34
20. Arm length, vertical	729	790	856	39	662	724	788	38
21. Downward grip reach	612	666	722	33	557	700	664	33
22. Chest depth	210	243	280	22	209	239	279	21
23. Abdominal depth, sitting	199	236	291	28	185	219	271	26
24. Buttock-knee depth, sitting	569	616	667	30	542	589	640	30
25. Buttock-popliteal depth, sitting	458	500	546	27	440	482	528	27
26. Shoulder breadth (biacromial)	367	397	426	18	333	363	391	17
27. Shoulder breadth (bideltoid) [12]	450	492	535	26	397	433	472	23
28. Hip breadth, sitting	329	367	412	25	343	385	432	27
29. Span	1693	1823	1960	82	1542	1672	1809	81
30. Elbow span	Na	Na	Na	Na	Na	Na	Na	Na
31. Head length	185	197	209	7	176	187	198	6
32. Head breadth	143	152	161	5	137	144	153	5
33. Hand length	179	194	211	10	165	181	197	10
34. Hand breadth	84	90	98	4	73	79	86	4
35. Foot length	249	270	292	13	224	244	265	12
36. Foot breadth	92	101	110	5	82	90	98	5
37. Weight (kg), estimated by Kroemer	58	78	99	13	39	62	85	14

From cad.fk.um.edu.my/ergonomics/anthoropometry/us.pdf

As a designer you are requested to arrange seating in an auditorium such that most individuals will have an unobstructed view of the speaker screen (see Figure 4.11). To do so you will need to

a. Determine the body dimensions that are critical in the design:
   i. Erect sitting height
   ii. The line of vision of people sitting
b. Select a design principle and the percentage of the population you plan to accommodate.
c. Determine the minimum to raise based on your design principle.
d. Measure the dimensions of your seat in the classroom. Redesign the seating to fit 95% of females. How would the design affect 50% of males?

10. Most houses have vents that open and close manually without any central control. Cities across the United States recommend the use of such vents in an effort to save energy. In most cases, household occupants do not use the entire house at the same time; the

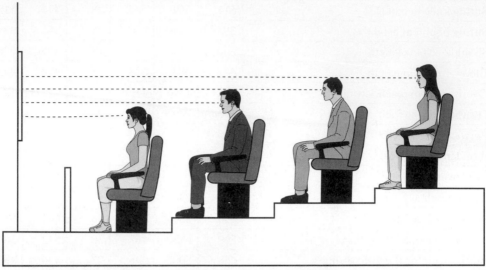

Based on METHODS, STANDARDS AND WORK DESIGN, 10/e by Benjamin Niebel and Andris Freivalds.

Figure 4.11: Design Criteria

tendency is to use certain rooms for long periods of time. For example, the living room and dining room may be used heavily, while the kitchen is used at certain hours of the day. To cool or heat a room, the vent system must work to cool or heat the entire house. A house can save energy if the vents of unused rooms are closed; this will push the hot or cold air to where it is needed most and will reduce the load on the conditioning system. Develop an objective tree to clarify the need statement, and prioritize the objectives laid out in the problem statement. You may add features that are not listed clearly in the problem statement but will give an additional advantage to the proposed design.

11. A young engineering graduate joined his father's business, which provides machining services for sheet metals work. Twelve months on the job, tragedy struck when his father died of a heart attack. As the only child, he has to take full charge of the business. Remembering that bigger margins can be earned only in product development, he wants to produce washing machines. The country he comes from has unique needs for washing machines, which many of the machines developed in other developed countries do not provide. Conduct a matrix analysis for him to use to identify the features provided by the washing machines currently on the market.

12. Devise and carry out an information-gathering exercise for a "reading assistant to hold and present multiple open books for cross referencing" using the three steps given in Section 4.6.

13. Devise and carry out an information-gathering exercise for a "single-stage scissor lift" using the resources listed in Section 4.6.3.

14. How is the direct search different from the indirect search?

15. Consider the issue outlined in Problem 6 as a need statement.

    A design team did a market analysis for this statement but forgot to fill in all of the information. Gather information from available sources and complete the following market analysis.

    i. The SIC codes that are associated with this design are
       a. Airflow controllers: air conditioning and refrigeration valve manufacturing (provide the codes)

    b. Air conditioning units: domestic and industrial manufacturing (provide the codes)

    c. Air ducts: sheet metal 5039

  ii. Materials used in the industry are (a) sheet metal (b) (provide the codes) (c) (provide the codes) (d) (provide the codes).

 iii. Major associations that deal with air conditioning are as follows: (a) Air Conditioning and Refrigeration Institute (b) (provide the codes) (c) (provide the codes) (d) (provide the codes)

 iv. Major companies (a) Trane (b) (provide the codes) (c) (provide the codes)

  v. Patent search (a) Sarazen et al., 4,493,456 (b) (provide the codes) (c) (provide the codes)

16. Develop a strategy with which your team will conduct the market analysis task.

17. Estimate the time required for your team to conduct a market analysis assignment, using any of the design statements in Problems 6–8.

18. Define *market analysis* and discuss how is it different from information gathering.

19. Perform a Web search and name three industries that change a product at a frequency of at least once each year.

20. Write a plan to conduct your market analysis.

21. List as many Web addresses as you can find that provide patent search. Name some differences among the sites.

22. Interview a librarian in your school. and identify research sources that are available for your use.

23. How many patents were filed during the past 10 years for coin sorters? List a few and identify their differences.

References

The Autobiography of Lincoln Steffens (foreword by Thomas Leonard). Berkeley, CA.: Heydey Books, 2005.

CROSS, N. *Engineering Design Methods: Strategies for Product Design*. New York: Wiley, 1994.

KING, W. J. *The Unwritten Laws of Engineering—Mechanical Engineering*. Vol. 66, No. 7. New York: ASME Press, 2001.

WEINER, S. "A New Look at Information Literacy at Massachusetts Institute of Technology." Presentation at the ASEE Conference, Chicago, IL, 1997.

http://www.bsigroup.com/
http://www.osha.gov/
http://www.ntis.gov/
http://www.census.gov/epcd/www/naicstab.htm
http://vancouver-webpages.com/global-sic/
http://www.thomasregister.com
http://www.tipcoeurope.com/
http://iml.umkc.edu/ndt/html/lexus.html
http://www.lexis-nexis.com/Incc/
http://www.google.com
http://www.briefing.com
http://www.briefing.com/
http://www.sme.org/
http://www.asme.org/
http://www.manufacturing.net/
http://stats.bls.gov/blshome.htm

http://stats.bls.gov/blshome.htm
http://www.uspto.gov/
http://www.uspto.gov/
http://www.european-patent-office.org/
http://www.european-patent-office.org/
http://www.nist.gov/
http://www.nist.gov/
http://www.entrepreneur.com/
http://www.entrepreneur.com/
http://www.tenonline.org/
http://www.tenonline.org/
http://www.tradedimensions.com/
http://www.tradedimensions.com
http://www.findex.com/
http://www.findex.com

CHAPTER 5

CUSTOMER REQUIREMENTS

"No matter how much you want it to be a technical problem, it's a people problem."

~Anonymous

A designer has to fully comprehend two views of the design task at hand: (1) what the product is, and (2) what it will do. We discussed "what the product is" in Chapter 4, and this is best described by the design brief. We will describe the second part, "what the product will do," in this chapter. Customers are the people who will use the product; therefore, they are the right people to decide what the product should do. However, a product will generally have several different types of customers. For example the lecturer, students, cleaners, scheduler, estate management, and finance controller are all customers for a lecture room in a university. They all, as a group, are called stakeholders. This chapter explains the derivation of stakeholder requirements. The output of this activity is a list of prioritized requirements with importance ratings, and optional Demand or Wish marking.

5.1 OBJECTIVES

By the end of this chapter, you will be able to

1. Define what is a customer requirement and differentiate it from a need.
2. Classify customer requirements into different types using different criteria.
3. Use data from interviews and focus groups to establish customer requirements.
4. Produce prioritized customer requirements.

5.2 CUSTOMER REQUIREMENTS

What is the difference between a *need* and a *requirement*? The answer is simple. Needs are a vague set of wishes that express what customers would like a product to do for them—for example, "Get me from point A to point B as quickly and safely as possible." Requirements, however, are the designers' detailed breakdown of what the product should do and achieve based on the customers' statements. They should carry out this translation *without* providing solutions. It is essentially an expanded and more organized form of the initial needs as expressed by the customers. This is the reason why they are still regarded as "customer requirements." Some customers may even provide sufficiently detailed statements of their needs that warrant these items to be moved unchanged to the requirements stage.

For example, a need from a customer could be "Something that will hold sufficient quantities of water, have the ability to heat the water efficiently, and have a way to pour this water into a mug or cup safely without spilling or burning." At first sight, you may immediately think of an electric kettle, but it is important at this stage not to jump to any conclusions or solutions. The next stage is to research the market and obtain more information on the customers of this product. This will enable the identification of the type, frequency, and quantity of usage. Indeed, you may realize, if this is for a commercial environment, that the quantity of water that is regarded as sufficient can be achieved only by a large dispenser-type machine. Even if the product was for domestic use, there are many other ways to provide energy to heat water, as well as vessels of varying shapes to hold and pour the water. An extract of our solution-neutral requirements may look like this:

- Holds varied quantities of water
- Heats varied quantities of water
- Boils water fast
- Is energy efficient
- Is easy to move around
- Provides safe handling during pouring
- Pours hot water without spilling
- Has esthetically pleasing surface
- Automatically switches off from the energy source or alerts user when the water is boiling

A customer requirement is an expression of a desired behavior of the product; it is about the responses or outputs of the product, not about the inputs. The customer is not the designer, but the customer does provide some essential data and helps the designer to design a product that will appeal to customers. The customer should help in the desired areas but not in every area.

For example, that the car should reach from 0 to 60 mph in 15 seconds is a legitimate customer requirement, while the size and layout of the cylinders in the engine are matters for the designer to deal with. The design team should decide the areas where they need inputs from the customers. Obtaining customer requirements is not a random activity; it is an activity that is carefully planned.

5.2.1 Types of Customer Requirements

There are many types of requirements such as functional performance and reliability. Design teams often use checklists of requirement types where inputs from the customer are valuable for designing a customer survey. A checklist is a document that contains typical customer-requirement items, usually developed by an experienced designer. The checklist ensures that no area will be missed. Ullman [1] has such a checklist, which is given here as Table 5.1.

Otto and Wood [2] classify customer requirements differently. In the first instance they classify them as direct and latent—the direct ones are easy for the customer to identify, whereas the latent ones are important but hard to identify as they relate to the system within which they operate. Viewing differently, they classify requirements as (1) constant ones that will be there always and (2) variable ones that may be discarded if an anticipated technological advancement takes place. Another classification divides them into *general* and *niche* requirements. These classifications are generic, but it is helpful to know that they exist.

Pahl and Beitz [3] classify the requirements as *demanded items* and *wish-to-have* items. They use this classification to identify the principal and secondary function carriers when they develop embodiment design.

Table 5.1: Types of Customer Requirements Listed by Ullman [1]

Functional Performance

- Flow of energy
- Flow of information
- Flow of material
- Operational steps
- Operation sequence

Human Factors

- Appearance
- Force and motion control
- Ease of controlling and sensing state

Physical Requirements

- Available spatial envelope
- Physical properties

Reliability

- Mean time between failures
- Safety hazard assessment

Life Cycle Concerns

- Distribution (shipping)
- Maintainability
- Disposability
- Repairability
- Cleanability
- Installability
- Retirement

Resource Concerns

- Time
- Cost
- Capital
- Unit
- Equipment
- Standard
- Environment

Manufacturing Requirements

- Materials
- Quantity
- Company capabilities

Based on Ulman, D. G. The Mechanical Design Process. New York: McGraw-Hill, 1992.

5.2.2 Characteristics of Well-Formulated Requirements

Ullman [1] enumerates the following characteristics of well-formulated requirements:

1. Requirements are discriminatory. (Ullman explains that a requirement that an ideal mate should have hair does not discriminate enough since almost everyone has hair.)
2. Requirements are measurable.
3. Requirements are orthogonal, meaning that each requirement addresses a unique feature in the design and there should not be any overlap. (He explains orthogonal properties by saying that (1) the product must give a smooth ride over rough roads and (2) the product should reduce shock from bumps, are not orthogonal requirements.)
4. Requirements are universal in that they are applicable to all the alternatives, not just for some of them.
5. Requirements are external; they should refer to the responses, not to the internal design parameters.

5.2.3 Customer Requirements and Level of Satisfaction

Identifying customer requirements is a process of satisfying the customer to such a level that he or she will be a loyal consumer of the product. This provision of function and level of satisfaction is analyzed at three levels. This is called the Kano model; it is shown by an illustrative graph in Figure 5.1.

The Kano model was developed by Dr. Noriaki Kano in the early 1980s. In the Kano model, there are three different types of product quality that yield customer satisfaction: basic quality, performance quality, and excitement quality. With basic quality, customers' requirements are not verbalized because they specify assumed functions of the device. The only time a customer will mention these requirements is if they are missing. If they are not fully

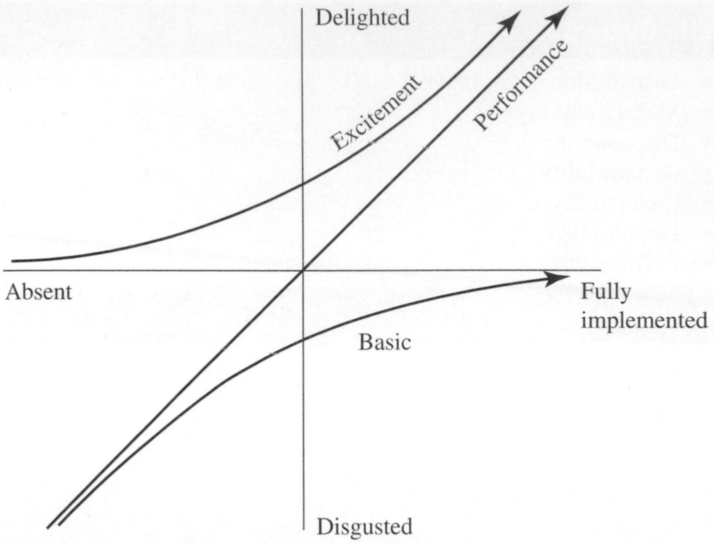

Figure 5.1: Kano Model for Customer Satisfaction

implemented in the final product, the customer will be disgusted with it. If they are included, the customer will be neutral. An example is the requirement that an adult bicycle should have brakes. The performance quality refers to customers' requirements that are verbalized in the form that indicates the better the performance, the better the product. The excitement quality involves requirements that are often unspoken because the customer does not expect them to be met in the product. However, if they are absent, customers are neutral. If the customers' reaction to the final product contains surprise and delight at the additional functions, the product's chance of success in the market is high.

5.3 CHOOSING CUSTOMERS

There can be several groups of customers or stakeholders for a product, and they all must be represented. Further there can be subgroups within each group. Explaining the choice of customers for a cordless drill, Ulrich and Eppinger [4] divide the market into three segments: (1) home-owners who make occasional use of the product; (2) handy people who make frequent use; and (3) professionals who make heavy-duty use. Within each group they identify lead users, users, retailers, and service centers. They propose a customer-selection matrix, which is given in Table 5.2.

Table 5.2: Customer-Selection Matrix for Cordless Drill

	Lead Users	Users	Retail or Sales Outlets	Service Centers
Home Owner (occasional use)	0	5	2	3
Handy Person (frequent use)	3	10		
Professional (heavy-duty use)	3	2	2	

Based on Ulrich K.T. and Eppinger S.D., Product Design and Development, Third Edition, Tata McGrawHill Publications New Delhi 2004.

Ullman [1] outlines different stakeholders such as the consumer, the design team's manager, the sales staff, the service-personnel and, a company's employees in the manufacturing, assembly, and shipping divisions. Customers to be surveyed should be chosen carefully, and they should represent the entire stakeholder group as a whole.

5.4 THE METHOD

Three methods have been advocated and are widely used as effective methods. They are

1. Interviews
2. Focus groups
3. Observing the product in use

5.4.1 Interviews

In this method a design team member discusses the requirements of a product with a single customer in the environment where the customer uses the product. The design team member records the customer's responses. This method does not work well if the team member simply shows up and asks seemingly random questions, which can make the team member appear ignorant. One way to reduce this effect is to send the topics or even questions in advance to the surveyed group. Dym and Little [5] describe a structured interview where the interviewer has a set of questions to ask an interviewee, who may or may not be informed of the questions in advance. This ensures direct answers to the questions while providing opportunities for additional questions and discussions. It also ensures that interesting side issues will not obscure the main issues the design team wants to discuss.

Ulrich and Eppinger [4] outline some helpful questions and prompts to follow after the initial introduction and explanation of the purpose of the interview in the following way:

- When and why do you use this type of product?
- Walk us through a typical session using the product.
- What do you like about the existing products?
- What do you dislike about the existing products?
- What issues do you consider when purchasing the product?
- What improvements would you make to the existing product?

Otto and Wood [2] call the method the like/dislike method and suggest using a table-like record that shows the findings (Table 5.3):

When the Ulrich and Eppinger [4] or Otto and Wood [2] methods are used, the collected requirements will often be mixed up. They must be grouped to form a hierarchical structure. The requirements should be organized using the *affinity diagram* method. The affinity diagram organizes a large number of requirements according to their natural relationships. However, if you use a checklist such as the one proposed by Ullman [1] for the interview, a hierarchical structure will result as the recorded document.

5.4.2 Focus Groups

In the focus group method a moderator facilitates or conducts a two-hour discussion with a group of 8 to 10 customers while the design team watches and videotapes the proceedings. Often questions such as Why? are discussed in these gatherings. The moderator is typically a professional market researcher. Again, the questions and discussions are structured and planned.

Table 5.3: Like/Dislike Method Data Collection Form

Customer Data: Project/Product Name Customer: Address: Willing to follow up? Type of User:		Interviewer(s): Date: Currently uses:	
Question	Customer Statement	Interpreted Requirement	Importance
Type of use			
Likes			
Dislikes			
Suggested improvements			

OTTO, KEVIN; WOOD, KRISTIN, PRODUCT DESIGN, 1st Ed., ©2001. Reprinted by permission of Pearson Education, Inc., New York, New York.

5.4.3 Observing the Product in Use

This method relies on the hypothesis that watching customers using the existing product to perform a task can reveal important requirements about the customer requirements.

5.5 ELICITING THE CUSTOMER REQUIREMENTS

The design team elicits customer requirements after gathering the customer comments as records of the interviews, focus group recordings, or observations of the product in use.

Ulrich and Eppinger [4] provide the following guidelines for eliciting customer requirements:

1. Express the requirement in terms of what the product should do, not in terms of how it might do it.
2. Express the requirement specifically.
3. Try and use positive rather than negative phrasing.
4. Express the requirement as an attribute of the product.
5. Avoid the words *must* and *should*.

5.6 PLANNING TO EXTRACT CUSTOMER REQUIREMENTS

The assertion "No plan is a good plan for disaster" is most applicable when it comes to extracting customer requirements. We have witnessed students engaged in product development projects sit together for a few minutes and scribble a few sentences or phrases, which they call customer requirements. They never use the list again except for its inclusion in the final report, and often the product has nothing to do with the listed customer requirements. We include this section to emphasize the importance of planning and using a systematic process to establish customer requirements. The sole purpose of getting the customer involved in product development is to understand and include the features that will make the customer feel happy and comfortable in using the product. The first and foremost thing, therefore, is an understanding of the product by both the customer and the designer. Since the bulk of the design work *is development design*, some form of the design is already available. In the case of a new designs, the designer should develop a sufficiently complete description of what the product is and does so that the customer feels comfortable with the product concept and can come up with ideas that lead to relevant customer requirements.

The only consolidated document about the product at this point of the process is the design brief, which essentially tells what the product is and what the company's objectives are. The starting point can be a reminder of this so that the customer can focus his thoughts on how the product is or is not achieving these objectives.

The requirements in general can be about performance, target production cost, manufacturing facilities, customers, service life, environment, size, weight, maintenance, appearance, quality and reliability, shipping, industry standards, safety, and packing. The strategy therefore should be to choose the potential areas for questioning so that valuable customer requirements could be extracted. Choose the areas you want to cover in advance as you prepare your questions. Normally 20 to 25 requirements are adequate for a medium-sized project.

Example 5.1: User's Requirements for a Single-Stage Scissor Lift

This example is based on an assignment to students at the United Arab Emirates University.

In this example the requirements are grouped according to the checklist classification.

Table 5.4: User's Requirements for a Single-Stage Scissor Lift (following Ullman's Checklist)

Functional Performance	Flow of energy	1. Works using single-phase AC, 230V sockets 2. Allow additional sockets for hand tools
	Flow of information	3. Press-on switch for moving up or down
	Flow of material	4. Payload of 400 kg (person + tools)
Human Factors	Appearance	5. Pleasing color paint
	Force and motion control	6. Easy to move and maneuver
	Ease of controlling and sensing state	7. Lockable in position
Requirements Physical	Available spatial envelope	8. Pass through domestic doors 9. Pass through domestic lifts
	Physical properties	10. Maximum height – 1000 mm 11. Stowed height – 400 mm or less 12. Platform area about 0.64 m^2
Reliability	Safety hazard assessment	13. Hand rails to hold when in motion 14. Can accommodate 100N toppling force by hand

Continued on next page

Lifecycle Concerns	Distribution (shipping)	15. Should be lockable at stowed height for transport
	Maintainability	16. Little or no maintenance
Resources	Time	17. Rise time about 20 seconds
	Cost	18. About $1500
	Standard	19. EN280
Manufacturing requirements	Materials	20. Mostly steel sheet metal and bars

Otto and Wood [2] give the derivation of requirements of an electric wok using the like/dislike method. It is reproduced here as Figures 5.2 and 5.3 with permission.

5.7 RELATIVE IMPORTANCE OF THE REQUIREMENTS

A customer is a representative of a group of people whom the company wants to be purchasers of the new product that is being developed. Typically we define *requirement* to mean something that is demanded or obligatory. Following this we can define *customer requirement* to mean "a particular characteristic of a good or service stipulated as obligation by customers." Since there can be several such customer requirements and all of them cannot have the same level of importance and all customers will not rate the requirement as important to the same level, deriving customer requirements for an identified need is a special activity.

The design team identifies customer requirements after having interacted with the customers. The most reliable method to establish the importance of customer requirements is to refer them back to the customer and conduct a new survey. Alternatively, some of the references in the original customer survey can be used as an indicator of the importance of the ratings. Another option is for the design team that conducted the survey to rate the requirements.

5.7.1 Prioritizing Customer Requirements

Once you have established a list of requirements, the next two steps are to prioritize and organize these requirements so that the designer is aware of the essential requirements as well as the ones that can be compromised due to conflict, cost, or other reasons. Conflicts may arise sometimes when a customer wishes for more than one feature that a given product cannot provide—for example, a portable travel kettle that can hold 10 liters of water at any one time. In this case the designer must identify whether the priority for the customer is the portability or the ability to hold 10 liters of water or whether there is a way to compromise and/or bias one requirement over the other.

Customer Data: Electric Wok Customer: John Doe Address: MIT Willing to do follow up? Y Type of user: College student		Interviewer(s): Date: Currently uses: Frying pan	
Question	**Customer Statement**	**Interpreted Need**	**Importance**
Typical uses	Stir-fry Steaming Frying/scrambling eggs Cooking pasta Cooking chili/stews [Everyday cooking]		
Likes	Non-stick surface Size Can stand on its own Temp. response rate high Aesthetically pleasing Depth of dish	Non-stick surface Compact Able to stand on its own Quick temp. response Aesthetically pleasing Deep sides	Good Good Should Good Good Should
Dislikes	Short cord Moves around too much when stirring (doesn't grip surface of table) Entire assembly is too high/tall Handles are hard to grip (esp. if oil splatters) Have to watch contantly Temp. adjustment gets too hot/also too low to read Sides don't get as hot; may overcook on bottom Afraid to get bottom wet	Long extension cord Can grip tabletop Compact (flat) elec. unit Handles are easy to grip Auto shut-off Temp. switch is insulated from heat Temp. switch is in an easily accessible/ readable spot Constant temp. distribution Bottom is watertight	Must Good Good Nice Should Should Good Should Good
Suggested Improvements	Retractable cord Better gripping bottom Hole through handles to grip Make heating element casing flatter Have clip for lid (to flip back & forth) that is also removable Deep-frying accessory (wire mesh shelf) Maybe has ears on both sides in case someone is lefty		

Figure 5.2: Portion of Customer Interview Data for Electric Wok Redesign: Customer 1A

Customer Data: Electric Wok Customer: John Doe Address: MIT Willing to do follow up? Y Type of user: College student		Interviewer(s): Date: Currently uses: Frying pan/pot	
Question	**Customer Statement**	**Interpreted Need**	**Importance**
Typical uses	Cooking stir-fry Chinese food Frying pan substitute Quick, easy dishes		
Likes	Bottom heats up quickly	Heats quickly	Must
	Little to no smoke emitted when cooking		
	Ears are covered in a type of insulating material	Ears remain cool (not hot) to touch	Good
	Non-stick surface	Non-stick surface	Should
	Fairly lightweight	Wok is lightweight	Nice
	Attractive design/color	Aesthetically pleasing	Nice
	Useful for college-age students (w/out easy access to stoves)		
	Easy to clean	Inside can be cleaned easily	Nice
Dislikes	Side slow to heat up	Sides heat up at same rate	Should
	Need to stir food around a lot to cook well	Smaller bottom area	Must
	Flat bottom: difficult to tilt wok/ move food	Rounded bottom to be able to tilt/ shake wok	Should
	No off button/switch	Include off switch	Should
	Cumbersome to wash	Heating unit is detachable	
	Afraid to get bottom (heating element) wet	Wok is watertight	Good
	Needed extension cord to cook on table	Long extension cord	Good
Suggested Improvements	One-sided handle (long) "Flip-out" handle to conserve shelf space Include off button Make bottom more round Notch on rim for spatula/stirring utensil to rest on Longer cord (or include extension?)		

Figure 5.3: Portion of Customer Interview Data for the Electric Wok Redesign: Customer 2A

To prioritize requirements, the designer assigns an importance rating for each requirement from 1 to 10, where 10 is the most important and 1 is the least important. It should also be determined whether a particular requirement for the product is essential or not. If it is considered essential, it should be classified it as a *Demand* denoted by the letter D. A Demand is always given the top rating value of 10. Other nonessential requirements are considered to be *Wishes* and are denoted with the letter W. The ratings for these Wishes range from 1 to 10 as described previously. Identifying the most important requirements will become very useful during the embodiment stage.

Usually the designer uses feedback from the customers and market research to determine the importance rating for each requirement. A group working on a design project can rank these wishes to conduct a survey in the event that the product the design group is working on will be beneficial to a group of customers who are easily identified and accessed. Companies desiring to conduct such an ordering of wishes use surveys. To distinguish numerically the importance of each wish, use a weighting factor. There are various methods for allocating weighting factors for these wishes.

a. *Absolute measure* (for each wish individually), where each wish is rated from 1 to 10. Wishes that have the same importance are assigned the same value. The total measure is irrelevant.

b. *Relative measure*, where a total of 100 is maintained, and each wish is given a value according to its importance. The weighing factor in this event could be a percentage figure totaling 100% when all values are added.

In many instances a product will have more than one customer. For example, the brake pads for a car involve drivers, mechanics, and salespeople as essential customers who may view and evaluate attributes differently. The attributes are assigned to three different columns; one column for each of the customers (drivers, mechanics and salespeople), according to their importance. All three customers will have to input a value for the weighting factor independently and as each see reasonable for each of the attributes. Students should be aware of this strategy; however, in many design situations where students compose the design team and the instructor or an industry representative is the customer, it is reasonable to ask the customer and the design team to grade the wishes.

We will subsequently use importance ratings to quantitatively assess how well the design of the product is progressing, employing a popular technique called quality function deployment (QFD), which is covered in more detail in Chapter 7.

 EXAMPLES

Example 5.2: Wheelchair Retrieval Unit

This example is based on an assignment given to students at Florida State University. Design a wheelchair retrieval unit to assist nurses in situations where a patient is taking a walk with the nurse, but after walking about 30 minutes the patient feels tired. The nurse must remain with the patient to provide support and should be able to use one hand to activate the wheelchair retrieval unit to bring the wheelchair to where the patient stands from where it was left when the patient started walking. The purpose is to bring the chair from where it was left by the nurse and patient when they started walking to where the nurse and patient reached.

Solution

Design the wheelchair retrieval unit with the following requirements:

- Unit dimension no larger than 30 cm^3
- Capability to reach the patient in 1 minute
- Responsive within 30 meters away from the nurse

Appearance is important. The finished product should be marketed within two years' time. Manufacturing cost should not exceed $50 per unit at a production rate of 1000 units per month. In the design process, after the designers create the objective tree and the function tree, it is important that the designers complete a table of specifications. Designers must decide on the level of generality of the specifications. The requirements are shown in Table 5.5.

This example is for a wheelchair retrieval unit that can be operated with one hand and help a nurse who is supporting a patient to retrieve the wheelchair. We've considerably expanded the initial statement. We've also taken into account the range of users as well as safety considerations. Now we have to distinguish the demands (D) from the wishes (W).

Table 5.5: Prioritized Requirements for a Wheelchair Retrieval Unit

D or W	Requirements	Importance (1–10)
D	Dimension 30 × 30 × 30 cm^3	10
D	Can reach patient within 1 min	10
D	Water resistant	10
D	Stable	10
D	Can support patient weight	10
D	Can provide transportation and seating	10
D	Easy to operate	10
D	Low maintenance	10
D	Durable	10
D	Safe	10
W	Stops quick	7
W	Comfortable	6
W	Easy to control	7
W	Low power drain when not in operation	5
W	Minimum number of parts	6
W	Lightweight	4
W	Compact/foldable	3
W	Low production cost	7
W	Can operate by itself	3
W	Small turning radius	5
W	Can maneuver through obstacles efficiently	8

Example 5.3: Automatic Aluminum Can Crusher—Requirements

Continuing the case study that we started in Section 4.7, this section considers the automatic aluminum can crusher. As a reminder, the following was the original need statement:

Design and build a device/machine that will crush aluminum cans. The device must be fully automatic (i.e., all the operator needs to do is load cans into the device; the device should switch on automatically). The device should automatically crush the can, eject the crushed can, and switch off (unless more cans are loaded). The following guidelines should be adhered to:

- The device must have a continuous can-feeding mechanism.
- Cans should be in good condition when supplied to the device (i.e., not dented, pressed, or slightly twisted).
- The can must be crushed to one-fifth of its original volume.
- The maximum dimensions of the device are not to exceed 1.5 × 1.5 × 1 feet.
- Performance will be based on the number of cans crushed in one minute.
- Elementary school children (K and up) must be able to operate the device safely.
- The device must be a stand-alone unit.
- The total cost of the device should not exceed the given budget ($200).

After completing the detailed market research, the engineering team meets and decides whether each requirement is a demand (D) or a wish (W) and allocates importance ratings. An example of the list of prioritized requirements is given in Table 5.6.

Table 5.6: Aluminum-Can–Crusher Requirements

List of Requirements in No Particular Order					
Pleasing to eye (stylish and fashionable)	W	5	Stand-alone unit	D	10
Internal parts totally enclosed; safe from ambient environment	D	10	Large capacity feeder and refuse container	W	4
Blends with surrounding	W	5	Cans automatically removed	D	10
Dimensions 1 × 1.5 × 1.5	D	10	Low peripheral force	W	8
Inconspicuous	W	5	Easy to start	W	9
Many colors available	W	1	Shock absorption	W	8
Built from polymer	W	2	Easy cleaning	W	3
Housing constructed from molded polymer	W	2	Durable refuse container	W	4
Paintable surface	W	3	Can counter	W	2
Ability to reset after kill switch has been used	W	9	Receiving containers on casters	W	3

Continued on next page.

Plexiglass window to view operation	W	1	Sealed bearings	W	7
Low noise	W	7	Ability to mount to various surfaces	D	10
Machine rendered inoperable when opened	W	10	Compact size	D	10
Operable by elementary students	D	10	Ability to crush various sizes of containers	W	3
Total cost < $200	D	10	Operator-free operation	W	6
Automatic kill switch and reset button within easy reach	W	10	Enough force to crush cans	D	10
Wiring kept away from moving parts	D	10	Starts immediately	W	7
Crushing mechanism inaccessible from feeder and dispenser	W	10	High-efficiency engine	W	9
Yellow light to indicate improper use of machine	W	3	Lightweight	W	3
Internal parts safe from liquid damage	W	8	Low loading height	W	7
Stops easily and immediately	W	9	Flip-open lid	W	1
Little heat produced	W	6	Solvents unable to hurt finish	W	6
No sharp corners	W	8	Limit number of light tolerances	W	7
Ability to stop in mid-operation	W	10	Many cans crushed per minute	W	8
Green light to indicate it is OK to load	W	3	Reduce volume by 80%	D	10
Extra wiring insulation	W	6	Can, glass, and plastic crushing	W	5
Operation and safety stickers	W	9	Runs on standard 110V outlet	W	9
Red light to indicate crushing mechanism is in operation	W	3	Long-running capability	W	9
No flying debris	W	10	Consolidates mechanical functions	W	5
No exhaust	W	8	Portable	W	4
Continuous	D	10	Easily accessible interior	W	7
Easy access to clear jams	W	6	Container to hold refuse liquid	W	2
Easy to maintain and disassemble	W	3	No or little service required	W	3
Low vibration	W	8	Utilizes ground to stablize	W	5
Variable length retractable cord	W	3	Small force required to depress switches	W	9
Stops easily and immediately	W	8	Stand-by mode	W	3
Utilizes gravity	W	2	Weatherproof	W	3
High material strength	W	9	Less than five assembly steps	W	5
Retails for < $50	W	3	Large storage of crushed cans	W	3

5.9 CHAPTER SUMMARY

1. Needs are a vague set of wishes that customers would like a product to perform whereas requirements are the designers' detailed breakdown of what the product should do and achieve based on the customers' statements.

2. A customer requirement is an expression of a desired behavior of the product; therefore, it is about the responses or outputs and not about the inputs.

3. The design team should decide the areas where they need inputs from the customers.

4. Design teams when planning a customer survey often use checklists of types where inputs from the customer are valuable.

5. Identifying customer requirements is aimed at satisfying customers to such an extent that they will keep purchasing the product.

6. Choosing customers for surveying is a very important activity.

7. Interviews, focus groups, and observing the product in use are the three methods for conducting a market survey.

8. In a structured interview, the interviewer has a set of questions to ask an interviewee, who may or may not be informed in advance.

9. Ulrich and Eppinger give a set of questions for the interview.

10. Otto and Wood give a form to be used in their Like/Dislike form.

11. In the focus group method, a moderator facilitates or conducts a two-hour discussion with a group of 8 to 10 customers while the proceedings are observed by the design team and a video is made.

12. The sole purpose of having the customer involved in product development is to understand the features that will make the customer feel happy and comfortable in using the product and to make sure those features are included.

13. A customer is a representative of a group of people whom the company wants to be purchasers of the new product that is being developed.

14. In order to prioritize requirements, the designer assigns an importance rating for each requirement from 1 to 10, where 10 is the most important and 1 is the least important. Another distinction that is usually made is whether a particular requirement for the product is essential or not. If it is considered essential, then it is classified as a Demand and denoted with a letter D wishes will need to be ranked. Demands have the same high priority and the letter "D" classifies that status, while wishes need to be rated/ranked.

5.10 PROBLEMS

1. In this activity each group—as shown in the following list—asks as many questions as they can to clarify the given design statement:
   a. Design a coffee maker.
   b. Design a safe ladder.
   c. Design a safe chair.
   d. Design a safe lawn mower.

2. Use the statement "Connect two bodies of water with no gradient." Develop a series of questions to better understand the need.

3. Use the following statement: "Ice is forming on the roof with density gradient." Develop a series of questions to better understand the need.

LAB 5: Kano Model Customer Needs Assessment

This lab deals with a customer needs assessment based on the Kano model. The Kano model defines customer needs based on customer satisfaction (see Figure L5.1).

As shown in Figure L5.1, there are five types of customer needs: (1) *must-be*, (2) *one-dimensional*, (3) *attractive*, (4) *indifferent*, and (5) *reverse*. A customer need in terms of a product attribute is considered *must-be* if its absence produces absolute dissatisfaction and its presence does not increase satisfaction. An attribute is considered *one-dimensional* if its fulfillment increases the satisfaction and vice versa. An attribute is considered *attractive* if it leads to a greater satisfaction—it is not expected to be in the product. An attribute is considered *indifferent* if its presence or absence does not contribute to the satisfaction. An attribute is considered *reverse* if its presence causes dissatisfaction and vice versa. Thus, to develop a customer-focused product, it is important to do the following:

a. Keep the must-be attributes.

b. Add a good number of one-dimensional and attractive attributes.

c. Avoid indifferent attributes as much as possible.

d. Avoid the reverse attributes.

To determine the preference of customers, the Kano model provides a questionnaire, as shown in Table L5.1. A customer selects one answer (Like, Must-be, Neutral, Live-with, or Dislike) from the *functional* side and the other from the *dysfunctional* side to assert his or her preference. Moreover, the Kano model provides a definition of consistency from the customer answers, as shown in Table L5.2. For example, if a customer selects Like from the functional side and Neutral from the dysfunctional side, it means that the attribute is "attractive" to customer needs.

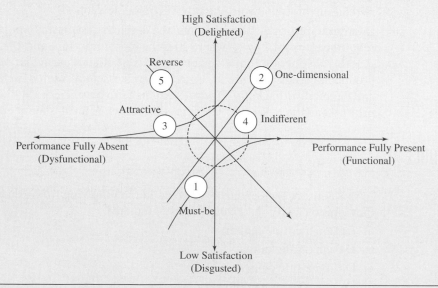

Figure L5.1 Definition of Customer Needs

Table L5.1 Example Questionnaire

Product: Room Heater *Attribute:* Steamer	*Functional:* Room Heater with a Steamer	*Dysfunctional:* Room Heater without a Steamer
	Like	Like
	Must-be	Must-be
	Neutral	Neutral
	Live-with	Live-with
	Dislike	Dislike

Table L5.2 Customer Needs Evaluation

	Dysfunctional				
	Like	**Must-be**	**Neutral**	**Live-with**	**Dislike**
Functional	Q	Λ	Λ	Λ	O
Must-be	R	I	I	I	M
Neutral	R	I	I	I	M
Live-with	R	I	I	I	M
Dislike	R	R	R	R	

Attractive (A), Indifferent (I), Must-be (M), One-dimensional (O), Questionable (Q), and Reverse (R)

Table L5.3 shows customer answers for three attributes of a mobile phone: (1) same-size keypad and display; (2) keypad small, display large; and (3) keypad-large, display small. The intention is to identify the customer preference regarding the relative size of the keypad and display of a mobile phone.

Figure L5.2 shows how the status of same-size keypad and display is determined from the answers of fifteen customers using Kano evaluation in Table L5.2. As seen Figure L5.2, this customer's attribute is "Indifferent" (does not help much to increase the customer's satisfaction). There are a relatively large number of customers that consider it to be "Reverse" (they do not want keypad and display to be equal in size). Therefore, the product developer should avoid this attribute and find solutions from the other two.

GROUP TASK Find out the status of the two other attributes in Table L5.3, and determine what the relative size of the keypad and display of mobile phone should be according to your overall evaluation.

Table L5.3 Customer Answers for Three Attributes of a Mobile Phone

Customer	Same-size keypad and display		Keypad small, display large		Keypad large, display small		Same-size keypad and display	
	Functional	Dysfunctional	Functional	Dysfunctional	Functional	Dysfunctional	Functional	Dysfunctional
1	Dislike	Must-be	Must-be	Dislike	Dislike	Dislike	Must-be
2	Live-with	Neutral	Like	Dislike	Dislike	Must-be
3	Dislike	Like	Like	Dislike	Dislike	Like
4	Live-with	Neutral	Must-be	Dislike	Dislike	Must-be
5	Neutral	Live-with	Like	Neutral	Dislike	Must-be
6	Must-be	Live-with	Like	Neutral	Dislike	Must-be
7	Dislike	Like	Like	Dislike	Dislike	Like
8	Neutral	Neutral	Must-be	Dislike	Dislike	Must-be
9	Like	Dislike	Must-be	Live-with	Dislike	Must-be
10	Neutral	Neutral	Must-be	Dislike	Dislike	Like
11	Dislike	Must-be	Must-be	Dislike	Dislike	Must-be
12	Must-be	Dislike	Must-be	Live-with	Dislike	Must-be
13	Neutral	Neutral	Dislike	Like	Dislike	Like
14	Like	Live-with	Like	Live-with	Live-with	Like
15	Like	Neutral	Like	Live-with	Live-with	Neutral

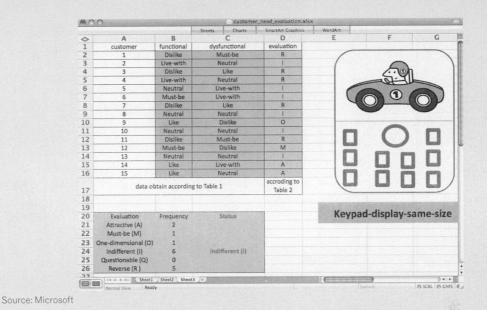

Source: Microsoft

Figure L5.2 Determining the Status of Same-Size Keypad and Display

References

[1] Ullman, D. G. *The Mechanical Design Process*. New York: McGraw-Hill, 1992.

[2] Otto K. N., and Wood K. L. *Product Design: Techniques in Reverse Engineering and New Product Development*. Upper Saddle River, NJ: Prentice Hall, 2001.

[3] Pahl, G., and Beitz, W. *Engineering Design: A Systematic Approach*. New York: Springer-Verlag, 1996.

[4] Ulrich K. T. and Eppinger S. D., *Product Design and Development* (3rd ed.), New Delhi: Tata McGraw-Hill Publications 2004.

In addition, the following books, articles, and websites were used in preparing this chapter:

CROSS, N. *Engineering Design Methods: Strategies for Product Design*. New York: Wiley, 1994.

DYM, C. L. *Engineering Design: A Synthesis of Views*. Cambridge, UK: Cambridge University Press, 1994.

HENSEL, E. "A Multi-Faceted Design Process for Multi-Disciplinary Capstone Design Projects." *Proceedings of the 2001 American Society for Engineering Education Annual Conference and Exposition*, Albuquerque, NM, 2001.

KARUPPOOR, S. S., BURGER, C. P., and CHONA, R. "A Way of Doing Design." *Proceedings of the 2001 American Society for Engineering Education Annual Conference and Exposition*. Albuquerque, NM, 2001.

SUH, N. P. *The Principles of Design*. New York: Oxford University Press, 1990.

PRODUCT CONCEPTS

CHAPTER 6

ESTABLISHING AND STRUCTURING FUNCTIONS

"Design is not just what it looks like and feels like. Design is how it works."

~Steve Jobs

In Part 2 we explained the *requirements stage*, identifying and clarifying the objectives for a design project to formulate the *design brief*. This is followed by the design team's understanding the project thoroughly and completely so that they can listen to the customers and record their input. The design team's understanding and translating the input as customer requirements (response characteristics of the design) followed. These needs were further analyzed on an individual basis and deployed as measurable metrics and units of measurement. Some of these would be mandatory while some would require input from the customers to determine the importance ratings. These inputs are referred back to the customers, and their importance ratings are obtained.

In Part 3, Chapter 6 describes the functions, or the purposive actions performed by the product. The text starts with an introduction to functions and the level of decomposition and abstraction, followed by an explanation of functional and physical representations. It then identifies the constituents of the functional representation of a device—namely, the goals, actions by the user, function, behavior, and structure and shows the origins of the various definitions for functions. It then explains purpose and action functions as explained by Deng [1], which can be used in conceptual design. The two main approaches to functional representation, *function tree* and *function structure of flows* are examined. The text concludes with a demonstration of function-based reverse engineering.

6.1 OBJECTIVES

By the end of this chapter you should be able to:

1. Describe a product in terms of its physical and functional elements.
2. Establish the five key concepts of functional representation—namely, goals, actions, functions, behaviors, and structures.
3. Understand and explain various definitions of functions.
4. Establish the *purpose* and *action* functions relating to a product.
5. Develop the function tree for a product.
6. Develop function structure in terms of flows for a product.

6.2 FUNCTIONS

The overall function of a product is the relationship between its input and output. We can further break down the function of the product to subfunctions, which identify purposive actions that the product is meant to perform. While requirements, as set by the customer, are "wish-lists" that describe what the product should do, functions are solution-neutral engineering actions that the product will perform. In this stage you will convert "wishes" into engineering terminology, which is more relevant to the design team. Remember that functions remain solution-neutral; hence, they are still a means for further ascertaining the problem. It may seem that a lot of time is spent at identifying and refining the problem and that it might seem to be an inefficient process. Some inexperienced designers become impatient and decide to short-cut or even skip this stage and try to suggest different concepts or solutions immediately. Remember that it is unlikely that the best solution can be provided to something where the best definition of the problem is not available.

Functions should consider *what* the product does (how it addresses the problem), not *how* it does it (how it addresses a solution). A function involves the following two components:

- An action verb
- A noun representing the object on which the action verb takes place

You can describe any given function with the combination of an action verb and a noun.

6.2.1 Decomposition and Abstraction

Products are designed, developed, and manufactured to perform intended functions. A product can be described in terms of (1) its physical elements or (2) its functional elements. The functional elements of a product are individual operations and transformations that contribute to the overall performance of the product. Functional elements are represented by verbs acting on nouns—for example, "transmit torque." Functional decomposition breaks down a system by functions into subsystems and sub-subsystems until the system is completely defined by its basic or action functions. Normally we stop the process of subdividing at some stage even though further division is possible. This is because the functions below that are easily understood at that level. This process of stopping without going ahead with decomposition is called *abstraction*. For example if "fill kettle" is the final function in the functional decomposition of "making coffee." It is understood that it consists of the functions "open kettle," "open faucet," "collect water," "close tap," and "close kettle." You can easily understand complex systems by using functional decomposition. The resulting description or structure of the

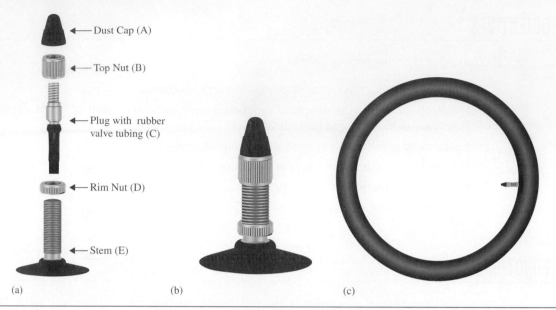

Figure 6.1: Inflating Valve of a Bicycle Tube

system in terms of its functions is called the *functional model* of the system. As an example consider the inflating valve for a bicycle tire shown in Figure 6.1.

Figure 6.1(a) shows the exploded view of the valve assembly of a bicycle tire, and Figure 6.1(b) shows the assembled view of the valve. Figure 6.1(c) shows the inflated tire fitted with the valve. The stem protrudes out of the tire, and the rubber foot is glued to the inner side of the tire. The tire is wrapped around the rim with the stem protruding through the special hole made for it in the rim. The rim nut secures the tire in the correct position on the rim. The stem houses the plug with the rubber valve tubing. The plug is secured to the stem by the top nut. On the threaded top of the plug, the dust cap is fitted to protect the air passage from blockage. When high-pressure air is connected to the plug, the air passes through the passage and reaches the inner side of the rubber valve tubing. The tubing expands and opens up for the air to reach the tube. When the air supply stops, the rubber tubing is pressed with the plug by the air inside the tube to close the inlet passage. The top nut locks the plug with the top part of the tubing tightly, and this prevents air escaping from the tire. Figure 6.2 describes the product in terms of the physical or functional elements.

6.3 DESCRIPTION OF DEVICES IN FUNCTION DOMAIN

Vermaas [2] identifies five key concepts related to functions that completely describe a device. They are

1. *Goals of device*: These define a state of affairs that a user wants to have or realize by using the device
2. *Action*: The behaviors of the users when they use the device
3. *Functions*: The desired roles played by the devices
4. *Behaviors*: The state changes the device exhibit
5. *Structure*: Materials and fields of devices, their configurations, and their interactions

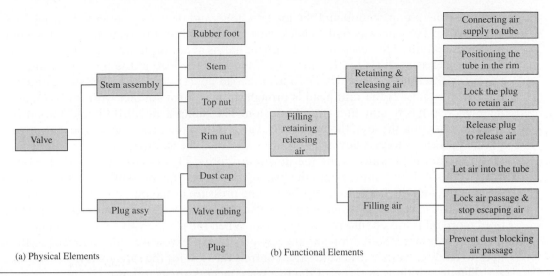

(a) Physical Elements   (b) Functional Elements

Figure 6.2: Functional and Physical Descriptions of the Valve

6.3.1 Example: A Hydraulic Jack for a Car

A hydraulic jack is used to lift a bigger load by applying a much smaller effort. Figure 6.3(a) shows the main components of the system. The reservoir is connected to the lower half through (1) the inlet check valve and (2) the needle valve. The oil leaves the reservoir through the inlet valve and returns through the needle valve. The inlet valve is a one-way valve that would permit oil to pass through, only to leave the reservoir. The lower half is divided into two

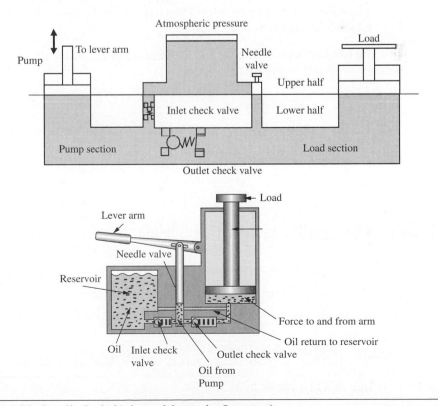

Figure 6.3: A Hydraulic Jack (Adapted from the Internet)

sections—the pump section and the load section—and they are connected through the outlet check valve. The one-way outlet check valve permits oil to flow from the pump section to the load section only. The reservoir or the upper half is at atmospheric pressure all the time.

The jack is used (1) to lift a load up, (2) to hold the load at the lifted position, and (3) to lower the lifted load safely down. To lift the load up, (1) the jack has to stand under the load on level ground, (2) the ram head is brought to have firm contact with the load, and (3) the needle valve has to stay in a closed position. The pump arm is lifted, and the oil flows into the pump section through the inlet check valve. When the arm reaches the maximum height, the pump arm is forced down. This closes the inlet check valve. Now the trapped oil in the pump section is pressurized by the downward force. This opens the outlet check valve and permits the high-pressure oil into the load section. This high pressure exerts force on the load-side piston, and the load moves up. When the pressure at the pump section becomes equal to that in the load section, the outlet check valve closes and the load remains in the same position until more oil is forced into the load section. When the needle valve opens, a contact between the high-pressure load section and the low-pressure reservoir is made. This makes the oil flow back to the reservoir. The needle valve controls the flow for the safe lowering of the lifted load.

The descriptions in terms of the five key concepts are as follows:

1. *Goals of the Device*: Raising a load by small increments of height, relieving the original support, using a much smaller effort, keeping the load at the elevated position as long as needed, and bringing it back to the original position safely when required.

2. *Action*: Setting the jack on stable ground, bringing the load and the jack ram in contact, closing the release valve and pumping the oil using the handle to lift the load, keeping the release valve closed for maintaining the load at an elevated position, and opening the release valve progressively for lowering the load safely. These are the actions the user must complete. We refer to these actions as *inputs* or *environment* as the case may be.

3. *Functions*: At a very high level you can group the functions into three main ones— namely, lift, hold, and release.

You can further decompose and arrange the goals, actions, and functions in a hierarchical manner as shown in Figure 6.4.

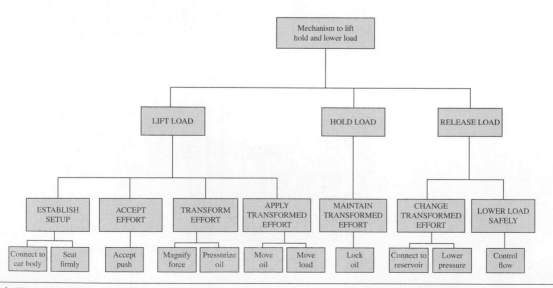

Figure 6.4: Function Tree of a Hydraulic Jack

4. *Behaviors*: Behaviors are the state changes exhibited by the device. When the effort is given at the handle bar, the lever mechanism magnifies the effort. The magnified effort pressurizes the oil. These are examples of behaviors of the structure or device.

5. *Structure*: The structure refers to the parts and subassemblies that make up the product. The hydraulic jack has 14 parts as shown in Figure 6.5. Some of these parts, like the top cap and oil plug, perform trivial functions, so we will not consider their functions at this stage but will consider them later at the detail design stage. The full parts list and their functions are given in Table 6.1.

In Figure 6.4 the purpose functions are shown by capital letters, and the action functions are shown by initial capital followed by lowercase letters.

6.3.2   Other Definitions in the Function Domain

Functional representation: Describing problems and solutions in terms of their functions and allowing reasoning about them

Functional reasoning: Functional reasoning is the process of reasoning (thinking deeply and deducing) at the functional level in order to generate concepts and evaluate them. In this context *functional synthesis* refers to the generation part of the solution. Functional reasoning

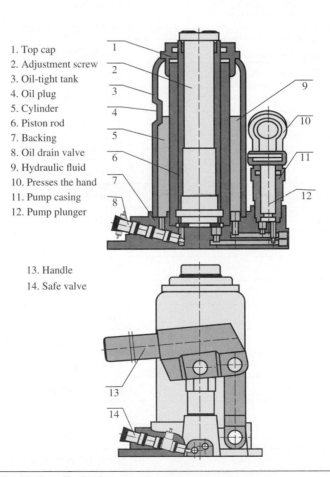

1. Top cap
2. Adjustment screw
3. Oil-tight tank
4. Oil plug
5. Cylinder
6. Piston rod
7. Backing
8. Oil drain valve
9. Hydraulic fluid
10. Presses the hand
11. Pump casing
12. Pump plunger

13. Handle
14. Safe valve

Figure 6.5: Parts of a Hydraulic Jack

Table 6.1: Parts and their Functions in a Hydraulic Jack

Part	Function
Top cap	Secure the tank and cylinder
	Guide the piston in z degree of freedom (dof)
	Stop all other degree of freedom to piston
	Stop accidental oil leak
Adjustable screw	Couple the vehicle to piston
	Translate in z direction
	Transmit load to piston Rod
Oil-tight tank	Store imported and return oil
	Supply oil to the pump
Oil plug	Import oil from outside
	Stop oil leak
	Indicate oil level in oil tank
Cylinder	Store high-pressure oil
	Guide piston in z direction
	Stop all other degree of freedom to piston
	Stabilize the pressure from oil
Piston rod	Transmit pressure force to vehicle (load)
	Position the force from piston to oppose the load
	Translate in z degree of freedom (dof)
Backing	Support the cylinder
	Support the load through the piston
	Transmit load to the ground
	Transport high-pressure oil to cylinder
	Stop high-pressure oil going to the oil tank
	Transport return oil to the oil tank
	Transport low-pressure oil to the pump
	Export very high pressure oil to environment
Oil-drain valve	Regulate return of oil to the oil tank
Hydraulic fluid	Sense the opening of the inlet valve
	Transfer into the pump
	Store energy from the pump
	Sense high-pressure valve opening

Part	Function
	Transfer to the cylinder
	Supply energy to the piston
	Sense opening of the release valve
Presses the Handle	Transmit energy from the handle to oil
(Pump plunger Input)	Change the magnitude of force
	Store high-pressure oil
Pump casing	Guide plunger in z direction
	Stop all other degree of freedom to plunger
	Stabilize the pressure from oil
Pump plunger	Extract low-pressure oil
	Transmit energy to oil
	Transport high-pressure oil to cylinder
Handle	Import human-effort force
	Change magnitude
	Transfer force to input to pump plunger
Safety valve	Export very high-pressure oil to environment

thus consists of two parts: (1) a functional representation of the object to be reasoned about and (2) a reasoning scheme.

Functional reasoning has several uses, some of which are listed as follows:

1. To fix the functional structure of a new design without referring to any solutions.
2. To describe and redesign existing products
3. To complement physical descriptions of artefacts by explaining the roles of their constituent parts
4. To explain and exchange ideas among design teams at different geographic locations
5. To archive existing designs.

6.3.3   Position of Functions

This section tries to establish a position for functions in the hierarchy. At the top level are the goals or the purpose of the product. Then the product has to be in the right environment and should be given the right inputs. If these are right, the product performs the functions. But functions are generated by the behavior of structures. It is worth noting here that the right inputs or environment are fundamental to the correct functioning of the product. The sequence described previously is correct from the design perspective. However, from the use perspective the structure is set up in the right environment, and the right inputs are given to the structure. This in turn generates the right behavior and hence the function. The performance of the intended function achieves the desired goal. This is schematically shown in Figure 6.6 as the hierarchy of function domain.

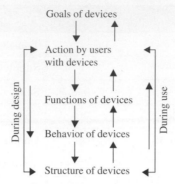

Figure 6.6: Hierarchy of Function Domain

The hierarchy shown by Figure 6.6 is overly detailed for simple products, and function is an abstract representation. Therefore, the functional representation of a product does not contain all the five elements (goals, actions, functions, behaviors, and structures). Often some are bypassed, by and hence we extend or adapt the definition of the function. This leads to (1) goals being described as functions and (2) behavior being described as functions, and this has produced various definitions and types for functions. Consider an illustration:

Figure 6.7 shows a single-stage gearbox connecting two parallel shafts, A and B. We can describe the function of the gearbox in the following ways:

1. *Function as the goal*: Transfer motion between parallel shafts. This is defined at a high level of abstraction and often is called the *purpose function*.

2. *Function as the desired behavior*: Transmit energy from Shaft A to Shaft B. This is the desired behavior from the device. This is defined at a moderate level of abstraction and can be called a Purpose Function or an Action Function since the device does something exhibiting behavior.

3. *Function as the desired effects of behavior of the structure*: Turn shaft B at lower speed and high torque. These are the action functions that the structure performs to provide the expected behavior.

Because of the possibility of defining functions from different perspectives there are several definitions for function, some of which are given in Section 6.4.

(6.4) DEFINITIONS OF FUNCTIONS

1. Functions fulfill the expectations of the purposes of the resulting artifact.
2. Function is an intended purpose of a device.

Shaft A

Shaft B

Figure 6.7: A Single-Stage Gear Box

3. Function is the logical flow of energy, material, or signal between objects or the change of states of an object due to flows.
4. Functions are the intended input/output relationships.
5. Function is the description of the behavior abstracted by users in order to make use of it.
6. Function is the fulfilment of a goal or purpose.
7. Function is the intended behavior.
8. Function is referred to at a higher level of abstraction and as the purpose of the product.
9. Functions convey what components do.
10. Function is the intended effect that a device has on its environment.
11. Function is an action of product on its inputs and outputs.

6.4.1 Definition of Function by Miles [3]

Lawrence D. Miles, the founder of *value analysis*, made the first definition of function in the *functive* format. A functive is a pair of words where a verb followed by a noun, as in "turn shaft." Although the word *functive* is not widely used, it is used exclusively in value analysis to emphasize that a function should be described by a verb followed by a noun. Function analysis translates the structure of a product into a structure of words.

Table 6.2 lists verbs and nouns recommended for use in defining functions in value analysis [4] . This listing is made on the supposition that any product will have a prime function. A product may have a secondary function in addition to the prime function. Within the context of this book, the purpose of defining functions is to use them in the conceptual design process. The functive format, where only a verb and noun are used to describe a function, is hard to achieve at the product-definition level; often a verb/noun phrase format is used in the early stages.

Table 6.2: List of Verbs and Nouns

Verb List				
Absorb	Control	Hide	Minimize	Rotate
Actuate	Convert	Hold	Modulate	Satisfy
Aid	Create	Ignite	Mount	Seal
Allow	Direct	Impart	Move	Secure
Amplify	Ease	Impede	Open	Shield
Apply	Emit	Induce	Position	Shorten
Assist	Emphasize	Inject	Preserve	Space
Assure	Enclose	Instruct	Prevent	Standardize
Avoid	Ensure	Insulate	Promulgate	Steer
Change	Establish	Interrupt	Protect	Support
Close	Exude	Limit	Receive	Suspend
Collect	Facilitate	Locate	Rectify	Time

Continued on next page.

Comfort	Fasten	Maintain	Reduce	Tolerate
Conduct	Filter	Maximize	Repel	Transfer
Contain	Guard	Mesh	Resist	Transmit

Noun List				
Access	Decoration	Flux	Noise	Task
Aesthetics	Density	Force	Odor	Time
Area	Dependability	Friction	Oxidation	Torque
Care	Deterioration	Heat	Pressure	Uniformity
Catalysis	Direction	Horsepower	Protection	User
Chromaticity	Dust	Image	Radiation	Variation
Color	Emissivity	Information	Repair	Vibration
Corrosion	Energy	Injury	Rust	Voltage
Current	Flow	Insulation	Stability	Volume
Damage	Fluid	Light	Status	Weight

Based on Fowler T. C. "Value Analysis in Design," Van Nostrand Reinhold, 1990.

6.5 FUNCTIONS FOR CONCEPTUAL DESIGN

Deng [1] proposes two function types, which are useful in conceptual design—*purpose function* and *action function*. To define the product concept, first produce a functional model of the product. This model should describe the goals, purpose functions (what is needed to achieve the goals), and the action functions (what is needed to achieve each of the purpose functions). The two functions relate to the different levels of a design hierarchy and of abstraction. Generally, the overall function and some of its subfunctions at the upper-level design hierarchy are expressed as a design intention, with the relevant concrete actions not known; hence, they are purpose functions. This overall function and the upper-level subfunctions may need to be implemented by some lowerlevel subfunctions via certain physical behaviors.

At the functional level, all the functional models give the desired output for the given specifications and constraints on the technical system. They are derived from the customer requirements and thus, together with the specifications, form a clear definition of the problem. Two main methods of functional modeling are in practice. They are

- Function tree modeling
- Function structure based on flows

6.6 FUNCTION TREE MODELING

A function tree provides an overview of the whole system in a hierarchical manner using conventional function analysis. It is based on functions of existing systems and gives increased capability to formulate problems from different views. If the function doesn't meet functional requirements at first, it usually has supporting and correcting functions. There are lower-level functions within its main function. Functions can help in defining subfunctions but may not relate directly to the primary function. In addition, very low-level functions cannot be seen in the function tree because of abstraction. The main difference between a function tree and function analysis is that the function tree illustrates the hierarchical structure of a product's functions. The function tree starts from the function analysis, however it extracts functions and changes these functions to function requirements and arrange those in a hierarchical structure.

6.6.1 Steps for Creating the Function Tree for an Existing System

STEP 1. Extract the Functional Requirements

Take a picture or make a sketch of the product and label its parts. This will ensure correct referencing of the parts during the construction of the function tree. Label at a sufficiently high level of abstraction to avoid cluttering with details. Consider each labeled item, and establish the purpose for its inclusion in the product from a functional point of view. Then list all the functions that part performs to achieve this purpose. Normally functions are described in the verb+object format. But for developing new products, the picture or sketch should give more details; therefore, it should be modified to the form of a function requirement with its own adjectives and descriptions. This will give you an opportunity to think more objectively as well as an opportunity to improve the system.

STEP 2. Arrange the Functions as a Function Tree

Establish a concise function for the product in the "verb, noun, and adjectives" format. This will form the root of the tree. Consider the functions list, and extract the functions describing the purpose for the inclusion of each unit. These form the next level purpose functions for the product. Now consider the remaining functions under each group. These are the action functions. Thus the leaves of the tree will be action functions, and all higher-level nodes will be purpose functions.

STEP 3. Analyze and Expand the Function Tree

Evaluation is the process of assessing the degree of fulfillment of the purpose for which the functions are created. A systematic scanning from the root to the leaves will identify the degree of fulfillment along with the causes for deviation. A function tree shows only the main philosophies of the system and focuses on the evolution of the system. It is not focused on the resources near the system, and it identifies only what function should be improved and what function does not act well so that they can be acted upon in new product development. You can include additional functions, for both purpose and action, at this stage.

6.6.2 Establishing the Function Tree for a Kick Step Stool

Extract Functional Requirements

The subsystems are standing platform, platform legs, stepping platform, conical drum, and wheels. The functions can be listed as shown in Figure 6.9.

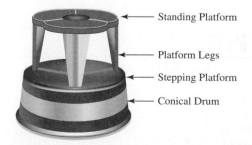

Figure 6.8: A Kick Step Stool

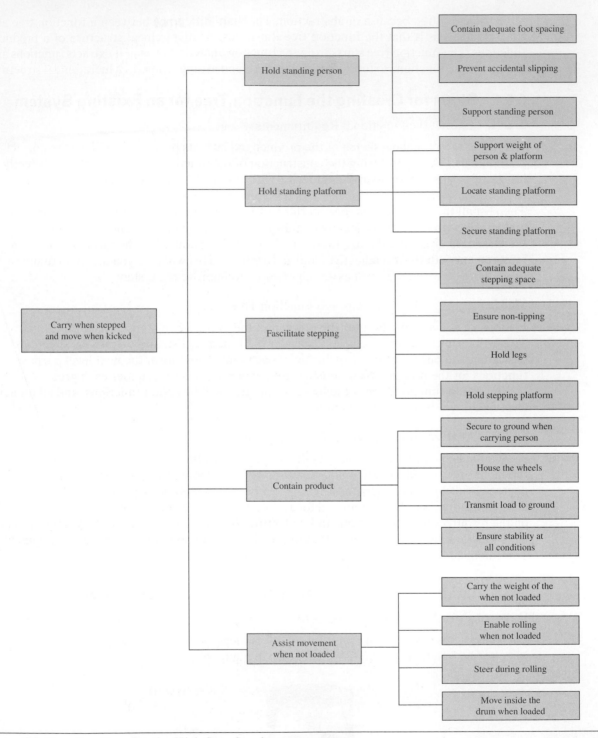

Figure 6.9: Function Tree for the Kick Step Stool

6.6.3 Function Tree Modeling for a New Product

You can establish function tree modeling for a new product using an affinity diagram, which is a graphic tool designed to help organize loose, unstructured ideas generated in brainstorming. In this method, collect and group disparate but related ideas into meaningful categories during the brainstorming process. Each team member writes down silently all the functions he or she can think of on individual cards or post-it stickers and posts them on a bulletin board or a wall. Once all team members completed writing down their ideas, the group discusses them so everyone understands what the others have written. The team then groups the related functions together, and a heading function or notation is given. The team continues this process of clustering and grouping under one title, which is continued until the entire hierarchy has been formed.

6.6.4 Function Tree Modeling for a Reading Assistant

The function list established for a reading assistant using the affinity diagram method is given here:

1.	Present	book
2.	Lock	Required page
3.	Locate	Chosen presenter
4.	Lock	Height (presenting)
5.	Allow	Required height
6.	Allow	Height adjustment
7.	Allow	Book area (single book)
8.	Allow	Six book presenters
9.	Allow	Reading angle adjustment
10.	Position	Chosen reading position (angle)
11.	Locate	Chosen presenter
12.	Position	Chosen presenter
13.	Use	Collapsible structural arrangement
14.	Allow	Quick opening up
15.	Allow	Quick shutting down
16.	Ensure	Minimal stowed space
17.	Allow	Reading light fixture
18.	Ensure	All-reading-position Stability
19.	Establish	Elegant look
20.	Provide	Mobility
21.	Allow	Pencil tray
22.	Allow	Coffee mug stand
23.	Use	Lockable base

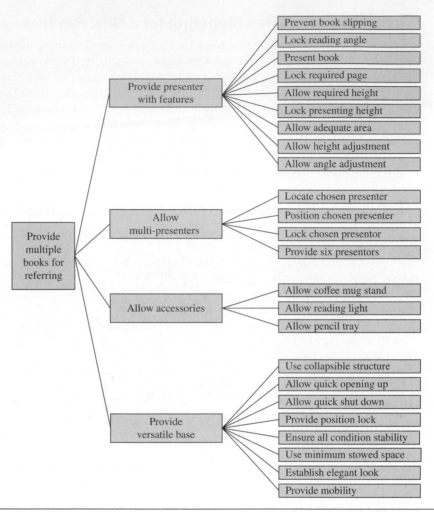

Figure 6.10: Function Tree for the Reading Assistant

These functions have been incorporated in a function tree, as shown in Figure 6.10.

6.7 FUNCTION STRUCTURE MODELING BASED ON FLOWS

The overall function of a product is the relationship between its inputs and outputs. The function of the product can be further broken down into subfunctions that identify purposive actions the product is meant to perform. A black-box model is an input/output model that is useful for figuring out the functions of engineering systems. In this model a rectangle called a black box represents the system under consideration while arrowheads on the left and right side of the rectangle mark inputs and outputs. The inputs and outputs can be of the three types: (1) material, (2) energy, and (3) signals. Figure 6.11 shows the representation. Enter all inputs and outputs to obtain a clear visualization of the system being considered.

A function structure breaks down the overall function as given in the "black-box diagram" into a hierarchy of subfunctions, represented by the function blocks and connected

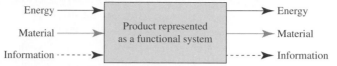

Based on Dym, C. L. Engineering Design: A Synthesis of Views. Cambridge, UK: Cambridge University Press, 1994.

Figure 6.11: Representation of a Product Function as Input and Output of Flows

by the flows. The flows are the material, energy, and signal. A function is the operation that the product performs on a flow or a set of flows to transform it from its input state to its output state.

1. Energy flow examples include electrical energy, potential energy, kinetic energy, heat, sound, and so forth. For example, a motor transforms electrical energy into kinetic energy.
2. Information flow is a signal provided to a device or set of data that the device acts on. For example an "on/off" switch provides a signal to a device.
3. Material flow is any physical entity that the device transforms. For example, a machine cuts the metal to shape.

Functional description: A functional description of an activity or a product gives a function (verb) acting on a flow (object) such as cut metal, hold book, present in position. Functional descriptions are solution neutral and do not refer to any particular solution.

Steps to construct a function structure

1. Understand and plan the functions the product performs.
2. Create a black-box model of the product describing the overall function of the product and showing the input and output flows.
3. One by one, trace the flows from input to output by asking the following two questions: (a) What happens to the input and the outputs inside the box? (b) Where does the output come from?
4. Assemble and combine the traces.

Example 6.1: Function Structure of a Radio (Adapted from Dym and Little [5])

- A radio receives a chosen airborne RF signal, amplifies it as required, and broadcasts it. Figure 6.12 shows the black box diagram for radio.
- The inputs are (1) electric power from the mains, (2) RF signal (another form of energy) as broadcast by the radio stations, and (3) control signals (e.g., volume, station, etc.).
- The outputs are (1) waste heat energy, (2) the desired signal as sound amplified to the required volume, and (3) various parameters that indicate the status of the outputs such as the volume and frequency.

Trace of Input 1: the Power

Common power outlets give power at 240V or 110V in the AC form. Often this has to be transformed to 12V in the DC form. In the process some power is wasted and becomes heat. This heat has to be considered as an output. There can be some waste heat from other units. All the heat must be collected to form the output "waste heat." Figure 6.13 shows this schematically.

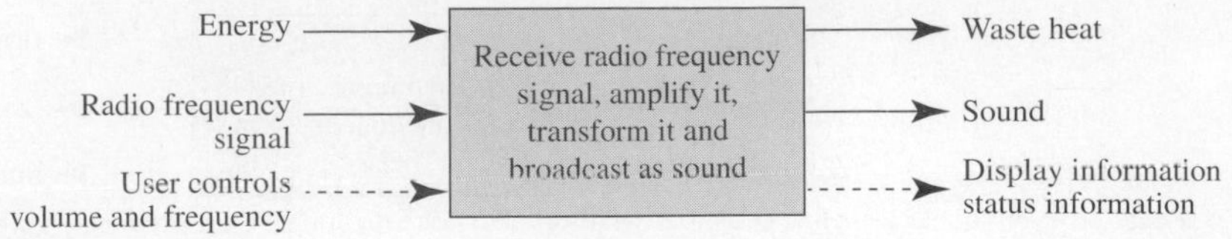

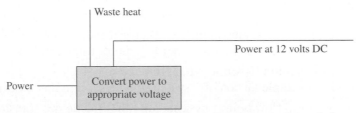

Based on Dym, C. L. Engineering Design: A Synthesis of Views. Cambridge, UK: Cambridge University Press, 1994.

Figure 6.12: Black-Box Diagram of a Radio

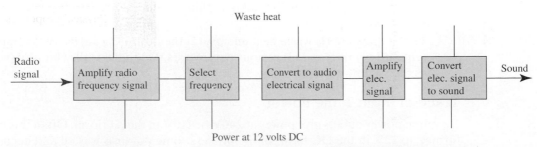

Based on Dym, C. L. Engineering Design: A Synthesis of Views. Cambridge, UK: Cambridge University Press, 1994.

Figure 6.13: Trace of Electric Power

Trace of Input 2: the RF Signal

The radio circuitry first amplifies the RF signal in the air to usable levels using the electric power. After amplification. the required frequency based on user input is selected. The chosen frequency is then converted into audio electric signal, after which the electric power amplifies the audio signal. Finally the amplified electrical signal is converted to sound and broadcast. Figure 6.14 shows the trace of the RF signal.

Trace of Input 3: User Control Signals

User controls have two types of inputs: (1) frequency or station selection and (2) volume selection. Output in this flow is the display of signal choice and volume, along with the status of indication. The display unit will take electrical energy for its operation and will waste a portion as heat.

Assemble and Integrate

Now we can integrate the three flows to create the function structure as shown in Figure 6.15.

Based on Dym, C. L. Engineering Design: A Synthesis of Views. Cambridge, UK: Cambridge University Press, 1994.

Figure 6.14: Trace of RF Signal

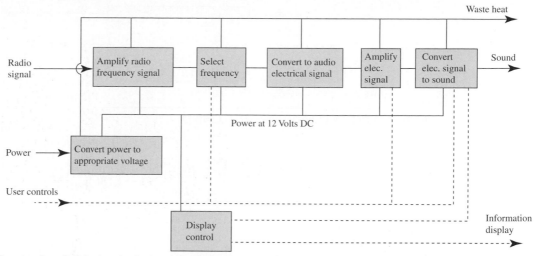

Based on Dym, C. L. Engineering Design: A Synthesis of Views. Cambridge, UK: Cambridge University Press, 1994.

Figure 6.15: Function Structure of a Radio

Example 6.2: Hand Held Nailer (Adapted from Ulrich and Eppinger [6])

The nail gun shown in Figure 6.16 uses compressed air to drive nails into various materials. Therefore, the functional description of the product is "drive nails." The input flows are compressed air (a form of energy), and nails (the material involved), and the triggering signal is information. The desired output flow from the product is a driven nail; other outputs are waste heat and noise. Figure 6.17 shows the black-box diagram.

Trace of Input 1: Nails

The nails are loaded into the gun in various quantities, and they are stored appropriately. On receiving the triggering signal one nail is isolated. That nail then receives the energy and is driven into the material. Figure 6.18 shows a representation.

Figure 6.16: A Nail Gun

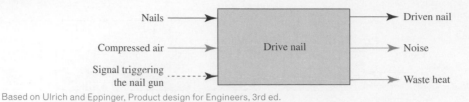

Based on Ulrich and Eppinger, Product design for Engineers, 3rd ed.

Figure 6.17: Black-Box Diagram of a Nail Gun

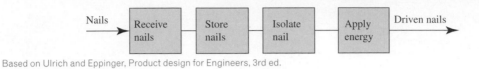

Based on Ulrich and Eppinger, Product design for Engineers, 3rd ed.

Figure 6.18: Trace of Nails

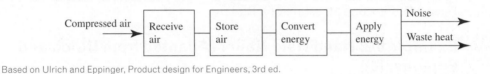

Based on Ulrich and Eppinger, Product design for Engineers, 3rd ed.

Figure 6.19: Trace of the Compressed Air Energy

Trace of Input 2: Compressed Air

Although compressed air is a material in this product, it carries energy; therefore, it is treated as energy. Figure 6.19 shows a representation.

Trace of Input 3: Trigger Signal

The user gives the trigger signal, and therefore a sensor is required to be in the product. The signal then branches into two paths so as (1) to isolate the nail and (2) to apply the energy. Figure 6.20 shows a representation.

Assembling and Integrating the Traces

The traces are then assembled and integrated, and in the integrating phase the signals and converted energy are connected to the trace of the nails. Figure 6.21 shows the assembled and integrated function structure.

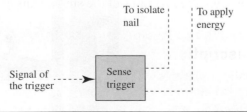

Figure 6.20: Trace of the Signal

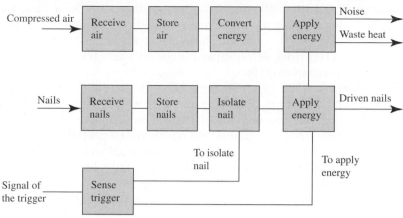

Based on Proceedings of Engineering Design Conference 2000, Published by the IMechE publishers

Figure 6.21: Integrated Function Structure

 ## 6.8 REVERSE ENGINEERING: ESTABLISHING THE FUNCTIONAL MODEL OF AN EXISTING PRODUCT

In many instances, the product already exists in the market. In this case, we can analyze an existing design by identifying its subsystems in generic terms to establish the product concept or specification in solution-neutral form; we can then generate a function tree based on our analysis. We can use the function tree and other data generated to identify the weak subsystems and replace them, or we can use latent technological developments when they prove to be advantageous. This process is commonly known as *reverse engineering* and essentially amounts to functional decomposition in reverse. This practice is important when a product is already available and the designers wish to benchmark, adapt, or improve the design.

Reverse engineering starts with the description of the process or the performing of the functions by the product. This is done to identify the underlying principles of the product and its subsystems. From the description of the process, it is possible to identify the subsystems and the constituent parts (a parts tree) of the product under consideration. This gives the *embodiment design* of the product. The process in carrying out reverse engineering can be summarized as follows:

1. Description of the process
2. Breaking down the product into subsystems
3. Establishing the functions of the subsystems

6.8.1 Reverse Engineering Example: Dishwasher Process Description

Dirt is usually a mixture of fatty and solid particulate material (although solids may be present without fat) and will usually contain any or all of the following: pigment, carbon, clay, iron oxide, protein, and fat. When particulate dirt is present on its own, it can be removed by simply washing with water. When this particulate dirt is mixed with fat, however, water is not sufficient to remove the dirt. The fat adheres to the surface, and a chemical is required to reduce this clinging force and remove the dirt. The chemical must also prevent the dirt from redepositing on the surface. These chemicals are the detergents or soaps that are used for washing. Over time, detergent manufacturers have tried to develop detergents that require less and less mechanical effort.

Detergents use two main processes to remove dirt. The first method increases the contact angle of the fat with respect to the surface to which it is attached. This causes the fat to roll up into a ball, which then can be removed easily. The second process uses the polar chemical characteristics of dirt. Intermolecular attraction between a part of the detergent known as the surfactant and the fat results in the fat being more attracted to the detergent solution than to the surface. Hence, the fat leaves the surface and is held in the solution. The first of these two processes takes place at higher temperatures than the second. For this reason, it is clear that the washing process must take place at very specific temperatures. When a program is set, the detergent is flushed by the flow of water. Then the dishes and the water mixed with the detergent are agitated and heated according to the pattern predetermined by the program selected. This ensures the removal of all dirt. The dirty water with the detergent is then pumped out, and fresh water fills the machine for rinsing purposes. Drying is then optionally applied by blowing hot air onto the dishes.

Functional Subsystems

From the description of the process, we have identified the following functional subsystems:

- Water addition system
- Detergent addition system
- Heating system
- Container and agitation system
- Dirty water removal system
- Rinsing system
- Drying system

Figure 6.22 illustrates the reverse engineering of the dishwasher, starting from the parts (embodiments) at the bottom of the chart and working up toward the functions. This demonstrates the power of remaining solution-neutral at the beginning of a design, as it leaves open the option to venture into adapted designs and different products (washing machine and dishwasher), still based on similar functions and concepts. In this case, many of the parts that perform similar functions in the two products can be standardized, which will lead you to a single core design and then to variants that differentiate between the two products. This is the basis for *design reuse*, which is commonly used by companies that want to diversify their product offerings while remaining within their own expertise and reducing costs by reusing many of their existing parts.

6.9 REDESIGN METHODOLOGY

In most cases, designers redesign by repeating the cycle and focusing only on the point at which the desired modification should be made. This practice often causes multiple constraint violations and at times overlooks the functional requirement. A redesigning approach should allow the designer to perform a stepwise refinement and to integrate the initial design concept to generate possible variants without affecting the overall functional requirement.

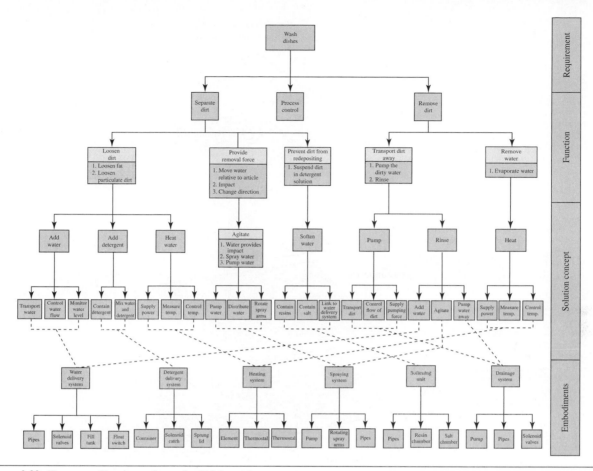

Figure 6.22: Reverse Engineering of a Dishwasher

In addition, the approach should enable the designer to allow change at any level, whether at a part level or at a system level.

The main aim for redesigning an object based on functional reasoning is to enable identification of possible design variants in order to fulfill a desired functional need. The present approach described by Hashim [7] includes both structure and function in the same schematic representation and involves:

- The structural and functional analysis of an existing design
- The production of abstract representations of design functions and design entities using conceptual graphs
- Design interrogation based on a functional tree approach

The functionality of each entity is deduced by observing the physical interactions between the entity and neighboring design entities. The analysis of structure and function is layer-based—that is, through hierarchical levels, which enables redesign to take place at any particular level of the hierarchy without having to fully develop the schematic representation. This approach also allows the overall function of the design to be understood in terms of the function or functions of individual design entities (see Figure 6.23).

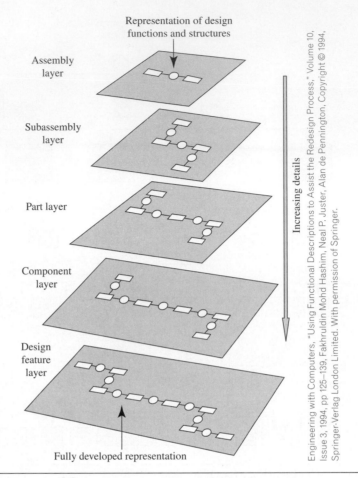

Engineering with Computers, "Using Functional Descriptions to Assist the Redesign Process," Volume 10, Issue 3, 1994, pp 125–139, Fakhruldin Mohd Hashim, Neal P. Juster, Alan de Pennington, Copyright © 1994, Springer-Verlag London Limited. With permission of Springer.

Figure 6.23: Design Hierarchy [7]

Conceptual graphs are logical forms that depict the relationship between concepts and conceptual relations. These concepts and their relations are represented by boxes and circles, respectively. In this approach concept nodes (represented by boxes) represent design entities, while the relation nodes (represented by circles) represent the functional relationship between two or more neighboring design entities. The entity that performs the function is called the *functor*, and the entity to which the function is being applied is called the *acceptor* (see Figure 6.24).

Design interrogation involves the process of deductive reasoning to generate possible design solutions that fulfill the required function.

Figure 6.24: Conceptual Representation of Functor, Function, and Acceptor

6.9.1 Screwdriver

The functional description of the screwdriver is made up of three elements: receive, transmit, and apply. The physical description also is made up of three elements: handle, shank, and tip. Figure 6.25 gives a functional and physical description of a screwdriver.

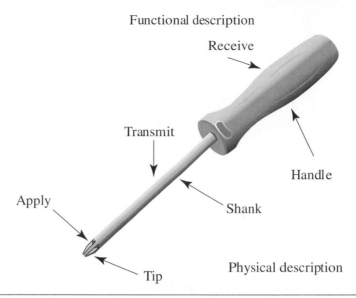

Functional description

Figure 6.25: Functional and Physical Description of a Screwdriver

A screwdriver receives a torque and positioning effort from a hand, transmits it to the tip or application unit, and applies it to the screw. This torque may be used to drive in or draw out screws.

Figure 6.26 shows a design interrogation for each of the screwdriver parts and their functions. The left side lists the functions with functional explanation along with the associated part of the screwdriver. Understanding the functions allows for redesigning the product while maintaining the functions.

Possible redesigns based on the interrogation are shown in Figure 6.27.

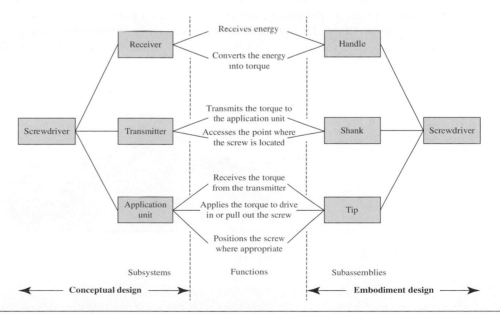

Figure 6.26: Design Interrogation of a Screwdriver

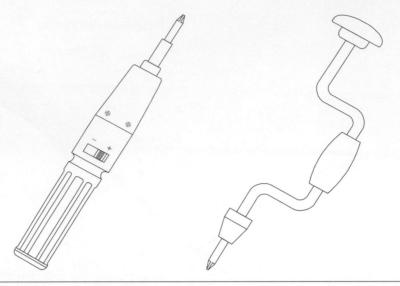

Figure 6.27: Redesign of a Screwdriver

6.9.2 Stapler

A stapler works by simply forcing a U-shaped piece of wire metal (staple) through an object, usually paper. Once the staple has penetrated the paper, the legs of the staple are bent to form a clasp. These days, many staplers have an anvil in the form of a "pinning" or "stapling" switch. This allows a choice between bending the staple legs either in or out. The outward-bent staples are easier to remove and are for temporary fastening or "pinning," The inward-bent staple, which is what has been used traditionally, is used for permanent fastening. For simplicity, the stapler taken in this example is a basic standard 10 size stapler.

As seen in Figure 6.28, the stapler consists of basically four parts: the head, base, magazine, and connecting pin. Figure 6.29 shows the structural decomposition diagram, Figure 6.30 shows which of the parts are connected to each other, and Figure 6.31 shows the functional diagram. Functionally, the head serves as a point of application of force when sheets of paper are stapled together. The base acts as an anchor and sits on a flat surface. The magazine holds the staples and is connected to the head and base by the connecting pin.

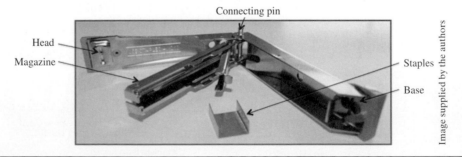

Figure 6.28: Basic Parts of a Stapler

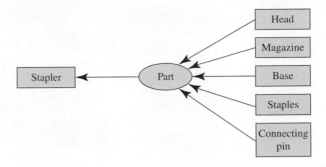

Figure 6.29: Structural Decomposition

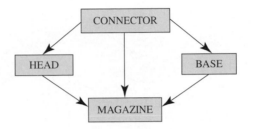

Figure 6.30: Physical Connectivity

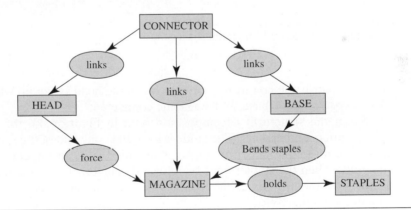

Figure 6.31: Functional Diagram

On the base there is an anvil, which is a metal plate that the staple is forced into. This makes the staples bend to hold the pages together.

On further decomposition, the main parts are broken down further, as shown in Figure 6.32. Functionally, as shown in Figure 6.33, the head houses the recoil flap that brings the head back to position once the staples have been inserted into the magazine. At the back

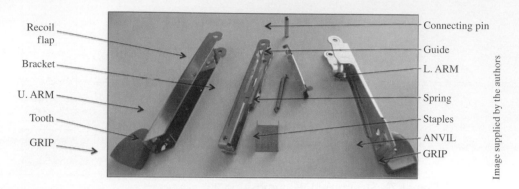

Figure 6.32: Parts (Exploded) of a Stapler

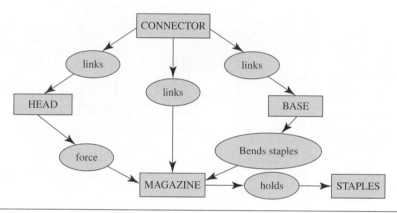

Figure 6.33: Functional Diagram for a Stapler

of the stapler is the connector pin, which lets the head swing up when staples are loaded and also keeps the head, base, and magazine connected.

From the structural decomposition seen in Figure 6.34, the magazine consists of the spring, guide, and bracket. The guide is located at one end of the spring and pushes the staples down the magazine once the stapler is reloaded. The spring can be pulled back along with the guide when the stapler must be reloaded. The bracket of the magazine is what houses all the parts of the magazine. Figure 6.35 shows the parts connectivity and figure 6.36 shows the functional connectivity.

In this example, the base consists of the lower arm, the anvil, and, as on the upper arm, a grip. The anvil forces the staple to bend inward when force is applied to the stapler head. The lower arm anchors the stapler when it is placed on a surface, whereas the grips provide for comfortable holding when the stapler is held in the user's hand.

The relation between the various parts of the stapler with their behaviors and functions are given in Table 6.3. Please refer to Figures 6.28 and 6.32 for the parts.

Design problems such as staples penetrating through papers can be easily identified for potential redesign of parts that contribute to the product failure. Producing possible redesigns is left as an assignment for students.

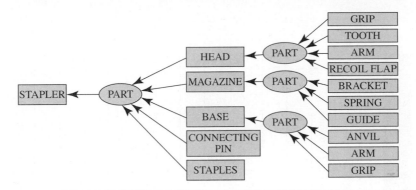

Figure 6.34: Structural Decomposition

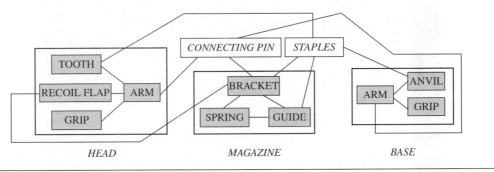

Figure 6.35: Physical Connectivity

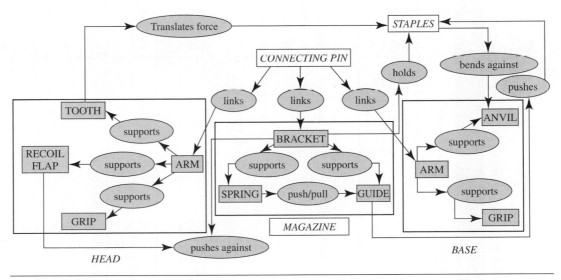

Figure 6.36: Functional Connectivity

Table 6.3: Relation between Stapler Parts and Their Behaviors and Functions

Part		Function	Behavior
1. HEAD	1.1 Tooth	Translates force to staple	Tooth presses staple
	1.2 Recoil Flap	Repositions arm after stapling	Provides a spring-like recoiling action
	1.3 Arm	Point of force application	Takes the external load applied on its surface
	1.4 Grip	Provides for comfortable grip	Seats fingers comfortably
2. MAGAZINE	2.1 Bracket	Supports spring / Supports guide / Holds staples	Has a hook for holding one end of spring / Has a notch for holding one end of spring / Walls provide support
	2.2 Spring	Eases movement of guide	Push/pull(s) guide
	2.3 Guide	Keeps staple in line with tooth	Pushes staples against bracket
3. BASE	3.1 Anvil	Bends staple	Grooves provide for bending action
	3.2 Arm	Acts as an anchor	Flat surface allows for anchoring
	3.3 Grip	Provides for comfortable grip	Seats fingers comfortably
4. CONNECTING PIN		Connects magazine with head and base / Allows the arm of the head to swing	Goes through grooves in all parts and is fixed at both ends / Acts as a pivot
5. STAPLES		Binds sheets of paper together	Pierces through the sheets of paper and bends inward at the ends

6.10 CHAPTER SUMMARY

1. The overall function of a product is the relationship between its inputs and output.

2. Functions are solution-neutral engineering actions that the product will perform.

3. The combination of an action verb and a noun are used to describe any given function.

4. Products are designed, developed, and manufactured to perform intended functions, and a product can be described in terms of its (1) physical elements or (2) functional elements.

5. Functional elements of a product are individual operations and transformations that contribute to the overall performance of the product.

6. The process of stopping without proceeding with decomposition is called *abstraction*.

7. The resulting description or structure of the system in terms of its functions is called the *functional model* of the system.

8. Five key concepts related to functions to completely describe a device are

   a. *Goals of device:* These are results that users want to have or realize by using the device.

   b. *Action:* Actions are the behaviors of the users when they use them.

   c. *Functions:* Functions are the desired roles played by the devices.

   d. *Behaviors:* Behaviors are the state changes exhibited by the device.

   e. *Structure:* Structure includes materials and fields of devices, their configurations, and their interactions.

9. In functional representation, problems and solutions are described in terms of their functions.

10. Functional reasoning is the process of reasoning (thinking deeply and deducing) at the functional level to generate concepts and evaluate them.

11. Functional reasoning thus consists of two parts: (1) a functional representation of the object to be reasoned about and (2) a reasoning scheme. For example, a function tree is a representation. The reasoning scheme says top-level functions are purpose functions, and the leaves of the tree are action functions.

12. A purpose function is a description of the designer's intention or the purpose of a design, and it is therefore abstract and subjective.

13. An action function is an abstraction of intended and useful behavior that an artifact exhibits.

14. Normally, a functional representation of a product does not contain all the five elements (goals, actions, functions, behaviors, and structures). This leads to several different definitions of functions.

15. Definitions of *function* include

   a. Functions fulfill the expectations of the purposes of the resulting artifact.

   b. Function is an intended purpose of a device.

   c. Function is the logical flow of energy, material, or signal between objects or the change of states of an object due to flows.

   d. Functions are the intended input/output relationships.

   e. Function is the description of the behavior abstracted by users in order to make use of the product.

   f. Function is the fulfillment of a goal or purpose.

   g. Function is the intended behavior

   h. Function is referred to at a higher level of abstraction and as the purpose of the product.

    i. Functions convey what components do.

    j. Function is the intended effect that a device has on its environment.

    k. Function is the action of a product on its inputs and outputs.

16. Lawrence D. Miles, the founder of value engineering, defined function first in the *functive* format, where a verb acting on a noun describes the function.

17. Miles recommends a list of verbs and nouns.

18. Identifying the purpose functions and action functions is useful for conceptual design.

19. A function tree provides an overview of the whole system in a hierarchical manner.

20. The procedure for creating a function tree for an existing system is as follows:

    a. Extract the functional requirements; Take a picture of the product, label it, and describe the functions performed by the various components.

    b. Arrange the functions in the form of a function tree. From the lower-level functions to the higher-level ones, arrange a hierarchy. Give descriptions for intermediate functions.

    c. Analyze, trim, and expand the function tree This is where the function tree is evaluated to see whether it actually represents the product.

21. An affinity diagram is used to create the function tree for a new product.

22. To produce the function structure of flows, a black-box model is used. A black-box model is an input/output model that is useful for determining the functions of engineering systems.

23. In a black-box model the inputs and outputs can be of the three flows: (1) material, (2) energy, and (3) signals.

24. A function structure breaks down the overall function as given in the black-box diagram into a hierarchy of subfunctions represented by the function blocks and connected by the flows.

25. The steps taken to construct a function structure are

    a. Understand and plan the functions the product performs.

    b. Create a black-box model of the product that describes the overall function of the product and shows the input and output flows.

    c. One by one, trace the flows from input to output by asking the following two questions: (1) What happens to the input and the outputs inside the box? (2) Where does the output come from?

    d. Assemble and combine the traces.

6.11 PROBLEMS

1. Discuss the following statement as it relates to the stage of the design process you have achieved thus far. The design process is the divergent–convergent process.

2. Why would it be wrong to state a mechanism of performing a function rather than to generalize and name the function in building up a function structure?

3. What is the difference between functions and subfunctions?

4. Would it be possible for different design teams to come up with different function structures for the same system or product? Why or why not?

5. As a team, choose one of the following products to take apart. Name the parts and what function they serve. Then build a functional structure for the product.

    a. Paper hole puncher

    b. Cellular phone

    c. Computer mouse

    d. Umbrella

    e. Tape dispenser

    f. Can opener

6. Develop a function tree to produce a device that is capable of converting wood chips into a fire log.

    a. Express the overall function for the system.

    b. Break down the overall function into a set of subfunctions.

    c. Draw a block diagram showing the interactions between the subfunctions.

    d. Draw the system boundary.

7. List the steps needed to generate the functional structure.

8. What is the difference between the functional structure and an objective tree?

9. Why can't you find a function for each objective you generate in the objective tree?

10. What function would you propose to satisfy the "must be safe" objective for the newspaper vending machine?

11. Generate a function tree for a coin-sorting device.

12. Generate a function tree for a leaf-removing device.

13. State whether the following statements are true or false.

Statements	T	F
1. Products are designed, developed, and manufactured to perform intended functions.	❏	❏
2. A product can be described in terms of (1) its physical elements or (2) its functional elements.	❏	❏
3. Functional elements of a product are individual operations and transformations that contribute to the overall performance of the product.	❏	❏
4. Functional decomposition breaks down a system by functions into subsystems and sub-subsystems	❏	❏
5. Abstraction is the process of stopping subdividing at some stage because the functions below that are easily understood at that level.	❏	❏
6. A functional model is the resulting description or structure of the system in terms of its functions.	❏	❏
7. The goals of a device are what the users want to have or realize by using the device.	❏	❏
8. Actions are the behaviors of the users when they use a device.	❏	❏
9. Behaviors are the state changes exhibited by the device.	❏	❏
10. Behavior is viewed as an observable characteristic of the function exhibited or manifested by the design structure.	❏	❏
11. Function can be defined as the desired goal—that is, the desired behavior or desired effect of the behavior of the structure.	❏	❏

Continued on next page.

Statements	T	F
12. A functive is a pair of words—one verb and one noun.	❏	❏
13. Miles defines functions in the functive format.	❏	❏
14. A function tree is the hierarchical representation of a product's functions and subfunctions.	❏	❏
15. The function tree of an existing product is derived by (1) extracting functions, (2) arranging the hierarchy, and (3) analyzing and expanding the tree.	❏	❏
16. The affinity diagram is a technique for defining the function tree.	❏	❏
17. Function structure is the representation of a function as a relationship between inputs and outputs using a black-box model.	❏	❏
18. Function structure is established by tracing the flows from input to output by asking the following two questions: (1) What happens to the input and the outputs inside the box? (2) Where does the output come from?	❏	❏

14. Establish the goal, action, function, behavior, and structure elements of a grocery cart.

15. Establish the function tree of a car scissor Jack.

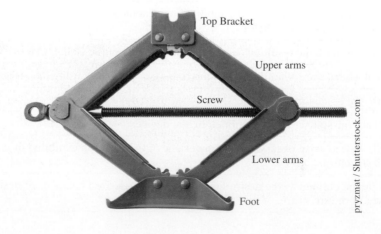

Top Bracket

Upper arms

Screw

Lower arms

Foot

pryzmat / Shutterstock.com

16. Establish the function structure of an electric kettle.

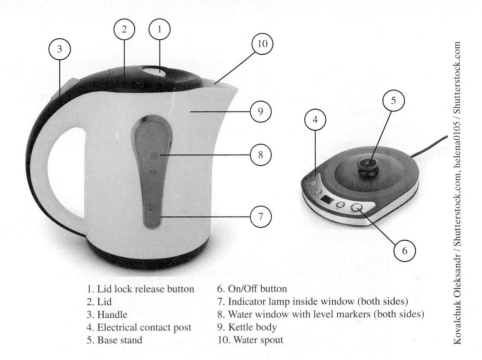

1. Lid lock release button
2. Lid
3. Handle
4. Electrical contact post
5. Base stand
6. On/Off button
7. Indicator lamp inside window (both sides)
8. Water window with level markers (both sides)
9. Kettle body
10. Water spout

Kovalchuk Oleksandr / Shutterstock.com, helena0105 / Shutterstock.com

17. Establish the purpose and action functions for a display table in exhibition halls.

Skalapendra / Shutterstock.com

18. Generate possible redesigns for the stapler based on the design integration example (Section 6.9.2).

LAB 6: Reverse Engineering

Purpose

This lab introduces concepts such as design for manufacturing, function analysis, and design for assembly. In this lab you will disassemble a drill to its essential components and then reassemble it to its original shape.[1] Before you assume responsibility for the drill, make sure that it is in good working condition. The following types of drills are available (we used three types; one type is enough for this activity): Black & Decker DR210K, Skil Twist 2106, and Skil Twist XTRA 2207. Students can choose another model if a model listed here is not available.

Procedure

Safety note: Never plug the cord into an electrical outlet while the drill housing is open.

1. Identify the type of drill, manufacturer, model number, and performance specifications.

2. Plug in the drill and run it. Listen to how it sounds and feels as it runs. Your senses are a useful qualitative measurement tool. The drill should sound and feel the same after you take it apart and put it back together again.

3. Before you disassemble the drill, list all of the functional requirements that the drill must satisfy. For example, one function could be "hold a drill bit without slipping."

4. Unplug the drill and disassemble it as far as possible. *Hint:* Carefully sketch as you disassemble. Put all parts in a bin.

5. Take images for the parts. Produce an exploded view of the drill (similar the one shown in Figure 6.31.

6. Produce a structural decomposition figure similar to Figure 6.32. Produce a physical connectivity figure similar to Figure 6.33.

7. Associate parts with functions. Produce a functional connectivity chart.

8. Determine the gear ratio. What types of gears are used? What are some other gearing alternatives that could be used in this assembly? Why are they not used?

9. Determine the type of bearing used. Why do you think this type was chosen?

10. Determine what type of motor is used. Why do you think this type was chosen? Describe how the motor in your drill works in terms of basic physical principles. Use illustrations to help your description.

11. Make a list of parts (also known as a bill of materials) of your drill. Identify the material used in each part. Also indicate whether the part is an off-the-shelf item or would need to be specifically manufactured for this product as a custom part. What percentage of the total number of parts can be purchased from a parts vendor? Explain your findings.

12. What is the total number of parts in your drill? Compare this number to other drills being dissected. Are there any parts that can be eliminated or combined into a simpler part?

13. Observe the tolerances in the assemblies. Why do you think these tolerances are needed?

---

[1]This lab is based on material presented at an NSF workshop, Central Michigan University, 1999.

14. Reassemble your drill. Be especially careful with the motor brushes.

15. What features of your drill make it easy to assemble? Is there anything that was particularly difficult to reassemble? How would you improve the design of your drill to eliminate this shortcoming? Are there any other improvements to the design that could improve the assembling of the drill? How long would it take for a factory worker to assemble the drill?

16. Once the drill is reassembled and you have no leftover parts, plug it in and try to run it. If it does not sound or feel the way it did before you disassembled the drill, or if a strange smell develops, *stop immediately.* Unplug the drill, and try to determine what went wrong with your assembly steps.

References

[1] Deng Y. M., "Function and Behavior Representation in Conceptual Mechanical Design," *Artificial Intelligence for Engineering Design, Analysis and Manufacturing*, Vol. 16, pp. 343–362, 2002.

[2] Vermaas P. E., "*Technical Functions: Towards Accepting Different Engineering Meanings with One Overall Account,*" *Proceedings of the 2010 TMCE Symposium*, I. Horvath, F. Mandorli, and Z. Rusak (Eds.), Ancona, Italy. 2010.

[3] Miles, L. D., *Techniques of Value Analysis and Engineering*, New York: McGraw-Hill, 1972.

[4] Fowler, T. C., *Value Analysis in Design*, Van Nostrand Reinhold, 1990.

[5] Dym, C. L. and Little, P. *Engineering Design: A Project-Based Introduction*, Hoboken, NJ: John Wiley & Sons, 2003.

[6] Ulrich, K. T., and Eppinger, S. D., *Product Design and Development*, 3rd ed., Tata McGraw Hill Publications. New Delhi 2004.

[7] Hashim, M., *Using Functional Descriptions to Assist the Redesign Process*, PhD Thesis, Department of Mechanical Engineering, University of Leeds, UK 1993.

In addition, the following books, articles, and websites were used in preparing this chapter:

BIRMINGHAM, R., CLELAND, G., DRIVER, R. and MAFFIN, D. *Understanding Engineering Design*. Upper Saddle River, NJ: Prentice Hall, 1997.

BURGHARDT, M. D. *Introduction to Engineering Design and Problem Solving.* New York: McGraw-Hill, 1999.

CHAKRABARTI, A. "Impact of Design Research on Practice: The IISc Experience," in Chakrabrati, A. and Lindemann, U. (Eds.) *Impact of Design Research on Industrial Practice*, New York: Springer, 2016.

CHANDRASEKARAN, B., and JOSEPHSON, J. "Function in Device Representation," *Engineering with Computers*, Vol. 16, pp. 162–177, 2000.

CHITTARO, L. and KUMAR, A., "Reasoning about Function and Its Applications to Engineering," *AI in Engineering*, Vol. 12, No. 4, pp. 331–336, 1998.

CROSS, N. *Engineering Design Methods.* New York: Wiley, 1994.

CROSS, N., CHRISTIAN, H, and DORST, K. *Analysing Design Activity.* New York: Wiley, 1996.

DYM, C. L. *Engineering Design: A Synthesis of Views.* Cambridge, UK: Cambridge University Press, 1994.

EDER, W. E. "*Problem Solving Is Necessary, But Not Sufficient.*" *American Society for Engineering Education (ASEE) Annual Conference*, Session 2330, Milwaukee, WI, 1997.

EDER, W. E. "*Teaching about Methods: Coordinating Theory-Based Explanation with Practice.*" *Proceedings of the American Society for Engineering Education (ASEE) Conference*, Session 3230, Washington, D.C., 1996.

EEKELS, J. and ROOZNBURG, N. F. M. "A Methodological Comparison of Structures of Scientific Research and Engineering Design: Their Similarities and Differences." *Design Studies*, Vol. 12, No. 4, pp. 197–203, 1991.

FULCHER, A. J. and HILLS, P. "Towards a Strategic Framework for Design Research." *Journal of Engineering Design*, Vol. 7, No. 2, pp. 183–194, 1996.

GERO, J. S., THAM, K. W, and Lee, H. S., "Behavior: A Link between Function and Structure in Design," in Brown D. C., Waldron M., and Yoshikawa H. (Eds.), *Intelligent Computer Aided Design IFIP*, pp. 193–225, North Holland: B-4, 1992.

HOLT, K. *"Brainstorming from Classics to Electronics." Proceedings of Workshop, EDC Engineering Design and Creativity*, W. E. Eder (Ed.), Zurich, Switzerland: pp. 113–118, 1996.

HUBKA, V., ANDREASEN, M. M., and EDER, W. E. *Practical Studies in Systematic Design*. London: Butterworths, 1988.

HUBKA, V. and EDER, W. E. *Theory of Technical Systems*. New York: Springer-Verlag, 1988.

HUBKA, V. and EDER, W. E. *Design Science: Introduction to the Needs, Scope and Organization of Engineering Design Knowledge*. New York: Springer-Verlag, 1996.

JANSSON, D. G., CONDOOR, S. S. and BROCK, H. R. "Cognition in Design: Viewing the Hidden Side of the Design Process." *Environment and Planning B, Planning and Design*, Vol. 19, pp. 257–271, 1993.

KEUNKE A. "Device Representation—The Significance of Functional Knowledge," *IEEE Expert*, Vol. 6, No. 2, pp. 22–25, 1991.

KUHN, T. S. *The Essential Tension: Selected Studies in Scientific Tradition and Change*. Chicago: University of Chicago Press, 1977.

PAHL, G. and BEITZ, W., *Engineering Design: A Systematic Approach*. New York: Springer-Verlag, 1996.

SIMON, H. A., *The Sciences of the Artificial*, Cambridge, MA: MIT Press, 1969.

SCHON, D. A. *The Reflective Practitioner: How Professionals Think in Action*. New York: Basic Books, 1983.

SUH, N. P. *Principles of Design*. Oxford University Press, 1989.

TUOMAALA, J. *"Creative Engineering Design." Proceedings of Workshop, EDC Engineering Design and Creativity*, W. E. Eder (Ed.), Zurich, Switzerland, pp. 23–33, 1996.

ULLMAN, D. G., *The Mechanical Design Process*. New York: McGraw-Hill, 1992.

UMEDA Y. and TOMIYAMA, T., "Functional Reasoning in Design," *IEEE Experimental Intelligent Systems and Their Applications*, Vol. 12, No. 2, pp. 42–49, 1997.

VIDOSIC, J. P., *Elements of Engineering Design*. New York: The Ronald Press Co., 1969.

WALTON, J. *Engineering Design: From Art to Practice*. New York: West Publishing Company, 1991.

WOOD, K. and GREER, J., "Function-Based Synthesis Methods in Engineering Design," in *Formal Engineering Design Synthesis* (Antonsson, E. and Cagan, J., Eds.), New York: Cambridge University Press USA, 2001.

CHAPTER 7

SPECIFICATIONS

"The beginning of wisdom is the definition of terms."

~ Unknown

Establishing the specifications is the last step in defining the problem before beginning to explore possible solutions. You have to further clarify the customer requirements at the specification stage of the design process. Specifications are a set of measurable parameters with values specific to the product that together as a whole define the product. Customer requirements and function diagrams are statements of what a design must achieve or do, but they are not normally set in terms of precise limits, which is what the specifications collectively do. This chapter (1) introduces and defines specifications, (2) explains a framework for the setting up and use of specifications, and (3) continues to describe four methods to establish specifications.

7.1 OBJECTIVES

By the end of this chapter, you should be able to

1. Define specifications and identify constituent elements.
2. Explain the use of specifications in the design and operation stages of a product.
3. Explain standards and identify relevant standards to incorporate in specifications.
4. Use one of four methods to establish specifications for a given product.

7.2 SPECIFICATIONS: WHAT AND WHY?

Products are designed to meet the customers' expectations expressed as customer requirements. While the requirements like "the product should be lightweight" are helpful in developing a clear sense of the issues of interest to customers, they provide little specific guidance about how to design and engineer the product. Different people can easily interpret the statement differently as they leave too much margin for subjective interpretation. A design must establish a set of specifications that spell out in precise, measurable detail about what the product is supposed to do. *Product design specifications* (PDS) is a set of statements of what the product should do, showing what the designer should be trying to achieve. Ullman [1] describes specifications as measurable behaviors of the product-to-be that will help later in the design process in determining its quality. They give the precise descriptions of the properties of the artifact to be designed. It is a specific goal statement that clearly and specifically lists the requirements, as well as the specific performance needed to meet each requirement. In short, this is the answer to the first challenge, "defining the problem well."

Consider sliced bread as an example. The specifications describe the product characteristics and what it can do (providing energy and nutritional elements in quantitative terms). Figure 7.1 shows the specifications of two brands of bread.

As can be seen from the example, specifications are lists of measurable parameters representing characteristics of bread and numeric values and units. Specifications in general spell out in precise and measurable detail what the product is and what it can do. In the case of the bread, specifications define the weight and size as well as the calories and nutritional details. They describe what the product is and the benefits it provides.

Examples of single specifications:

- *A kettle* boils a maximum amount of 5 cups of water and a minimum of 1 cup.
- *An access platform* should have an outreach of 20 meters.
- *A single specification* consists of a metric and a value with units. For example, "Average time to assemble" is a metric and "75 seconds" is a value with units. The value may take different forms: a number, a range, an inequality, or even membership in a set.
- *Together*: "A metric and a value with units form a single specification."
- *Product specifications* are a collection of individual specifications.
  Specifications specify the product and thus form an important part of the product concept.

Specifications of Warburtons Extra Thick Sliced White Bread and Klosterman 100% Whole Wheat Bread		
Specification	**Product 1 Warburtons Thick Sliced**	**Product 2 Klosterman Whole Wheat**
Picture of the Product	© studiomode / Alamy Stock Photo	Enlightened Media / Shutterstock.com
Product Description	Warburtons extra-thick sliced white bread	24 oz. 100% whole wheat bread
Product Code	W9011	
Unit Weight	800 g	680 g
Calories	234 per 100 gram	70 per 31 gram slice
Slice Thickness		15.9 mm
Pack Size	12 slices + 2 crust	20 slices + 2 crust
Recommended Storage	Cool dry place	Fresh or Frozen
Product Length		298.5 mm
Product Width		108 mm
Product Height		108 mm
Nutritional Details	Yes	Yes
Ingredients	Yes	Not given

Figure 7.1: Specification of Sliced Bread

7.3 A FRAMEWORK FOR SPECIFICATION, DESIGN, AND TESTING

Specifications form the basis of a contract between the designer and the client. If the specifications of a kettle states that a kettle will boil 5 cups of water at 20°C in 2 minutes, the designer should design the kettle so that it is capable of containing 5 cups of water and of boiling it from 20°C in 2 minutes. This will be tested by the standards agency before it is labeled as meeting a standard—even more so when medical and pharmaceutical products are involved.

In the pharmaceutical industry, validation is the process of producing documented evidence that provides a high degree of assurance that all parts of the facility works consistently and correctly when brought into use. Traditionally validation consists of three parts:

1. *Installation qualification*: documented verification ensuring that the system is installed as specified in the design.
2. *Operational qualification*: documented verification ensuring that the system or equipment is operating as intended and specified.
3. *Performance qualification*: Documented verification ensuring that the system in its normal operating environment produces acceptable quality products as expected and specified.

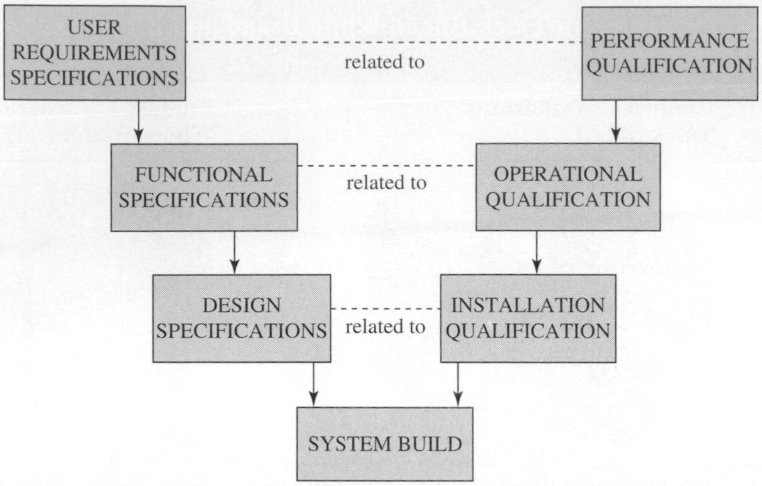

Figure 7.2: Framework for Specification Design and Testing

This process is implemented through a framework called a "V" model, which is shown in Figure 7.2.

In the "V" model the qualification process occupy the right arm. It refers to various specification elements, and the "V" model houses these elements in its left arm. It is the definition of the product with complete details. Good automated manufacturing practice (GAMP) describes the definitive and binding statements at various levels as specifications and defines the various specifications as follows:

1. *User requirements specifications*: the user describes what the system is supposed to do. The statement should include all essential requirements, and if possible, should include a prioritized set of desirable requirements.

2. *Functional specifications*: The supplier writes these in conjunction with the user, providing a detailed description of the functions of the equipment or system. This is a design output.

3. *Design specifications*: the supplier provides a complete definition of the equipment in sufficient detail so that it can be built. This is a design output.

This framework provides a balanced view between the design and operation of the system. The left arm shows the design activities in a linear way. In the design process, the user requirement specifications are taken as the input to develop the functional specification, and the functional specification is taken as the input for the design specification. The equipment is built according to the design specification and qualification as tests begin. While the design process is the focus of this book, you should keep in mind that the outputs should be verifiable with qualifications. This means that each one of the specifications should be measurable.

There are four widely used techniques for generating engineering specifications. They are (1) customer-based, (2) QFD-based, (3) performance-based, and (4) theme-based. The following sections explain them.

7.4 FOUR CUSTOMER-BASED METHODS BY ULRICH AND EPPINGER [2]

In the customer-based method, you obtain customer statements in their original form. The design team then analyzes the statements and rewrites them as needed while providing focus to deploy them in the product. To ensure that a need is properly deployed, the design team must identify a precise and measurable characteristic of the product that will reflect the degree of satisfaction of that particular need. This particular characteristic of that need embodies a unit such as a meter, a Newton, or a second, and can be measured easily. This unit is the metric of that need. Each need is considered individually, and each need requires a metric. Thus the customer statements are translated into easily measurable metric units. Each metric is assigned a numeric value to form a specification, and the full set of specifications derived from the full set of needs is called the *specifications of the product*. Assume that the translation from customer needs to a set of precise, measurable metrics is possible and that meeting specifications will lead to customer satisfaction. Chapter 5 dealt with identifying the relative importance of the needs; therefore, translating the prioritized needs (customer requirements) into metrics is used as the starting point.

7.4.1   Establishing the Metrics

This is the most crucial part of defining the specifications and thus of defining the design problem as well. It should define the customer need precisely and accurately.

Consider the water-level indicator in a lake that conserves water. Let the customer statement be "must be able to read the amount of water in the reservoir." The customer may use this information (1) to determine the average depth of water in the lake, (2) to determine the quantity of water available in the reservoir for consumption, or (3) to estimate the fullness of the reservoir. These details must be extracted either when establishing the needs or when establishing the metrics so that the specifications precisely and accurately fulfill the customer requirements.

It is a good practice to consider each need and look for a precise, measurable characteristic of the product that will deploy that need. As in the example there may be more than one metric needed to meet a requirement or need (see Figure 7.3).

Soonthorn Wongsaita / Shutterstock.com

Figure 7.3: Different Metrics to Define the Need

Characteristics of Good Metrics (Adapted from Ulrich and Eppinger [2])

1. Metrics should be complete as far as possible—that is, one metric should be sufficient to deploy the need completely.

2. Metrics should be dependent, not independent, variables. Metrics are needed to measure how well the product performs, so they are responses.

3. Metrics should be practical. They must be easily and economically measurable.

4. Some needs cannot easily be translated into quantifiable metrics. Techniques must be devised to measure them indirectly through easily and economically measurable quantities.

You can easily figure out your metrics by filling out the chart in Figure 7.4

7.4.2 Competitive Benchmarking Information

To position your product relative to existing products, you have to know the details of your competitors' competing products. Knowing the details of their products will help to identify the realistic values for various characteristics. Record your competitors' product details in the same metrics and units as those in your product's need statements. This will facilitate comparison and the fixing of values for the various specifications for the product under development. A typical benchmarking information chart shown in Figure 7.5.

Metrics	Metric 1 (Units)	Metric 2 (Units)	Metric 3 (Units)	Metric 4 (Units)	Metric 5 (Units)	Metric 6 (Units)	Metric 7 (Units)	Metric 8 (Units)
Need 1	*							
Need 2		*						
Need 3			*					
Need 4				*				
Need 5					*			
Need 6						*		
Need 7							*	
Need 8								*

Figure 7.4: Need Metric Mapping

Metric Description	Importance	Units	Competitor Companies			
			A	B	C	D
Metric 1						
Metric 2						
Metric 3						
Metric 4						
Metric 5						
Metric 6						
Metric 7						

Figure 7.5: A Typical Benchmarking Chart

7.4.3 Setting the Target Values

With the knowledge of the originating need(s), their importance, and the values of the competing products, the design team can set a *target value* for each metric. Ulrich and Eppinger [2] suggest setting up two values—(1) acceptable values and (2) ideal values. These values can be given in five ways: (1) exactly *x*, (2) at most *x*, (3) at least *x*, (4) between *x* and *y*, and (5) one of a set of discrete values.

So far the specification we've developed is called a *target specification,* meaning that in the eyes of the design team if the product meets these specifications, it will fully satisfy the customer. However, this has to be checked with reality. The design team has to revisit the values set from both a technical and a cost point of view. These may require trade-offs and compromises. Once you've considered these constraints, firm up the values for the metrics to give the final values for the specifications. Figure 7.6 shows a typical set of specifications.

Title: Specifications for the --------					
Drawn by: ------------					
Need No.	Metric No.	Specification	Importance	Units	Value

Figure 7.6: Typical Specification Document

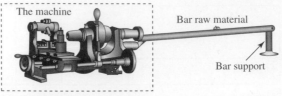

The machine

Bar raw material

Bar support

Based on http://www.lathes.co.uk/meriden/

Figure 7.7: General-Purpose Bar Support

Table 7.1: Customer Input, Needs, and Importance Ratings for a Bar Support

Input	Need	Importance
I want the support to be strong.	Strong material	7
I prefer the operation to be easy.	Easy operation	9
I prefer it to look good.	Good looking	6
It should be manually operated.	Manual operation	9
I want it to be maintenance free.	Limited or no maintenance	7
It should be usable by different machines.	Height adjustable	9
It should cost no more than $75.	Cost of the unit	6
There should be no moving or vibrating in use.	Firm seating during use	9
It should be durable.	Durable material	7
It should be safe	Stable	9

7.4.4 An Example: General-Purpose Bar Support

In this example the objective is to design and make a portable, compact device with a height-adjusting facility so that long bars, beams, and shafts can be supported at one end while being machined at the other end. Figure 7.7 shows a typical bar support in use. The customer verbatim, the needs derived from them, and the importance ratings of needs are given in Table 7.1. Establish these specifications.

Establishing the specifications starts with the prioritized needs as the input. The first activity is the establishment of the need-metric matrix. Figure 7.8 shows the need metric matrix for the bar support.

Attaching a value for each of the metric defines a single specification. Thus, the specification for the bar support can be similar to the one shown in Table 7.2.

 ## 7.5 QUALITY FUNCTION DEPLOYMENT-BASED METHOD FOR WRITING SPECIFICATIONS

This section first describes the quality function deployment (QFD) technique and then describes how the first chart in QFD (called the house of quality) is used to develop specifications.

The quality function deployment (QFD) method was first developed in Japan in the mid-1970s and was used in the United States in the late 1980s. Through the use of the QFD method, Toyota reduced the cost of bringing a new car model to the market by more than 60% and reduced the time by about 33%. Quality function deployment **extracts the demanded quality items from the customer and deploys them or realizes them through measurable design**

	Importance rating	Yield strength	Number of steps	Score on a 1–5 scale	Power tool required	MTBF	Height range	Cost of unit	Contact area while in use	Protective painting	Moment to topple
1. Strong material	7	x									
2. Easy operation	9		x								
3. Good looking	6			x							
4. Manual operation	9				x						
5. Limited or no maintenance	7					x					
6. Height adjustable	9						x				
7. Cost of the unit	6							x			
8. Firm seating during use	9								x		
9. Durable material	7									x	
10. Stable	9										x

Figure 7.8: Need Metric Matrix

Table 7.2: Specifications for the Bar Support

Specification	Quantity and Units
Material yield strength	>210 MPa
Number of steps for use	<4
Appearance rating	>3 on a 1 to 5 scale
Number of external power tools	0
Mean time between failure and maintenance	>1000 hrs
Height range	300 mm to 1200 mm
Price	<$75
Paint	Yes; choice from list
Toppling moment	>60 Nm

elements or attributes called quality functions, at the product planning, part planning, materials and manufacturing planning, and production planning stages (Vidosic [3]). It employs the principle of deployment where (1) product quality can be assured through the quality of subsystems, (2) subsystem quality can be assured by the quality of parts, (3) parts quality can be assured through the quality of process elements, and (4) quality of process elements can be assured through production methods and plans. For example, the qualities of an electric bulb (brightness, sharpness, life, etc.) can be assured by the quality of the subsystems glass container and filament system;

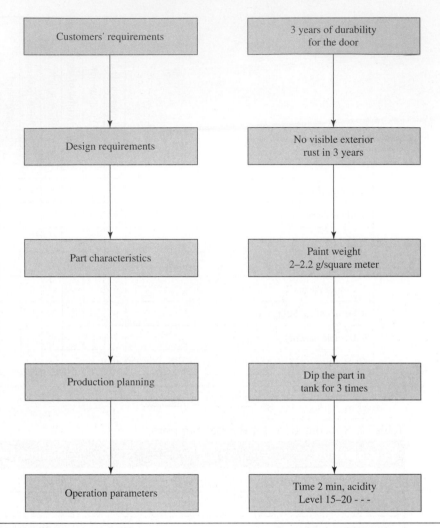

Figure 7.9: Durability of Car Door

the quality of the filament system can be assured by the characteristics of the parts element and conductor and so on. The vacuum inside the glass container is assured by the emptying process. The *principle of deployment* is a key feature in QFD technique. Each requirement is considered separately in the deployment process. An example is shown in Figure 7.9 where the requirement of three-year durability for a car door is taken through the QFD process.

The deployment is expressed as a matrix with rows representing the "what is to be deployed" and the columns representing the "methods (or hows) of deployment." Figure 7.10 shows the deployment implemented for a portable belt conveyor.

Each stage in the QFD process can be represented by a chart, and the QFD process is represented by a four-chart cascade as shown in Figure 7.11.

The cascade shows how in chart 1 the design requirements deploy the customer requirements, in chart 2 the part characteristics deploy the chosen design requirements, in chart 3 the manufacturing process deploys the chosen part characteristics, and in chart 4 the process parameters deploy the chosen processes. The last three stages were very familiar to the engineers. However the first stage where the customer requirements shape the design elements or quality functions was relatively new and was embraced and developed with enthusiasm. The developed stage 1 chart is the house of quality, which is described in Section 7.5.2.

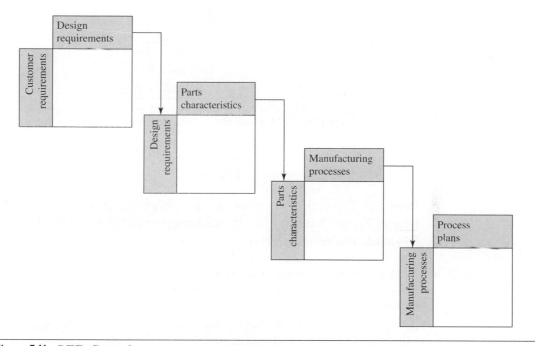

Whats	Design elements	Lightweight	Handle	Wheels	Plug power connection	Belt texture	No trap points	Limited torque	Rubber linings	Low drop height	Low center of gravity	Weight	Leg spacing	Floor fixing
Easily moved														
No component Damage														
Stability														

Figure 7.10: Deployment of Requirements for Quality Functions

Figure 7.11: QFD Cascade

7.5.1 House of Quality

House of quality is concerned with two items—(1) customer requirements and (2) quality functions or design characteristics. In the deployment process in the matrix form, the rows represent the requirements, and the columns represent quality functions or design characteristics. Thus additional boxes to the matrix in the horizontal direction will add details to requirements, and additional boxes in the vertical direction will add details to the design characteristics. The details thus added are called rooms, and the whole chart is called the house of quality.

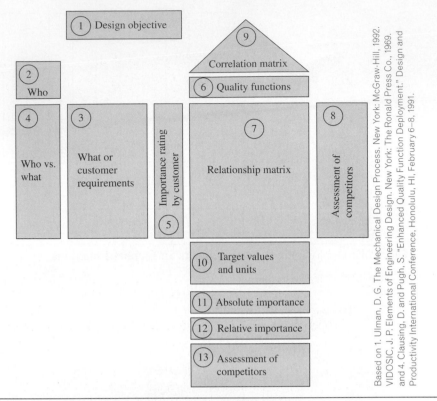

Based on 1. Ulman, D. G. The Mechanical Design Process. New York: McGraw-Hill, 1992. VIDOSIC, J. P. Elements of Engineering Design. New York: The Ronald Press Co., 1969. and 4. Clausing, D. and Pugh, S. "Enhanced Quality Function Deployment." Design and Productivity International Conference, Honolulu, HI, February 6–8, 1991.

Figure 7.12: Stage 1 QFD Chart Called the House of Quality Chart

Figure 7.12 shows a representation of the house of quality which includes:

1. **Region 1.** This region describes what the design objective is.
2. **Region 2.** This first entry lists the customers or stakeholders for the product.
3. **Region 3.** The requirements established in Chapter 6 are listed as rows. These are the first of the two main entries in the house of quality chart.
4. **Region 4. Who vs. What:** This entry records the source of the requirement or the stakeholder who raised this requirement. This is important to relate the constraints and mandatory requirements.
5. **Region 5. Importance Rating by the Customer:** Entries in this region record the importance of each requirement on a linear 1–10 scale.
6. **Region 6. Engineering Metrics or Quality Functions:** These are the second of the two main entries in the house of quality chart. This one records what must be achieved in order to deliver the customer requirements, taken one by one. The quality functions must be quantifiable and measurable.
7. **Region 7. Relationship Matrix:** This region records the answers to the question "If the quality function in region 6 is successfully achieved, will the customer need be satisfied and to what degree?" In this context 9 = strong correlation, 3 = medium correlation, 1 = weak correlation, and a blank means no correlation.
8. **Region 8. Assessment of Competitors or Existing Design:** This rates the customer's perception of the degree of attainment of the identified need by competitive solutions or by the existing design.

9. **Region 9. Correlation of Quality Functions:** This region indicates how the quality functions reinforce or oppose each other. It is important that the opposing quality functions be reconciled.

10. **Region 10. Target Values and Units:** It has been pointed out earlier that the quality functions should be quantifiable and measurable. This region records the target quantity and units of each of the quality functions.

11. **Region 11. Absolute Importance of Quality Functions:** This is the importance of each quality function based on the strength of relationship and the importance of corresponding customer requirements. It is calculated by the inner product (sum product) of the column containing the importance of the customer requirements and the relationship matrix column corresponding to individual quality function.

12. **Region 12. Relative Importance:** This region records the absolute importance scores normalized on a 1 to 10 scale or as percentages.

13. **Region 13. Assessment of Competitors or Existing Design:** This region involves assessment of product performance relative to competitive solutions on each quality function.

Example 7.1: House of Quality for the Splashguard of a Bicycle

Splashguards

Sirapob / Shutterstock.com

Figure 7.13: Splashguard of a Bicycle

Figure 7.13 shows a bicycle with a splashguard.

1. **Region 1. Design Objective:** An attachment prevents water from hitting the rider when riding on a wet road.

2. **Region 2. Who the Customers and Stakeholders Are:** Rider, mechanic, and marketing are identified as the customers.

3. **Region 3. Requirements:** They are identified by individual interviews from the three categories.

    Riders' Requirements

    Keep water off the rider

    Won't mar the bicycle

    Won't catch water, mud, and debris

    Won't rattle

    Won't wobble or bend

    Has a long life

    Won't rub on wheel

Is lightweight
Mechanics' Requirements
Is easy to attach
Is easy to detach
Is quick to attach
Is quick to detach
Won't interfere with rack, panniers, or brakes
Marketing Requirements
Fits universally
Is attractive
Costs less than $3

4. **Region 4. Who Wants What?:** This is a revised format of the information given previously.

5. **Region 5. Importance Ratings:** These rate the importance of the requirements obtained from customers.

6. **Region 6. Engineering Metrics or Quality Functions:** The design problem is stated in terms of parameters that can be measured and have target values. These quality functions are a translation of the voice of the customer into the voice of the engineer. They serve as a vision of the ideal product and are used as criteria for design decisions. In this step, develop parameters that tell if customers' requirements have been met. Start by finding as many engineering parameters as possible that indicate a level of achieving customers' requirements. Make every effort to find as many ways as possible to measure customers' requirements. This step is similar to the need metric mapping in the method discussed in Section 7.4. For the list of requirements listed for the splashguard, the quality functions can be as follows:

Requirements	Quality Functions
Keep water off the rider	Percentage of water hitting the rider
Won't mar the bicycle	Percentage of consumers who find the look of the splashguard appealing
Won't catch water, mud, and debris	Number of sharp bends and catches
Won't rattle	Number of attaching points
Won't wobble or bend	Number of attaching points
Has a long life	Lifetime
Won't rub on wheel	Clearance from wheel
Is lightweight	Weight of the splashguard and fixings
Is easy to attach	Number of steps
Is easy to detach	Number of steps
Is quick to attach	Time to attach
Is quick to detach	Time to detach
Won't interfere with rack, panniers, or brakes	Minimum clearance distance
Fits universally	Standardized sizes or the bikes it fits
Is attractive	Color combinations
Cost less than $3	Selling price

The house of quality chart is concerned with two inputs: (1) requirements and (2) quality functions. All other details are derivatives of these two information. Therefore, given this information, the house of quality chart can be built.

Some Observations

1. When starting to generate the list of quality functions, consider each requirement separately one at a time. Ideally this will give a one-to-one relationship, but occasionally there may be a quality function that would deploy two needs. On the other hand, occasionally one requirement may require two quality functions. A one-to-one relationship assures that a condition stipulating the value of the quality function can meet the originating requirement. This provides the designer with the ability to control (controllability) the level of achievement together with the ability to observe (observability) the level of achievement.

2. If something like "looks good" has to be measured, transform it into a testable measure such as "high score on 5-point attractiveness scale by >65% of riders." This means the setting up of a 5-point attractiveness scale (units = "points") such as 1 = ugly, 2 = tolerable, 3 = acceptable, 4 = attractive, 5 = captivating).

3. If the requirement is "attractive color," establish a list of attractive colors and the specific color should be a member in the list.

7. **Region 7. The Relationship Matrix:** Entries answer the question "If the quality function is successfully achieved, will the customer need be satisfied and to what degree"? for each quality function established in Region 6. Because of this, they are filled on a column-by-column basis. For example, if the quality function "attaching steps" is considered, it has a medium relationship with the requirement "quick to attach," and weak relationships with the requirement "won't interfere with rack, etc." and "fits universally," when answering the stipulated question.

   Weights are attached to these relationships depending on whether a strong or moderate discrimination is desired. A geometric progression of weights in the form of 1, 3, and 9 would give a strong discrimination whereas an arithmetic progression of weights in the form of 1, 3, and 5 would give a moderate discrimination. To avoid the confusion, graphic symbols are used in the example.

8. **Region 8. Assessment of Competitors:** This activity has not been carried out in this example.

9. **Region 9: Correlation Matrix:** Entries in this region are important for identifying conflicts.

10. **Region 10. Target Values and Units:** It has been said from inception that quality functions are measurable quantities. Entries in this region specify the values targeted for optimal performance together with their units. The targeted value for the quality function "lifetime" can be 2000 hours where 2000 is the value and an hour is the unit.

11. **Region 11. Absolute Importance:** Entries in this region show the importance of each quality function. They are calculated by the inner product (sum product) of the column containing the importance of the customer requirements and the relationship matrix column corresponding to individual quality function.

    Calculate the inner product as described in the explanation of Region 11 in Section 7.5.2. For this example, the absolute importance for the quality function "Attaching Steps" is $= 7 \times 9 + 5 \times 3 + 7 \times 1 + 9 \times 1 = 94$.

12. **Region 12. Relative Importance:** Entries in this region show the importance of each quality function in comparison to others. It is expressed as a number between 1 and 10 or as a percentage. Since the importance ratings for requirements are given on a 1 to 10 scale, the same basis is adopted here. The completed chart is shown in Figure 7.14.

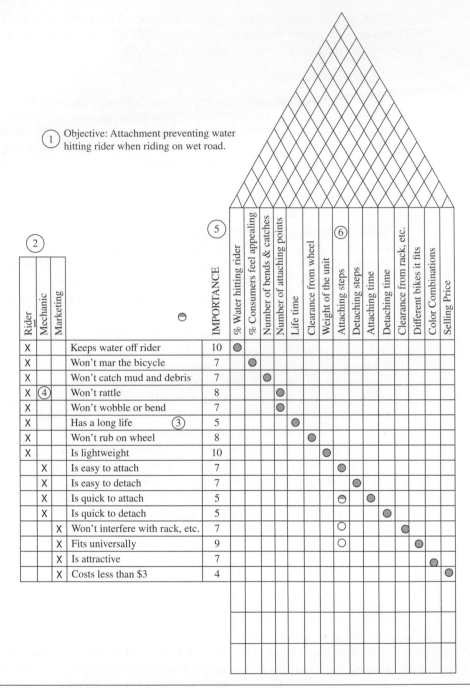

Figure 7.14: House of Quality Chart for the Splashguard (incomplete)

Quality function deployment provides a structured way to analyze customer requirements and quality functions. Different researchers have added several additional analyses. For example, Region 8 in the model described in Section 7.5.1 analyzes the competitors' products from the customers' perspectives. You can carry out a similar benchmarking from the designers' perspectives as shown in Region 13.

7.5.3 Specifications from the House of Quality Chart

Region 6 in the house of quality chart details the quality functions that deploy the customer requirements. Region 10 details their target values and units for these quality functions. Region 12 details the relative importance of the quality functions. By evaluating these three groups of information the design team can determine that some of the quality functions and their originating needs are only marginally important and hence can be ignored. Alternatively, they may decide that all quality functions are important and the product should provide them.

Chosen quality functions, their target values and units, and their relative importance ratings form the set of specifications in the QFD-based method for writing specifications.

7.6 FUNCTION-BASED METHOD BY DYM AND LITTLE

Dym and Little [5] *define performance specifications* as a description of how well the designed product must do something and *functional specification* as what functions the product must perform to realize its objectives. But functional specifications are incomplete if they fail to specify the performance. They classify performance specifications into three categories: (1) prescriptive or definition-based performance specifications, (2) interface performance specifications, and (3) detailed design performance specifications.

7.6.1 Performance-Specification Method

In the *performance-specification method*, the specification defines the required *performance* instead of the required *product*. While it describes the performance that a design solution has to achieve, it does not define any particular physical component that may constitute a means of achieving that performance.

Here's how to use the performance specifications method to produce a set of specifications.

STEP 1. *Consider how general your solution can be.*

Specifications that are too general may allow inappropriate solutions to be suggested. However, specifications that are too tight may remove all of the designer's freedom and creativity for the range of acceptable solutions. This level of limitation may also be connected with the definition of the customer you are dealing with, because the customer may also limit the class and the set of specifications. For example, designing a space shuttle puts the designer on a tighter limit of specification. Another example that demonstrates this point involves the design of an air jet for civilian use. Such a design has a different set of requirements from designing an air jet for military use. Recognizing the customer's needs is an important factor in defining the range for the set of requirements. The level of generality could be reserved through the development of the objective tree.

Different types of generality can be listed from the most general to least general as

a. Product alternatives
b. Product types
c. Product features

Consider this example to illustrate these levels. Suppose that the product in question is a domestic aluminum-can disposal device. At the highest level of generality, the designer would be free to propose alternative ways of disposing of aluminum cans, such as crushing, melting, shredding, and chemical dissolution. There might even be freedom to move from the concept of single-can to multi-can, or from a domestic product to a product for factory use. At the intermediate level, the designer would have less freedom and might be concerned only with different types of can crushers, such as foot operated or automatic. At the lowest level of generality, the designer would be constrained to consider only different features within a particular type of crusher, such as if it hangs on the wall or is self-standing, processes one or many cans in one step, and so on. At this level of developing specifications, designers need to review the need statement and objective tree to analyze where the objective tree stands with respect to the level of generality.

STEP 2. *Determine the level of generality at which the product operates.*

In general, the customer or client determines the level of generality. For example, the customer may ask the designer to design a safe, reliable aluminum-can crusher. Here, the customer has already established the type of product and placed the constraints on the type of aluminum can disposal unit: It must be a crusher. The set of requirements will follow for a can crusher, not a can-disposal unit. Sometimes customers may state a solution to a product such as "can crusher," whereas in fact they do not really care whether the cans are crushed. What they actually need is a product that disposes of cans efficiently. Communicate with the customer to ensure that the apparent limits are really necessary as they may have simply not considered other alternatives. This further emphasizes the need to accurately define the problem before suggesting possible solutions.

STEP 3. *Identify the required performance attributes.*

For clarifying the customer statement, use objective trees, function analysis, and market analysis. Objective trees and functional analysis determine the performance attributes of the product. Some of the requirements in the customer statement are "must meet" (called demands) and others are "desirable" (called wishes). Solutions that do not meet the demands are not acceptable solutions, whereas wishes are requirements that should be taken into consideration whenever possible. Pahl and Beitz [6] developed a checklist that can be used to derive attributes along with objective trees and function analysis (Table 7.3). List the attributes, distinguish whether the attribute is a demand or a wish, and then tabulate the requirements.

Table 7.3: Checklist for Drawing Up a Requirement List.[1]

Main Headings	Examples
Geometry	Size, height, breadth, length, diameter, space requirement, number, arrangement, connection, extension
Kinematics	Type of motion, direction of motion, velocity, acceleration
Forces	Direction of force, magnitude of force, frequency, weight, load, deformation, stiffness, elasticity, stability, resonance
Energy	Output, efficiency, loss, friction, ventilation, state, pressure, temperature, heating, cooling, supply, storage, capacity, conversion
Materials	Physical and chemical properties of the initial and final product, auxiliary materials, prescribed materials (food regulations, etc.)
Signals	Inputs and outputs, form, display, control equipment
Safety	Direct safety principles, protective systems, operational, operator and environmental safety
Ergonomics	The man–machine relationship, type of operation, clearness of layout, lighting, aesthetics
Production	Factory limitations, maximum possible dimensions, preferred production methods, means of production, achievable quality and tolerances
Quality	Control possibilities of testing and measuring, application of special regulations and standards

Main Headings	Examples
Assembly	Special regulations, installation, siting, foundation, transport limitations due to lifting gear, clearance, means of transport (height and weight), nature and conditions of dispatch
Operation	Quietness, wear, special uses, marketing area, destination (for example, sulfurous atmosphere, tropical conditions)
Maintenance	Servicing intervals (if any), inspection, exchange and repair, painting, cleaning
Recycling	Reuse, reprocessing, waste disposal, storage
Costs	Maximum permissible manufacturing costs; cost of tooling, investment, and depreciation
Schedule	End date of development, project planning and control, delivery date

[1] Based on ENGINEERING DESIGN: A SYSTEMATIC APPROACH by G. Pahl and W. Beitz, translated by Ken Wallace, Lucienne Blessing and Frank Bauert, Edited by Ken Wallace. Copyright © Springer-Verlag London Limited 1996.

STEP 4.   *State succinct and precise performance requirements for each attribute.*

Identify the attributes that can be written in quantified terms; then attach the performance limit to the attribute. The limits may be set by the customer statement or by federal and government agency standards. These attributes may include, but are not limited to, maximum weight, power output, size, and volume flow rate. For presenting the specifications, use a metric-value combination (e.g., the size must be less than 1 m^3). For wishes and demands that do not have defined value attached to the metric, we use the quality-function-deployment method, which is described in Section 7.5.

7.6.2   Case Study Specification Table: Automatic Can Crusher

Continuing our can crusher example from previous chapters, the next stage is to draw up a specifications table. Review the prioritized requirements and market analysis results at this time (see Chapters 4 and 5). From the need statement for the aluminum can crusher in this scenario, the specifications that *must* be met are

1. The design of the crushing mechanism is not to exceed 20 × 20 × 10 cm in total size.
2. The can crusher must have a continuous feed mechanism.
3. The can must be crushed to 1/5 of its original volume.
4. The device must operate safely; children will use it.
5. The device must be fully automatic.

   Table 7.4 shows an example specification table that includes a metric and a value for this case study.

Table 7.4: Specification Table for Automatic Can Crusher.

Metric	Value
Dimensions	20 × 20 × 10 cm
Cans crushed	1/5 original volume
Weight	< 10 kg
Sales price	< \$50
Number of parts	< 100

Continued on next page.

Metric	Value
People able to use	≥ 5 yrs
Probability of injury	$< 0.1\%$
Manufacturing cost	$< \$200$
Steps to operate	1
Maintenance cost	$< \$10$ annually
Efficiency rating	> 95 percentile
Internal parts enclosed	100%
Storage of crushed cans	60
Loader capacity	>30 cans
Crush cans	≥ 15 cans/min
Crush cans Crush cans	$\geq 1.2 \times 10^{-2}\, \text{m}^3/\text{min}$ ≥ 0.57 kg/min
Noise output	> 30 dB
Starts	<10 sec
Runs	>2 hours at a time
Stops	<5 sec
Vibration magnitude	<5 mm
Vibrations	$<$ sec
Withstand	... 250 N
Maintenance	> 4 hrs/yr
Number of colors	
Lifespan	> 4000 hrs

7.7 THEME-BASED OR HEURISTIC METHOD FOR WRITING SPECIFICATIONS

The definition of heuristic refers to techniques, activities, or lessons that allow people to discover something for themselves or by finding solutions through experiments or loosely defined rules. Consider an equipment design and manufacturing company for a cement plant or pharmaceutical plant. The company would have been involved in the design of several such plants and would have gone through the process several times. This would have given the company unique insights that customers would not have. It will be in a position to identify specific areas (or themes) from which specifications arise. It would even know the alternatives that the customer can choose from and the merits and demerits of each choice. In such companies, their insights and experiences become routine procedures that must be followed whenever a design is initiated at one of those companies. Starting the systematic process from scratch in these situations is a waste of time. Understanding the themes and following the procedures would produce better specifications.

Similar situations can exist in connection with smaller products. People who write specifications for products as a routine can identify specific areas from which specifications arise. The guidelines in these situations are given as checklists. A typical checklist, shown in Table 7.3, is provided by Pahl and Beitz [6] in which they identify 16 themes as used in Table 7.3. Another such list of themes appears in Table 7.5.

Table 7.5: Themes from Which Specifications Arise

Performance	Quantity	Competition
Size	Materials	Finish
Shipping or container	Quality testing	Service
Economy	Manufacturing	Design time
Cost	Safety	Packing
Weight	Lifespan	Environment
Maintenance	Ergonomics	Subsystems
Standards	Market	–

Based on Pahl, G. and Beitz, W. Engineering Design: A Systematic Approach. New York: Springer-Verlag, 1996.

7.7.1   Development of Specifications for a Power Train

The requirement is a power train that would rotate a vertical shaft at about 2 to 3 rpm. The power required is estimated to be about 0.5 kw. Additional requirements are that the power can be from a battery that is rechargeable from the mains or from a single-phase domestic supply. Only one unit is required; therefore, it is preferable to use commercially available components as much as possible.

Themes from which specifications arise can be as follows:

1. Types of motor, power ratings and speeds
2. Available types and dimensions of pulleys and lengths for V belts
3. Types of gearboxes, power ratings, and reductions available
4. Available chain sizes and corresponding sprockets

The specifications can be as shown in Table 7.6.

Table 7.6: Specifications of a Power Train

Theme	Specification	Value and Units
Types of motor, power ratings, and speeds	Power Type Speed	0.5 to 0.75 kw AC or DC 1400 rpm for AC Speed controllable DC
Types and dimensions of pulleys available for V belts	Number of grooves Pulley ratios Belt lengths available	1 or 2 2 and 2.5 780 mm or 1250 mm
Types of gearboxes, power ratings, and reductions available	Spur or helical gears Number of stages Worm and wheel	1:10 2 or 3 1:75 to 1:99
Sprockets for chain	Sprocket sizes Number of teeth	75 mm and 150 mm 19 and 38

7.8 CHAPTER SUMMARY

1. Specifications are a set of measurable parameters with values specific to the product, which together as a whole define the product.

2. Specifications spell out in precise, measurable detail the product's characteristics and what it is expected to do.

3. Together a metric and a value with units form a single specification, and product specifications are a collection of individual specifications.

4. Validation is the process of producing documented evidence, which provides a high degree of assurance that all parts of the facility work consistently and correctly when brought into use.

5. In the V model the qualification process occupies the right arm, and the various specification elements or design are housed in the left arm.

6. There are four widely used techniques for generating engineering specifications. They are (1) customer-based, (2) QFD-based, (3) performance-based, and (4) theme-based.

7. In the customer-based method customer statements are obtained in their original form and rewritten as needed; they are then translated into metrics, providing the required focus to deploy them in the product.

8. To ensure that a need is properly deployed, a precise and measurable characteristic of the product called *metric* that will reflect the degree of satisfaction with that particular need is identified.

9. Specifications are drawn, based on these metrics.

10. In the customer-based method the customer statements are translated into easily measurable metrics with units, each metric is assigned a numeric value to form a specification, and the full set of specifications derived from the full set of needs is called the specifications of the product.

11. Each metric is set a target value, and these values can be given in five ways: (1) exactly x, (2) at most x, (3) at least x, (4) between x and y, and (5) one of a set of discrete values.

12. The specifications thus far developed are called *target specifications*, meaning that in the eyes of the design team if the product meets these specifications it will fully satisfy the customer.

13. Quality function deployment extracts the demanded quality items from the customer requirements and deploys them or brings them to effective life through measurable design elements or attributes called *quality functions*. This happens at the product-planning, part-planning, materials and manufacturing–planning, and production-planning stages.

14. In the first stage of QFD, quality functions are derived from the customer requirements, and the developed stage 1 chart is called the *house of quality*

15. House of quality is concerned with two items: (1) customer requirements and (2) quality functions or design characteristics. In the deployment process in the matrix form, the rows represent the requirements and the columns represent quality functions or design characteristics.

16. The relationship matrix records the answers to the question "If the quality function in Region 6 is successfully achieved, will the customer need be satisfied and to what degree?" In this context 9 = strong correlation, 3 = medium correlation, 1 = weak correlation, and a blank means no correlation.

17. Quality function deployment provides a structured way to analyze customer requirements and quality functions.

18. The design team draws the specification considering Rooms 6, 10, and 12 in the house of quality.

19. Chosen quality functions, their target values and units, and their relative importance ratings form the set of specifications in the QFD-based method for writing specifications.

20. Performance specifications describe how well the designed product is expected to do something, and functional specifications define what functions must be performed to realize the objectives.

21. Specifications are drawn from a checklist in the performance-based and the theme-based methods.

7.9 PROBLEMS

1. A set of measurable parameters with values specific to the product that together define the product is called the *specifications*. These parameters are either physical properties of the product or the capabilities of the product. Log on to the Internet to extract the specifications of a product and to identify these two classes of specifications.

2. Given the need statement "Design a domestic heating appliance," generate different levels of generality of solutions.
   a. Alternatives
   b. Types
   c. Features

3. Generate engineering characteristics for the following:
   a. Easy to hold
   b. Cost-effective
   c. Safe for the environment
   d. Easy to maintain

4. Consider the statement you worked on previously during an in-class team activity. By looking at your objective tree, generate a list of specifications and a house of quality for the following statement:

   If you drive in the state of Florida, you notice some traffic congestion due to trees being trimmed on state roads and freeways. Assume that the Florida Department of Transportation is the client. Usually, branches 15 cm or less in diameter are trimmed. The allowable horizontal distance from the edge of the road to the branches is 2 m. The material removed from the trees must be collected and taken away from the roadside. To reduce the cost of trimming, a maximum of two workers can be assigned for each machine. The overall cost, which includes equipment, labor, and other things, must be reduced by at least 25% from present cost. The state claims that the demand for your machines will follow the price reduction (i.e., if you are able to reduce the cost by 40%, the demand will increase by 40%). Allowable working hours depend on daylight and weather conditions.

5. Develop a list of specifications for the following objectives. The table should include metrics, the corresponding values (assume a value), and units.
   a. Easy to install
   b. Reduces vibration felt by the hand
   c. Is not contaminated by water
   d. Lasts a long time
   e. Is safe in a crash
   f. Allows easy replacement of worn parts

6. List a set of metrics corresponding to the need that a pen should write smoothly.

7. Devise a metric and a corresponding set for the need that a roofing material should last many years.

8. How might you establish precise and measurable specifications for intangible needs, such as "The front suspension should look great"?

9. Can poor performance relative to one specification always be compensated for with high performance on the other specifications? Give examples.

10. Develop a house of quality for the following need statement.

    We have a mountain-sized pile of wood chips that we want processed into home-use fire logs. We are located in Tallahassee, Florida. We can have a continuous supply of wood chips throughout the year. We should be able to produce 50 fire logs per minute. The shareholders require us to have a large profit margin, and our prices should be lower than our competition for the same size logs. Needless to say, the logs should produce enough heat to keep our customers faithful and have environmentally clean exhaust.

11. Use a Web search engine to identify a list of companies that utilize QFD. Describe what they apply QFD for, and list differences and similarities in the steps developing QFD.

References

[1] Ullman, D. G. *The Mechanical Design Process*. New York: McGraw-Hill, 1992.

[2] Ulrich, K. T. and Eppinger, S. D., *Product Design and Development* (3rd ed.), New Delhi: Tata McGraw-Hill Publications 2004.

[3] Vidosic, J. P. *Elements of Engineering Design*. New York: The Ronald Press Co., 1969.

[4] Clausing, D. and Pugh, S. *"Enhanced Quality Function Deployment." Design and Productivity International Conference,*Honolulu, HI, February 6–8, 1991.

[5] Dym, C. L., Dym, C. L. and Little, P. *Engineering Design: A Project-Based Introduction,"* Hoboken, NJ: John Wiley & Sons, 2003.

[6] Pahl, G. and Beitz, W. *Engineering Design: A Systematic Approach*. New York: Springer-Verlag, 1996.

In addition, the following books, articles, and websites were used in preparing this chapter:

AKAO, Y. *Quality Function Deployment: Integrating Customer Requirements into Product Design.* (Translated by Glenn Mazur). New York: Productivity Press, 1990.

COHEN, L. *Quality Function Deployment: How to Make QFD Work for You*. Reading, MA: Addison-Wesley, 1995.

DAETZ, D. *Customer Integration: The Quality Function Deployment (QFD) Leader's Guide for Decision Making.* New York: Wiley, 1995.

DAY, R. G. *Quality Function Deployment: Linking a Company with Its Customers*. ASQC Quality Press, 1993.

DEAN, E. B. "Quality Function Deployment for Large Systems." *Proceedings of the 1992 International Engineering Management Conference*, Eatontown, NJ, October 25–28, 1992.

HAUSER, J. R. and CLAUSING, D. "The House of Quality." *The Harvard Business Review*, May–June, No. 3, pp. 63–73, 1988.

HUBKA, V. *Principles of Engineering Design*. London: Butterworth Scientific, 1982.

KING, B. *Better Designs in Half the Time: Implementing Quality Function Deployment in America.* Methuen, MA: GOAL/QPC, 1989.

MIZUNO, S. and AKAO, Y. *QFD: The Customer-Driven Approach to Quality Planning and Deployment.* (Translated by Glenn Mazur). Asian Productivity Organization, 1994.

MOEN, R., NOLAN, T., and PROVOST, L. P. *Improving Quality Through Planned Experimentation.* New York: McGraw-Hill, 1991.

NAKUI, S. *"Comprehensive QFD." Transactions of the Third Symposium on Quality Function Deployment*. Novi, MI, June 24–25, pp. 137–152, 1991.

PUGH, S. *Total Design*. Reading, MA: Addison-Wesley, 1990.

SHILLITO, M. L. *Advanced QFD: Linking Technology to Market and Company Needs*. New York: Wiley-Interscience, 1994. http://www.klostermanbakery.com/products/whole-grain-rich/24-oz-100-pct-whole-wheat-bread#

SOLUTION CONCEPT

CHAPTER 8

CONCEPTUAL DESIGN

> **"*You can't solve a problem on the same level that it was created. You have to rise above it to the next level.*"**
>
> ~Paraphrase of Albert Einstein

Conceptual design is the entry point to the *solution concept* stage of the design process model. As stated earlier, the challenges of engineering design are two-fold; (1) defining the problem and (2) identifying the solution. We took up the first challenge in Part 3 in the form of the *function diagram* and *design specifications*. The second challenge starts here.

The objective of conceptual design is to produce these three outputs: (1) A sketch of the overall system describing the scheme; (2) sketches of the important subsystems; and (3) rough dimensions of the proposed product.

This chapter starts with the definition of conceptual design as a scheme and analyzes the conceptual design of a power train for a scissor lift as an example that provides (1) the description of the scheme as a sketch, (2) a list of components, and (3) the description of a layout. We will then explain morphological analysis for conceptual design in detail and name eight other methods used for idea generation. With these in the background, we will analyze the approaches to conceptual design and design thinking. Conceptual design thinking can be described as the "harmonious integration of function providers delivering the functions set out in the functional description outlined by the function tree or function structure."

8.1 OBJECTIVES

By the end of this chapter, you should be able to

1. Explain conceptual design as an integrated scheme of function providers.
2. Generate a morphological chart and derive conceptual designs from it systematically.
3. Use systematic methods to generate function-based conceptual designs.
4. Analyze and evaluate conceptual design approaches.

8.2 A TYPICAL CONCEPTUAL DESIGN

Conceptual design is the stage where outline solutions are identified. French [1] identifies this solution in the form of schemes:

> By a scheme is meant an outline solution to a design problem, carried to a point where the means of performing each major function has been fixed, as have the spatial and structural relationships of the principal components. A scheme should have been sufficiently worked out in detail for it to be possible to supply approximate costs, weights, and overall dimensions, and the feasibility should have been assured as far as circumstances allow. A scheme should be relatively explicit about special features or components but need not go into much detail over established practice.

The task as a designer is to develop and outline solutions that are extensions of the product concept, or more specifically, extended or developed from the function structure and to meet the specifications. It is a transition between (1) a set of required functions, (2) a set of behaviors that fulfil the functions, and (3) a preliminary arrangement of components or structures that make the behaviors.

A typical conceptual design will have three kinds of documentation. They are

1. A sketch (hand-or computer-made) of the overall system with necessary notes
2. Important subsystems with sketches as necessary
3. Rough dimensions of the proposed product

As an example consider the power train for the single-stage scissor lift shown in Figure 8.1. The goal is to draw electric power from the main supply and use it to produce ascending and descending motions to the platform carrying the payload. This is achieved by providing a "to- and- from motion" to the free end of the scissor mechanism.

The scheme for the power train may be described as follows:

1. The motor rotates the motor-shaft.
2. The motor-shaft rotates coupling 1.
3. Coupling 1 rotates the input shaft of the gearbox.
4. The input shaft rotates the intermediate shaft inside the gearbox.
5. The intermediate shaft rotates the output shaft of the gearbox.
6. The output shaft rotates coupling 2.
7. Coupling 2 rotates the power screw.
8. The power screw translates the nut.
9. The nut shaft translates the free end of the scissor.
10. The scissor lifts or lowers the platform.

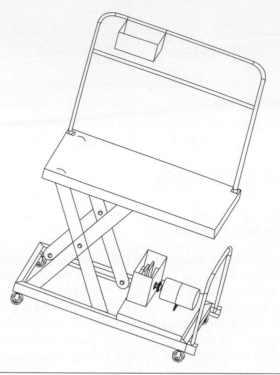

Figure 8.1: A Single-Stage Scissor Lift

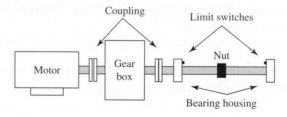

Figure 8.2: Sketch of the Proposed Power Train

The main components or subsystems of the power train (Figure 8.2) are

1. Motor and motor shaft
2. Coupling 1
3. Input shaft of the two-stage gear box
4. Intermediate shaft of the gear box
5. Output shaft of the gear box
6. Coupling 2
7. Power screw
8. Power- screw nut
9. Nut shaft

Rough dimensions and layout can be as shown in Figure 8.3.

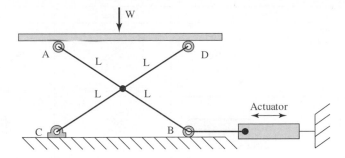

Figure 8.3: Layout of the Power Train with Rough Dimensions

8.3 CONCEPTUAL DESIGN USING MORPHOLOGICAL ANALYSIS

General morphological analysis is a method for identifying and investigating the total set of possible configurations contained in a given problem. The approach begins by identifying and defining the parameters of the problem to be investigated and assigning each parameter a range of relevant "values" or conditions. You then construct the morphological chart by listing the parameters in the left-most column. The cells in the row are filled with one particular "value" or condition of the corresponding parameter.

In the case of conceptual design, the parameters are the purpose functions. Because each function can be achieved in a number of ways, each way is entered in the rows of the chart. Figure 8.4 shows the schematic of a morphological chart; each subfunction is shown in the left column and the many possible solutions appear in the corresponding rows.

Figure 8.5 shows an example of this for a mechanical vegetable collection system. The parameters are concepts of devices that can provide specific functions. The general idea is to identify as many means possible to achieve the same functional requirement. The parameters under consideration are as follows:

1. Vegetable-picking device
2. Vegetable-placing device

Possible Solution/Options Sub function		1	2	...	j	...	m
1	F_1	O_{11}	O_{12}		O_{1j}		O_{1m}
2	F_2	O_{21}	O_{22}		O_{2j}		O_{2m}
:		:	:		:		:
i	F_i	O_{i1}	O_{i2}		O_{ij}		O_{im}
:							
n	F_a	O_{n1}	O_{n2}		O_{nj}		O_{nm}

F = Function of sub function
O = Options
ij = i,j solution/or option

Figure 8.4: A Generic Morphological Chart

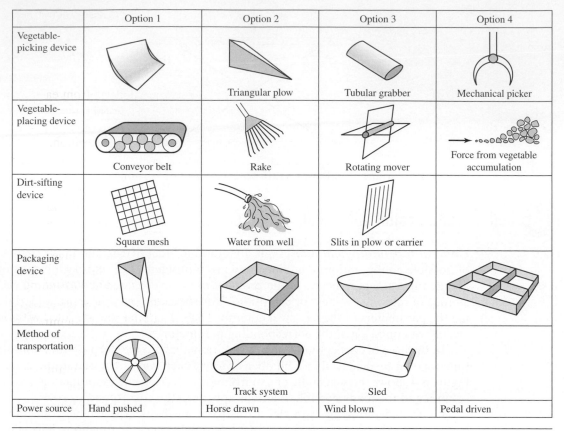

	Option 1	Option 2	Option 3	Option 4
Vegetable-picking device		Triangular plow	Tubular grabber	Mechanical picker
Vegetable-placing device	Conveyor belt	Rake	Rotating mover	Force from vegetable accumulation
Dirt-sifting device	Square mesh	Water from well	Slits in plow or carrier	
Packaging device				
Method of transportation		Track system	Sled	
Power source	Hand pushed	Horse drawn	Wind blown	Pedal driven

Figure 8.5: Morphological Chart for a Vegetable Collection System

3. Dirt-sifting device
4. Packaging device
5. Transportation method
6. Power source

8.3.1   Steps to Develop Concepts Using a Function-Based Morphological Chart

The strategy is to generate alternatives for each of the functional subsystems necessary and then choose compatible combinations.

STEP 1.   *Extract all purpose functions:*

These are the high-level functions that the product has to perform. In the function tree, they will be near the root. In a function structure of flows, these may be representations of subsets of functions inside the black box. In the morphological chart, these occupy the first column in the left.

STEP 2.   *Develop concepts for each function:*

The goal of this step is to generate as many concepts as possible for each of the functions identified in Step 1. In the previous example, vegetable placing could be achieved using a conveyor belt, a rake, a rotating mover, or simply by accumulating the vegetables. If there is a function that has only one conceptual idea, the definition of this function should be reexamined. Keep the level of abstraction and level of detail at the same degree.

STEP 3.   *For each function or subfunction, list all means or methods to be used:*

In this step, repeat the activities specified in Step 2 for each purpose function. These secondary lists are the individual subsolutions that, when you combine from each list, will form the overall design solution. You can express the subsolutions in rather general terms, but it is better if you can identify them as real devices or subcomponents. For example, to lift something you can use a ladder, a screw, a hydraulic piston, or a rack and pinion. The list of solutions should contain both the known components and any new or unconventional methods or solutions for achieving the specific task.

You can use many brainstorming or lateral thinking concepts to arrive at the solution. Techniques to carry this out are discussed in the following section. Remember that many possibilities are necessary for each function. Also, when trying to identify a provider for a given function, only that *specific* function or subfunction should be the criterion, not the overall function. You will take care of additional eliminations at a different stage in the design process.

STEP 4.   *Draw up the chart containing all of the possible subsolutions:*

This chart is called the *morphological chart*. The morphological chart is constructed from the functional list. Start with a grid of empty squares. The left column lists each function identified in the functional diagram. Across each row, list all alternative means (methods) for achieving the function. In the left column, enter in each method in a square. Enter all the solutions identified in Step 3. Once finished, the morphological chart contains the complete range of all theoretically possible solution forms for the product. You create the complete range of solutions by selecting one subsolution at a time from each row. The total number of combinations is rather large, so keep the list of methods to about five.

STEP 5.   *Identify feasible combinations:*

The next step is to consider the feasibility of the combinations formed by combining the individual concepts into complete conceptual designs. A large proportion will be incompatible and can be scrapped immediately. Among the compatible ones, some (probably a small number) will be existing solutions, and some will be new solutions.

Sometimes it is helpful to pick combinations based on "themes" for each potential solution—for example, the economical one, the environmental one, the fancy one, and so forth. Be careful not to narrow your choices with inappropriate or limited themes. It is also essential to maintain an open mind at this point by assuming that all selected solutions should seem feasible. Identify as many potential solutions as possible for further consideration and evaluation. Evaluating these solutions is the subject of Chapter 9. Section 8.4 discusses how brainstorming can help you generate ideas for morphological charts as well as potential solutions for the complete product.

8.3.2   Morphological Solution Space

Any design problem will have numerous solutions: viable and not viable, efficient and not so efficient, possible but not producible, and so on. The design method at the conceptual design stage should scan through the solution space and identify the set that has the near-optimal and the optimal solutions as its members. The situation was shown earlier as a schematic in Figure 1.1, which is reproduced here as Figure 8.6.

In functional representation, a design solution can be represented by the combination of one solution for each purpose function at a certain level in the hierarchy. If there are four functional requirements A, B, C, and D, a design solution is the combination of means for each one of A, B, C, and D.

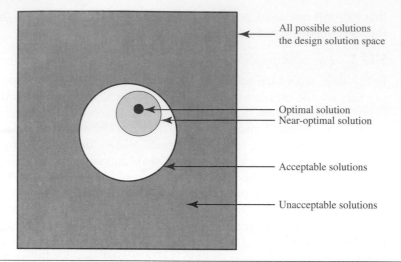

Figure 8.6: Design Solution Space

We can describe the means for finding the requirements for A, B, C, and D by stating

1. Let the means for A be given by A_1, A_2.
2. Let the means for B be given by B_1, B_2, B_3.
3 Let the means for C be given by C_1, C_2.
4. Let the means for D is given by D_1, D_2, D_3.

Then a design solution can be described as $A_i B_j C_k D_l$.

Consider a situation where there are four purpose functions A, B, C, and D. The morphological chart for that situation is given in Table 8.1.

There are only two means to provide function A, three means to provide function B, two means to provide function C, and three means to provide function D. Then there can be $2 \times 3 \times 2 \times 3 = 36$ design combinations. In other words, there are 36 design solutions in the solution space. Possible solutions are

$$\begin{bmatrix} A_1B_1C_1D_1 & A_1B_1C_1D_2 & A_1B_1C_1D_3 & A_1B_1C_2D_1 & A_1B_1C_2D_2 & A_1B_1C_2D_3 \\ A_1B_2C_1D_1 & A_1B_2C_1D_2 & A_1B_2C_1D_3 & A_1B_2C_2D_1 & A_1B_2C_2D_2 & A_1B_2C_2D_3 \\ A_1B_3C_1D_1 & A_1B_3C_1D_2 & A_1B_3C_1D_3 & A_1B_3C_2D_1 & A_1B_3C_2D_2 & A_1B_3C_2D_3 \\ A_2B_1C_1D_1 & A_2B_1C_1D_2 & A_2B_1C_1D_3 & A_2B_1C_2D_1 & A_2B_1C_2D_2 & A_2B_1C_2D_3 \\ A_2B_2C_1D_1 & A_2B_2C_1D_2 & A_2B_2C_1D_3 & A_2B_2C_2D_1 & A_2B_2C_2D_3 & A_2B_2C_2D_3 \\ A_2B_3C_1D_1 & A_1B_3C_1D_2 & A_2B_3C_1D_3 & A_2B_3C_2D_1 & A_2B_3C_2D_2 & A_2B_3C_2D_3 \end{bmatrix}$$

Table 8.1: Morphological Chart for Design Problem X

Purpose Function	Means 1	Means 2	Means 3
A	A1	A2	
B	B1	B2	B3
C	C1	C2	
D	D1	D2	D3

In a tree format, this could be the one shown in Figure 8.7. It is constructed by drawing the options for A from the root as the starting point. Each end of the branches (A1 and A2) is then connected to the options of B. Once that is completed the end of the branches B are connected to options of C. The process continues until all options for the purpose functions are exhausted. You can easily find viable solutions using the tree.

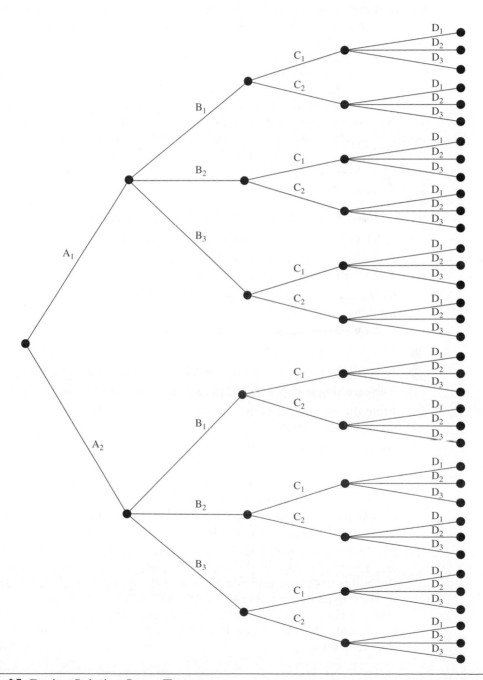

Figure 8.7: Design Solution Space Tree

8.3.3 Identifying the Viable Solutions

The condition for viability is that each subsolution or means should be compatible with others. In the tree, if any one connection is not viable, all branches beyond are not viable. If A_1B_1 is not viable, the entire first row in the solution space matrix and the corresponding branches in the tree are not viable even if others beyond it are viable. The possible viable solutions are those made up from compatible joints at every node in the tree. With this in mind, start the tree from left to right. The first node is the starting point for all solutions. Consider the branches A1 and A2. Both are viable, so mark a "Y" for "yes" on the link. Proceed to the next node where the means for B are considered. If A1 and B1 are not compatible, mark an "N" for "no" on that link. You do not need to consider the branches beyond it. Then consider the branches A1 and B2. Again, if A1 and B2 are not compatible, mark an "N" on that link; if they are compatible, mark a "Y" and proceed to the next node to consider B2 and C1. In this way proceed with all branches. This will eliminate a substantial part of the tree. Consider the means that are marked "Y" from start to finish. These are possible solutions unless they are incompatible with the two other means proposed. Because their number is small, they should be handled easily. This method is easy if the tree is already available. If you have to construct the tree, you can draw a tree showing only the compatible branches. This will substantially reduce the number of branches to be drawn.

8.3.4 Procedure to Define Viable Solutions

1. Draw the list of purpose functions that will completely define the product.
2. Consider each function individually and identify the means for providing these functions. Call them the number of options.
3. Create a matrix with (maximum number of means +1) columns and (number of purpose functions +1) rows.
4. In the first row starting from the second column, write means 1, means 2, in each cell.
5. In the first column starting from the second row, write the purpose functions, one per cell.
6. Now fill the matrix row by row by filling a single option in each cell.
7. The resulting matrix is called the *morphological chart* or the *morphological matrix*.
8. From the chart identify the number of options available for each purpose function and arrange them in ascending order.
9. Draw a vertical dotted line for each purpose function and a root node away from these lines.
10. Now start constructing the tree. On the line representing the function for which the smallest number of options is found (first line) mark off points representing the number of options. Draw lines connecting the root node to these points. Label the lines with the options' names—for example, A1, A2, and so forth.
11. Now move to the next line. Mark off points representing the number of options available for that function corresponding to each point on the first vertical line. Connect the points to the corresponding point on the previous line and label the lines with the options' names—for example, B1, B2, B3, and so forth. Now consider whether the nodes at the ends of the line segment are compatible. If they are, write "Y" on the line; if not, write "N" on the line. Repeat this procedure for all points between the first and the second line.

12. Identify the nodes on the second vertical line, which are the end points of lines marked with a "Y." Mark off points on the third vertical line corresponding to each one of them. Connect them, carry out the compatibility test, and label each one "Y" or "N." Do not do anything to the nodes connecting "N" lines.

13. Repeat the procedure with all the vertical lines.

14. The combinations with all "Y" labels are potentially viable concepts. It will be a small number, and further viability checks can easily be made.

The process can be stated as an algorithm that eliminates unacceptable solutions at early stages and converges towards near-optimal and optimal solutions. The process is enhanced by having the functions with the smaller number of options near the root of the tree since an "N" label there would cut down a substantial amount of the solution space.

8.3.5 Illustrative Case Study

An adjustable shower curtain rail is to be designed. The function tree for the product is given in Figure 8.8. It has four purpose functions.

Let the functions be labeled A, B, C, and D and the solutions or means proposed are (A1, A2), (B1, B2), (C1, C2), and (D1, D2). For simplicity, only two means per function have been considered. We've used steps 1 to 7 to prepare the morphological chart given in Table 8.2.

Possible solutions are

$$
\begin{bmatrix}
A_1B_1C_1D_1 & A_1B_1C_1D_2 & A_1B_1C_2D_1 & A_1B_1C_2D_2 \\
A_1B_2C_1D_1 & A_1B_2C_1D_2 & A_1B_2C_2D_1 & A_1B_2C_2D_2 \\
A_2B_1C_1D_1 & A_2B_1C_1D_2 & A_2B_1C_2D_1 & A_2B_1C_2D_2 \\
A_2B_2C_1D_1 & A_2B_2C_1D_2 & A_2B_2C_2D_1 & A_2B_2C_2D_2
\end{bmatrix}
$$

Figure 8.9 shows the complete picture of the solution space tree and the compatibility check. It is made up of 16 solutions with 30 lines connecting nodes. However if A1 and A2 are considered, there are two "N" branches, and they need not be considered any further. This eliminates two big branches each containing 7 lines. This reduces the number of lines by 14.

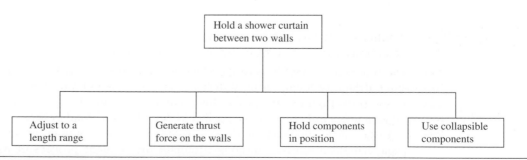

Figure 8.8: Function Tree of a Shower Curtain

Table 8.2: Morphological Chart

Function		Solution 1	Solution 2
A	Adjust to length ranges	Coupling Adjustment A1	Telescopic Adjustment A2 Valeriy Lebedev / Shutterstock.com
B	Generate force on shoes	Spring Compression B1	Threaded Coupling B2
C	Hold components in position	Spring Washer C1	Thread Friction C2
D	Use collapsible components	Two-Piece Connection D1	Three-Piece Connection D2

When B is considered again, the two branches with 3 lines with "N" notations don't need to be considered. Only 10 out of the 30 lines have to be drawn. From the analysis, viable solutions are $A_1B_2C_2D_1$, $A_1B_2C_2D_2$, and $A_2B_1C_1D_1$. If no further incompatibilities are detected, these solutions should be used for further development.

8.4 METHODS FOR IDEA GENERATION

The creative part of conceptual design is idea generation where new ideas are being generated to solve the design problem. Several design methods or techniques have been reported with evidences showing their usefulness. In this context a technique is a procedure, skill, or art used to complete a task. Shah [2] has categorized them into two broad groups: (1) intuitive and (2) logical. Most of the available methods are divided into these groupings as shown in Figure 8.10. He argues that the intuitive methods stimulate the thought process of human mind and, though the results are unpredictable, have the potential to produce novel solutions. Logical methods on the other hand systematically decompose and analyze the problem. The most suitable method for a given problem depends on the nature of the problem. This section describes the classification and structuring of the methods as arranged by Shah and describes some of those methods with examples.

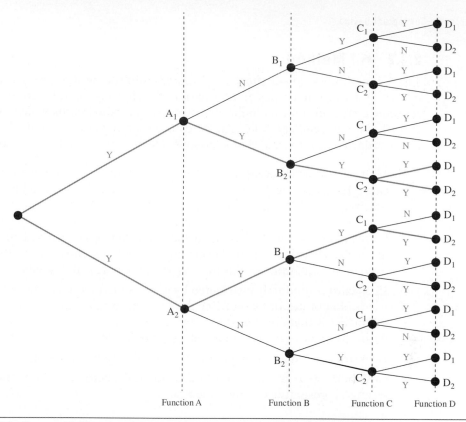

Figure 8.9: Compatibility Analysis within the Complete Solution Tree

8.4.1 Shah's Classification of Idea-Generating Methods

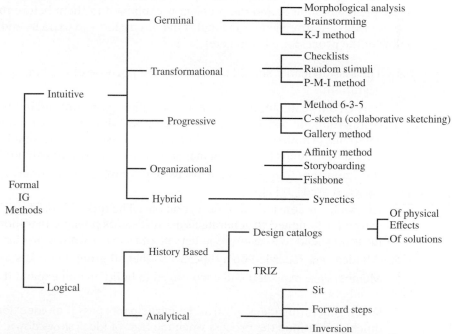

Shah J. J., Kulkarni S. V., Vargas-Henandez N. "Evaluation of Idea Generation Methods for Conceptual Design: Effectiveness Metrics and Design of Experiments," Transactions of the ASME, Vol. 122, pp. 378–384, 2000.

Figure 8.10: Classification of Idea-Generating Methods

8.4.2 KJ Method

The KJ method is named after the Japanese anthropologist, Jiro Kawakita. It establishes an orderly system from a chaos of information. When using the KJ method to generate ideas, all relevant facts and information are written on individual cards which are collated, shuffled, spread out, and read carefully. The cards are then reviewed, classified, and sorted based on idea similarity, affinity, and characteristics. The procedure in KJ method is as follows:

1. Multiple people write customer needs, product ideas, and comments onto 4 × 5-inch cards. Each card can be brief with only a title, or it can provide additional details such as a description and illustration. People can quickly sketch and handwrite ideas, needs, descriptions, and illustrations onto a 4 × 5 card. They can use a note's contents to search the web for related images and information and to create new notes based on search results. Images and web page content capture experiences, emotions, and concepts that may be hard to express through words or illustrations. People can reveal this data by several means.

2. Each card is randomly distributed by either shuffling the cards before handing them to each collaborator or by shuffling the cards around a table and having each person work on the cards that are closest to them. They then group similar ideas into piles. People naturally explore the grouped items by moving related notes next to each other.

3. Piles are labeled according to the need or idea they represent.

4. Collaborators relate the notes by drawing links between groups and creating meta-groups representing common themes.

8.4.3 Brainstorming

Brainstorming aims to produce a large number of ideas through the interaction of a group of people in a short time. The underlying principle is that when there are many ideas, the chance that there will be a better idea among them is higher than when there are few ideas. A group of people is assembled, and the problem is explained to them before they are asked to come up with ideas. A group leader facilitates the discussion and extracts and records ideas. Salient points of the process are listed here:

1. The session leader should clearly state the purpose of the brainstorming session at the beginning.

2. Participants call out one idea at a time, either going around in turn or at random. Going round in order structures the process, while going in random may favor greater creativity. Another option is to begin the brainstorming session by going in turn, and after a few rounds the session may be opened up to all to call out ideas at random.

3. It is important to refrain from discussing, complimenting, or criticizing ideas as they are presented.

4. Every idea is considered to be a good one. The quantity of ideas is what matters. The method is designed to generate as many ideas as possible in a short time. Discussing ideas may lead to premature judgment and slow down the process.

5. All ideas are recorded on a flipchart so that all group members can see them.

6. Members are expected and encouraged to build on and expand the ideas of other group members.

7. When generating ideas in turns, let participants pass if an idea does not come to mind quickly. Continue the process when the flow of ideas slows down. This will enable creating as long a list as possible and reaching for less obvious ideas.

8. After all ideas have been listed, clarify each one and eliminate exact duplicates.

8.4.4 Ideation: Asking Structured Questions

Ideation is a process of generating ideas for a design solution by asking a set of structured questions. It relies on the fact that when there are several ideas, the chance for there to be a good one is higher than when the number of ideas is small. There are many techniques and different ways to structure questions. The following is one such example. In this set, 15 classic questions aimed to guide thinking in particular directions are included to help generate new ideas for a product.

1. *What is wrong with it?* Make a list of all things that you feel are wrong with the present product, idea, or task.

2. *How can I improve it?* Forgetting feasibility, list all of the ways you would improve the present product, idea, or task.

3. *What other uses does it have in its present form?* What other uses can there be if the idea is modified? Can it perform a function that was not originally intended?

4. *Can it be modified?* For example, change of trim, shape, description, weight, sound, form, or contours, and so forth.

5. *Can it be magnified?* Can it be made larger, higher, longer, wider, heavier, or stronger.

6. *Can it be "minified"?* Can it be made smaller, shorter, narrower, lighter, or miniaturized.

7. *Is there something similar we can adopt?* What can be copied? Can it be associated with something else? Is there something in stock or surplus that can be used?

8. *What if we reverse it?* Try turning it upside down, turning it around, rearranging it, making an opposite pattern or an opposite sequence, and so forth.

9. *Can it have a new look?* Can we change the color, form, or style; can we streamline it or use a new package or a new cover?

10. *Can it be based on an old look?* Can we use a period or antique look, parallel a previous winner, or look for prestigious features Can we trade on "They don't build them like that anymore"?

11. *Can it be rearranged?* Can we try a different order, interchange components, piece them together differently, or change places.

12. *Can it be substituted?* What can take its place? Can we exchange plastic for metal, metal for plastic, make it light instead of dark or round instead of square? What other process, principle, theory, or method can be used?

13. *Can the ideas, principles, methods, groups, components, hardware, or issues be combined?*

14. *Can we simplify?* Can we make it easier to reach, disposable, simple to use, or having faster performance?

15. *Can it be made safer?* What devices, properties, controls, or sensors can be added to prevent injury, accident, or explosion?

The number of new ideas that are generated using such a simple technique will be surprising; try it: Pick an existing product, go through the following questions thoroughly, and allow sufficient time to consider each point. See how many new ideas can arise.

8.4.5 Checklist Method

The checklist method is a transformational method that is very popular. Ford [3] identifies three aspects—(1) attribute listing, (2) wishful thinking, and (3) demerit listing—as constituent parts of the process.

Attribute listing is the identification of all the important characteristics of the systems and the relevant people. It is expected to result in a list of all the relevant attributes that can be integrated in problem solving or in the development of the new idea.

Wishful thinking involves the unlimited imagination that focuses on ideals that can be achieved in resolving the current problem. The ideals from the list are to be compared with the available attributes and with the best way to combine the attributes outlined.

Demerit listing involves studying the current situation to identify all the possible ways in which it can be altered to achieve better results. In short, attribute listing explores personal and physical attributes, wishful thinking explores ideals, and demerit listing explores possibilities to improve the status quo. It aims at developing a new idea by thinking through a series of related issues or suggestions (ideals). The checklist may take any form and length. An example of a general checklist follows:

1. *Other uses:* Are there new ways of using the product, or can it be modified to use in other ways?

2. *Adapt:* What else is like this? What other ideas does this suggest? What could be copied? What could be emulated?

3. *Modify:* Try adding a new twist; try changing meaning, color, odor, or form, and so forth.

4. *Magnify:* Make it larger, increase the frequency, increase its strength, increase its thickness, or add extra value.

5. *Minify:* Make it smaller.

6. *Substitute:* What else can be used instead; what other ingredient or what other material?

7. *Rearrange:* Interchange components, change the layout, or change the order.

Checklists have to be made for each project to make it more effective

8.4.6 C-Sketch

Collaborative sketching or C-sketch was proposed originally as an idea-generation method in 1993 at Arizona State University under the name of 5-1-4 G. It originated as an extension of Method 6-3-5 in which six designers generate three ideas at each of five passes. In method 5-1-4 G, five designers work on a single design at a time in four passes. The designers work on sketches—hence the letter G (graphic) in the name. The method was renamed C-sketch in an attempt to provide a more descriptive name. The method works in the following manner:

1. Designers work independently, developing a sketch of their proposed solution to the problem for a predetermined length of time (cycle time). When the time has expired, the sketch is passed to the next designer.

2. This designer may then add, modify, or delete aspects of the proposed design solution. The fundamental limitation to changes in the sketches is that the entire design may not be erased.

3. In this manner, the sketches are passed sequentially through the design team. Designers add their own imprint on the design sketches.

4. At the conclusion of the exercise, a set of solutions will be available, whose number equals the number of designers participating in the method.

5. A secondary constraint is that sketching is the only allowed mode of communication among design team members.

Figure 8.11 illustrates the pattern for passing the sketches in a group of designers. The sketch from the first designer is passed to the second, and so on until all designers have worked on the same sketch.

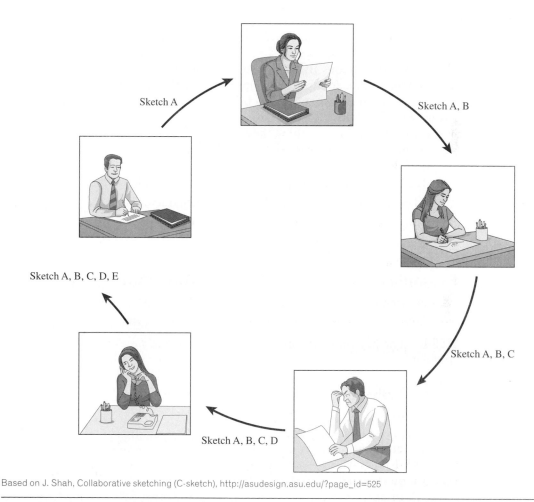

Based on J. Shah, Collaborative sketching (C-sketch), http://asudesign.asu.edu/?page_id=525

Figure 8.11: Illustration of C-Sketch Method.

8.4.7   Critiquing

Critiquing refers to receiving input on current design ideas. This could be collaborative, such as receiving a design critique from a colleague or individuals critiquing their own ideas (either systematically or intrinsically). This technique often spurs new thought by finding solutions to design flaws within current concepts.

8.4.8 Design by Analogy

Analogy can be defined as the similarity among features of two things, on which a comparison may be based. To understand the analogy better, consider the example of Dym and Little [4], which states "studying engineering is like drinking water from a fire hose." The properties of the water from the fire hose are (1) continuous supply, (2) large volume flow rate, and (3) high pressure. The statement says that engineering study also handles a vast syllabus, delivered at a rapid pace and with high rigor. Gentner [5] decomposes analogical learning into five subprocesses: (1) accessing the base system; (2) performing the mapping between base and target; (c) evaluating the match; (d) storing inferences in the target; and sometimes (e) extracting the commonalities.

Design-by-analogy is the use of examples to spark ideas as to what a solution to a problem might be. It is a process that identifies an analogous product that is known to a great extent, understands and establishes its key characteristics, and maps a relational structure from a base or known product to the target (the product to be designed). To illustrate the process consider the example that follow.

Brunel, the famous British engineer during Queen Victoria's time, observed that a ship-worm formed a tube for itself as it bored through timber. This is said to have led him to the idea of a caisson for underwater construction when he was building an underwater tunnel under the River Thames [6]. In a similar fashion Dym and Little [4] speak of the pesky little plant burrs that seem to stick to everything they are blown on to have led the way to the development of Velcro, the widely used fastener. Steps for design by analogy can be listed as follows:

1. Understand the design problem, and identify the key characteristic needed.
2. Identify a base system that exhibits that characteristic.
3. Carry out a mapping between the base and the target.
4. Evaluate and consolidate the design.

Example 8.1: Design of a Reading Assistant

The requirement is a compact and elegant reading assistant that will present a maximum of six books in open condition to a reader.

STEP 1. The key characteristic needed is *simple and flexible support for multiple surfaces.*

STEP 2. A rubber plant with leaves attached to the main trunk was identified as the base.

STEP 3. The base product consisting of (1) part in the pot, 920 main trunk, (3) stem of the leaf, and (4) the whole leaf, were mapped to corresponding (1) base, (2) column, (3) support arm, and (4) the board.

STEP 4. The interface between the column and the support arm were given rotational freedom to assist the user.

The base and the product are shown in Figure 8.12.

8.4.9 Attribute Listing

Attribute listing refers to taking an existing product or system, breaking it into parts, and then recombining these to identify new forms of the product or system. Since a large number of designs are development designs, this method is very useful. The method can be best explained by the example that follows.

Tamara Kulikova / Shutterstock.com

Photo provided by the authors

Figure 8.12: Base and Product of Design-by-Analogy

Paper Bicycle: Adopted from Design Studies

The requirement was to design and build a bicycle made up of paper with minimum weight to carry client passengers through specified distances. Other materials could be used, but there would be a weight penalty. The use of glue was permitted.

The ideation process started with listing the constituent parts of a BMX bicycle. Six subsystems—(1) frame, (2) drive mechanism, (3) front and back wheels, (4) tube interfaces, (5) dynamic interfaces, and (6) fork and steering—were identified. The problem has now transformed to designing equivalent subsystems or parts using paper. A morphological chart lent itself to the task, and the resulting chart is given in Figure 8.13. The bicycle built from this project is shown in Figure 8.14.

8.4.10 Basic Machine Method

The basic machine method was covered in Section 1.6; however, it is discussed here for completeness. Students are encouraged to study all idea-generation methods. The authors encourage students to utilize the morphological chart because of its simplicity and close connection with the previous design steps (functions and specification).

Prater [7] describes a machine as any device that helps to do work. It may help by changing the amount of force or the speed of action, by transforming energy like a generator, by transferring energy from one place to another like a drive shaft, by multiplying force like a press, by multiplying speed, or by changing the direction of a force. There are only six simple machines: the lever, the block, the wheel and axle, the inclined plane, the screw, and the gear. Table 8.3 analyzes these basic machines.

Complex machines are merely combinations of two or more simple machines. Consider a wheelbarrow as shown in Figure 8.15. It is a class-2 lever with a variation of the wheel and axle attached at the fulcrum. Functionally the lever helps to lift a larger load with a smaller effort, and the turning of the wheel with the help of friction provides the motion.

The conceptual design of the wheelbarrow can be considered as a scheme to combine the functions of the two basic machines. Now consider the barrow and hand mixer shown in Figure 8.16. It is a wheelbarrow with a provision to rotate the drum using the wheel and axle

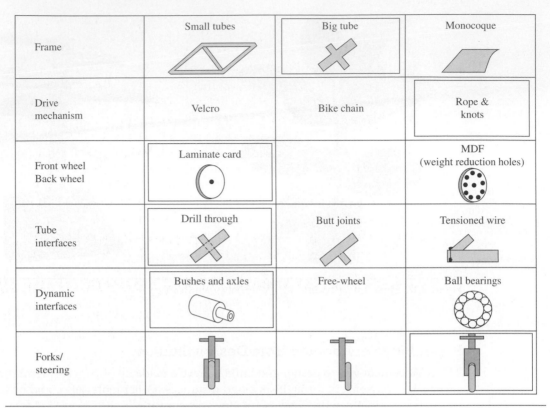

Frame	Small tubes	Big tube	Monocoque
Drive mechanism	Velcro	Bike chain	Rope & knots
Front wheel Back wheel	Laminate card		MDF (weight reduction holes)
Tube interfaces	Drill through	Butt joints	Tensioned wire
Dynamic interfaces	Bushes and axles	Free-wheel	Ball bearings
Forks/ steering			

Figure 8.13: Morphological Chart for a Paper Bicycle

Photo provided by the authors

Figure 8.14: Paper Bicycle Developed with Attribute Listing

where the drum is the axle and the hand crank is the device causing the rim of the wheel to rotate.

Students are advised to review Section 1.6. Table 8.15 and Figures 8.15 and 8.16 are represented here for convenience. Hence, in this method complex machines can be thought of as a harmonious integration of function providers, typically simple machines, delivering the function outlined by the function tree or the function structure.

Table 8.3: Basic Machines

Machine	Description	Function
Lever	A lever consists of the fulcrum (F), a force or effort (E), and a resistance (R). Depending on their relative positions, they are classified as Class 1 RFE (as shown), class 2 FRE and class 3 FER.	Levers of the first and second class help in overcoming large resistances with a relatively small effort by reducing the speed of the resistance. Third-class levers speed up the movement to overcome the resistance with the use of a large effort.
The Block or Pulley	A pulley is a wheel with a groove along its edge, in which a rope or cable can be placed.	A single pulley simply changes the direction of the effort. A combination of pulleys can reduce the effort needed to lift a resistant object.
Wheel and Axle	The wheel-and-axle machine consists of a wheel or crank rigidly attached to the axle, which turns with the wheel.	The force on the rim of the wheel causes the axle to rotate. Wheel and axle magnify the effort or speed it up. Variations are possible with separation of the axle from the wheel.
Inclined Plane	A plank where one end is resting at a lower level and the other at a higher level	Raise or lower heavy objects by applying a small force over a long distance.

Paul Matthew Photography / Shutterstock.com

Continued on next page.

Machine	Description	Function
Screw	An inclined plane wrapped around a cylinder	Reduce large amounts of circular motion to very small amounts of straight-line motion.
Gears 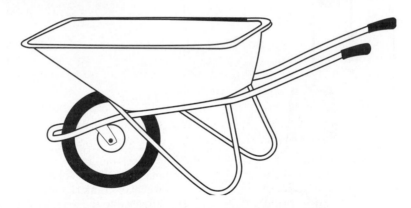	Meshing wheels.	Can change the direction of motion, increase or decrease the speed of the applied motion, and magnify or reduce the applied force. Gears also give a positive drive.

Figure 8.15: Wheelbarrow as the Combination of Two Basic Machines

Figure 8.16: Barrow and Hand Mixer as the Combination of Three Basic Machines

8.5 APPROACHES TO CONCEPTUAL DESIGN

Conceptual design problems can be approached in several ways. Four such ways are given here:

1. Partitioning the solution space
2. Dividing and conquering—the traditional approach
3. The function, behavior, and structure method
4. The adaptive design approach

8.5.1 Partitioning the Solution Space

Every design problem will have several feasible solutions, and there will be one optimal and a few near-optimal solutions in the solution space. The designer knows none of the solutions. The task is to partition the solution space into two parts; the optimal solution and the near-optimal solutions will be in one part, and inefficient solutions will be in the other. The designer can neglect the latter part and focus on the former part. This will enable the designer to arrive at the near-optimal or optimal solution quickly. The approach is to decompose the design problem and solution into manageable chunks, find solutions, and then integrate them into a single feasible solution.

Consider the story of a road construction. It is called a story because there is no evidence in hand for its occurrence. In a third-world country, a new prime minister came to power and scheduled a visit in a month's time to a particular district. The local member of Parliament was the cabinet minister in charge of irrigation, power, and highways. The minister wanted a new 30-mile road to be constructed within the present month so that the prime minister could ceremonially open it during his visit. The construction work involved clearing jungle and doing some cutting and filling before laying the foundation and bitumen. Traditionally the work starts at one end and continues to the other end using the newly laid road as the access road to the work site. The task in hand is the design and execution of a work system. The minister called his largest department, the River Valleys Development Board (RVDB), and asked the chairman whether he could undertake that job. The chairman considered the task and exclaimed "One mile a day! Impossible." The technical director also came up with the same answer. Finally the minister looked to the chief engineer, who said that he could do it provided he could use the stockpile of stones and bitumen from the RVDB and personnel and machinery from RVDB and that there should not be any financial restrictions. The minister agreed. That afternoon the chief engineer called 30 contractors and gave them each one mile to be laid in two weeks starting one week after that day. During the one week, he cleared the jungle and set the 30 work segments. He laid an access road by the side of the road to be constructed for machinery bringing supplies. He cleared enough land on the other side of the road for construction offices of the contractors. He finished the entire construction work in 24 days. The work-system design thought out by the chairman and the technical director was the single-entry type. The chief engineer's work system design was on the other half of the solution space, the multiple-entry type. The chief engineer left the inefficient part and searched for solutions in the efficient half. Thus partitioning and leaving out the inefficient part and searching in the other part is an efficient approach in certain cases.

Example 8.2:

A reading assistant named "ART OF READING" was developed in accordance with a design model having five stages: (1) requirements; (2) product concept; (3) solution concept; (4) embodiment design; and (5) detail design. The design brief was written, societal needs were identified, and the requirements were established using a systematic process of recording the customer statements, the needs (the technical interpretation of the statements), and finally the metrics and units. The designers' priorities were obtained by referring them to the customer once again. Following these steps, the function structure of the product was proposed. With these the product concept was drawn. The end product of this stage is the function structure and a set of specifications.

The product's main objective is to provide the reader with extra space to keep books open when reading and taking notes. The starting point was an ordinary table with sufficient space, as shown in Figure 8.17(a). It was envisaged that half of the table could be used for the books while the remaining half could be used as the desk for a computer and writing. However, this was not a satisfactory solution because it involved a single surface, requiring effort from the user to access books. This eliminated the portion of the solution space where the solutions were based on a single-space solution. Multi-layered tables were then considered with the refined criterion of *easy access height-wise*. On evaluation, the tables with large surface areas as shown in Figure 8.17(b) were ruled out. Thus another portion of the solution space was eliminated. Then multi-layered tables with segmentation, such as the "L" tables and the one shown in Figure 8.17(c), were considered and eliminated, given the improved criterion *easy and compact access*. Elimination of these removed a set of concepts based on compactness assessment. The fourth set of concepts considered the provision of several small surfaces as a separate unit and as integrated unit. Provision as a separate unit as shown in Figure 8.15 (d) was ruled out on the grounds of compactness. The fifth set considered the integrated approach, and two concepts—(1) the rotating dinner stand (lazy Susan style) as shown in Figure 8.17(e) and (2) the rotating holders on a shaft as shown in Figure 8.17(f)—were considered. On the grounds of easy access and compactness the concept shown in Figure 8.17(f) was chosen for further development. Figure 8.17(g) schematically shows the narrowing down process applied to the solution space. Figure 8.17(h) shows the final product.

8.5.2 Dividing and Conquering: The Traditional Approach

As explained in Section 8.3, in this approach the purpose functions are taken one by one, and ways to provide that function on its own are considered using morphological analysis. The success of this approach depends on choosing the purpose functions at the right level. If the level of abstraction is high, there will be very few functions, and it will be difficult to propose means. On the other hand if there are too many functions, there will be a combinatorial explosion and no meaningful solution can be obtained from them.

8.5.3 Function, Behavior, and Structure Method by Gero [8]

In this approach three aspects of the concept—function, behavior, and structure—are taken for development. Function variables describe what the object is designed for, the behavior variables describe the attributes that are expected or derived from the structure variables of the object, and the structure variables describe the components and their relationships. Behavior variables are grouped into two types: (1) the expected ones and (2) the derived ones or what the structure delivers. Within these Gero identifies eight processes: (1) formulation, (2) synthesis, (3) analysis, (4) evaluation, (5) documentation, (6) re-formulation type 1, (7) re-formulation type 2, and (8) re-formulation type 3. They are schematically shown in Figure 8.18.

In formulation, the requirements expressed in terms of function variables are transformed into behavior variables that are expected to deliver the functions. Synthesis defines a solution structure that is expected to deliver the expected behavior variables. Analysis derives the actual behavior from the synthesized structure. Ideally this will be the same as the expected behavior, but in reality there would be some differences. They provide a better insight into the problem and solution arena. If the derived behavior is satisfactory, the design is documented, and the concept is committed for further development.

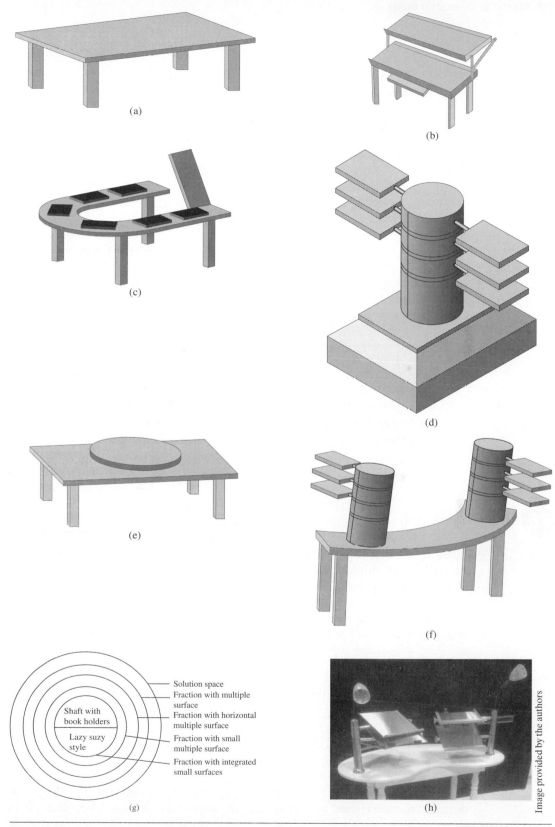

The labels in image (g) read:

Solution space
Fraction with multiple surface
Fraction with horizontal multiple surface
Fraction with small multiple surface
Fraction with integrated small surfaces

Shaft with book holders
Lazy suzy style

Figure 8.17: Conceptual Designs, Design Space Partitions, and the Finished Product

Image provided by the authors

249

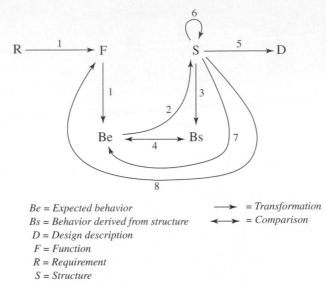

Be = Expected behavior
Bs = Behavior derived from structure
D = Design description
F = Function
R = Requirement
S = Structure

⟶ = Transformation
⟷ = Comparison

Based on Gero J. S., Tham K. W., Lee H. S., "Behavior: a link between function and structure in design," in Brown D. C., Waldron M. Yoshikawa H. (eds), Intelligent Computer Aided Design IFIP pp. 193–225, North Holland: B-4, 1992.

Figure 8.18: Gero's FBS Framework

1. Re-formulation type 1 administers changes in structure variables to adjust the unsatisfactory delivered behavior variable.

2. Re-formulation type 2 administers changes in expected behavior and structure variables to adjust the unsatisfactory delivered behavior variable.

3. Re-formulation type 3 administers changes in functions or their ranges if the derived behavior is unsatisfactory.

8.6 DESIGN THINKING

After discussing the methods and examples of conceptual design, we now look at the conceptual design thinking in an objective way. Design thinking is the combination of (1) understanding the context of the problem, (2) generating insights and solutions, and (3) analyzing and fitting various solutions to the problem context. Typical steps in design thinking may be described as (1) define, (2) research and ideate, (3) prototype (sketch or solid model), (4) test (physical model or analytical model), and (5) choose and implement. Within this context consider the scissor-type elevator. Let the starting point be the scissor mechanism. Then the problem is defined as providing to-and-fro motion to the free end at high force and low speed to lift and drop the platform. On the ideation step, the motor gives a rotational motion, and this has to be converted into linear motion.

There are several mechanisms, including the nut and screw, that can do this job, and one of them must be selected. Again, the motor delivers power at high speed and low torque, which has to be converted to low speed and high torque. Once again, there are several mechanical choices (e.g., gears, pulleys and belts, and chains and wheels) for doing this job. The chosen pieces must be assembled to form the concept in the prototype phase. The assembled concept is then tested analytically or otherwise to find out and confirm that the lift will provide the desired characteristics: maximum height, stowed height, and rise time. At this point there can and will be more than one viable concept. One of them should be chosen and developed further.

8.6.1 Supporting Design Thinking

Gero [8] states that design is the process of moving from function or intended purpose to structure; the components and their relationships that go to make up a design are mental models of past designs. They can be pictures of designs, schemes as described in Section 8.2, or prototypes. They act as a basis for design inasmuch as the designer can retrieve the models and utilize the knowledge encapsulated therein as they become relevant in the design process. It is convenient if these models are structured as functions, behavior, with structure and knowledge relating them. In this context *function* reveals the purpose, *structure* describes the components and their interconnections, and *behavior* spells out how the structure of the artifact achieves its functions. It therefore follows that in evaluating designs only those behaviors associated with function-related performance of the artifact are singled out for consideration.

8.6.2 Creativity

In the systematic design process, creativity is utilized in all steps. The need for group creative problem solving is utilized in constructing the objective tree, deducing information from market analysis, developing a function analysis, structuring a house of quality, and developing concepts. As such, many design books emphasize a strong connection between design and creativity. Some research concludes that design is a creative process by nature. In this section, a definition of creativity in relation to design is introduced.

Ned Herrmann [9], author of *The Creative Brain*, defines creativity this way:

"Creativity in its fullest sense involves both generating an idea and manifesting it— making something happen as a result. To strengthen creative ability, you need to apply the idea in some form that enables both the experience itself and your own reaction and others' to reinforce your performance. As you and others applaud your creative endeavors, you are likely to become more creative."

Lumsdaine [10] defines creativity this way:

"Playing with imagination and possibilities, leading to new and meaningful connections and outcomes while interacting with ideas, people and environment."

Kemper [11] has this to say about creativity:

"The word creativity is a rather curious state: Nearly everyone can recognize creativity instantly, yet no one can define the word in a fully acceptable manner. A companion word, invention, causes even more trouble. Every layperson is instantly confident of the word invention and knows exactly what it means; such a person can think of countless inventions: The electric light, the safety pin, even the atomic bomb. It is only Patent Office examiners and United States Supreme Court justices who believe there is a problem in defining what an invention is."

In these definitions, we see that creativity is associated with generating ideas. We need to ask whether creativity is an individual quality or can be associated with a group or team.

Furthermore, the more important question is whether creativity is natural or can be taught. If creativity is natural, then it can be categorized in each individual as either present or absent. However, research has found that there is a degree of creativity in each individual that depends on the cognitive style, personality, and the creative outcome; psychologists suggest that the level of creativity can be determined either statistically or assessed by experts.

Research has revealed that groups perform better on creative problem-solving tasks. It has been said that two heads are better than one, which also may be applied to creativity. Brainstorming is the best-known and most widely used technique for idea generation in groups. However, for brainstorming sessions to outperform individuals in generating ideas, participants must

1. Have some social relationships (e.g., as students you identified with each other the first day you walked into the classroom).
2. Have used some of the idea generated (you have done this through the objective tree and the function tree);
3. Have some technical experience pertinent to the problem (this is not as important but is preferable)
4. Have worked on some tasks interdependently.

8.6.3 How to Increase the Level of Creativity

The author McCabe [12] once said, "Ideas are like rabbits. You get a couple and learn how to handle them, and pretty soon you have a dozen." The following points will help you become more creative.

Know your thinking style: Since early history, humans have been trying to understand how the brain works. Herrmann [9] developed a metaphorical model of the brain that consists of four quadrants. This is covered in more detail in Lab 2 of this book. In teamwork, you may encounter different thinking styles. A good team is one that represents a full brain. You can train yourself to have all quadrants function at the same power or increase activities toward utilizing more of a specific quadrant.

1. Identify the weakness.
2. Attack problems that require the use of a weak quadrant.

Use visual imagery: Einstein asserted that imagination is more important than knowledge; knowledge is finite whereas imagination is infinite. Bernard Shaw [13] said, "You see things and say why? But I dream things that never were and I say why not?" The inner imagery of the mind's eye has played a central role in the thought processes of many creative individuals. Albert Einstein ascribed his thinking ability to imagining abstract objects.

Kekule [14] discovered the all-important structure of the benzene ring in a dream. Most visual thinkers clarify and develop their thinking with sketches. Drawing not only helps them to bring vague inner images into focus but also provides a record of the advancing thought stream. Sketches expand the ability to hold and compare more information than is possible from simple recall from the brain. Sketches are usually extensions of thought, and they provide a conference table for talking to oneself. Distinguishing between the state of sleep and awakeness and between visions and perceptions has been studied since ancient history. According to researchers in visual imagery, there are three states of visual imagery:

a. Dharana, in which you simply focus on a given place or object.

b. Dhydana, in which you strengthen the focus of attention.

c. Samadhi, in which you experience a union or fusion with the object and cannot distinguish between self and object.

As students, you have been practicing memorization since kindergarten; visualization will help enhance your memory as each image stores much more information than words. The recollection of these images will enhance your idea generation.

Reframing: Einstein asserted that problems cannot be solved by thinking within the framework in which the problems were created. Reframing involves taking problems out of their frame and seeing them in a different context. It allows consideration of potentially valuable ideas outside current frames. The most common habits that limit the ability to change mental frames are as follows:

a. *Pursuit of perfection:* Many people are under the false impression that working long and hard is sufficient to develop the perfect answer to the problem.

b. *Fear of failure:* You develop resistance to change because others may perceive you as incompetent.

c. *Delusion of already knowing the answer:* Once you find an answer, you do not look for others. Why look for another answer if this one works?

d. *Terminal seriousness:* Many people are under the illusion that humor and serious idea generation do not mix.

Humor: It has been said that "men of humor are always in some degree men of genius." One dictionary definition of humor is "the mental faculty of discovering, expressing or appreciating ludicrous or absurdly incongruous elements in ideas, situation, happenings, or acts." If you were to take the words *ludicrous* or *absurdly* out of the definition of humor, you would have a definition of creativity.

Information gathering: In market analysis (Chapter 4), you were introduced to the power of gathering information on the problem statement. Gathering information will also enhance creativity; it will allow you to view ideas generated by other creative minds (written brainstorming). Patents can serve as an excellent source of ideas. However, it is difficult to identify the specific patent that contains the idea that you are looking for.

There are two main types of patents: utility patents and design patents. Utility patents deal with how the idea works for a specific function. Design patents cover only the look or form of the idea. Hence, utility patents are very useful, since they cover how the device works, not how it looks. Since there are over 8 million utility patents, it is important to use some strategies to hone in on the one that you may be able to use. Use a Web search, such as that provided by the Patent Office. Keyword searches (as well as patent numbers, inventors, classes, or subclasses) are available.

The next place to find information is in reference books and trade journals in the relevant area. Another information source is to consult experts. If designing in a new domain, you have two choices for gaining the knowledge sufficient for generating concepts. You either find someone with expertise, or you spend time gaining experience on your own. Experts are those who work long and hard in a domain, performing many calculations and experiments themselves to find out what works and what does not. If you cannot find or afford an expert, the next best source is the manufacturer's catalog or the manufacturer's representative.

8.7 DEVELOPING CONCEPTS: MORPHOLOGICAL ANALYSIS SAMPLES

In this section you will be exposed to three examples to demonstrate the morphological analysis of three different devices.

Example 8.3: Mechanical Vent

Consider the case involving a design to automate the opening and closing of the air conditioning vents in a centralized location in the house. Design teams have developed a function analysis wherein the functions are

1. Select vent.
2. Send signal.
3. Receive signal.
4. Convert signal.
5. Open/close vent.

Figure 8.18 shows a morphological chart for this problem. The five functions are entered in the left-hand column of the chart. Each row represents a specific function. For example, row 5 is the function "open/close vent," and we need to find means to achieve the function of opening and closing the vent. The chart shows five different means or methods to achieve that function using gears, belt, electric field, cable, or impact plate. The methods are entered using words and diagrams in each of the five squares following "open/close vent." The chart in Figure 8.19 thus contains the five functions and the different (5, 5, 5, 6, 5) ways of achieving each function. Theoretically, there are 3750 possible different open/close vent machines, combining the various means. However, many of them are not viable. At least five or six of such combinations will be identified for building the system. Figure 8.19 shows the morphological chart for this unit. It consists of six different options for five functions.

Example 8.4: Wheelchair Retrieval Unit

A design team is required to design a retrieval unit for wheelchairs to assist nurses who perform walking activities with patients. In most cases, a single nurse is in charge of assisting patients during their walking exercise. However, the nurse cannot assist the patient while dragging the wheelchair. The design team developed a function analysis for the unit as follows:

1. Align wheelchair to patient and nurse.
2. Move wheelchair.
3. Steer wheelchair.
4. Stop wheelchair.

Figure 8.20 shows the morphological chart for this unit. It consists of six options for each of the four functions; thus, you will have 64 possible solutions. You may use the tree to select one viable option; ultimately, five or six different viable options will be selected out of these solutions. For example, one possible combination is rail, track, rail, reverse power. (that is, 2, 2, 3, 1).

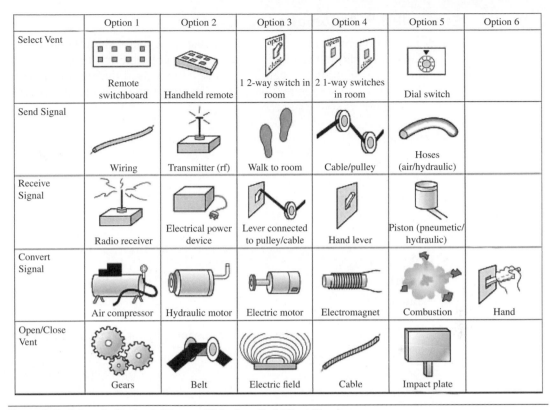

Figure 8.19: Morphological Chart of Mechanical Vent Device

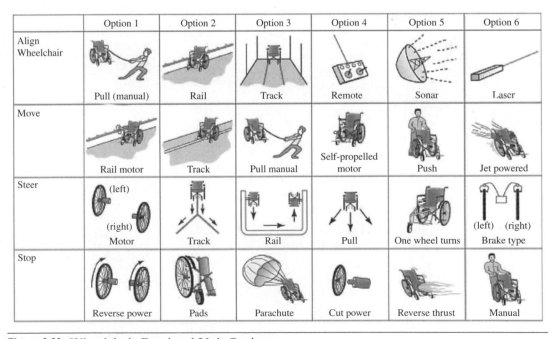

Figure 8.20: Wheelchair Retrieval Unit Options

Example 8.5: Automatic Can Crusher

The design team lists all the functions that must be accomplished. Then team members generate a matrix that shows the functions in the right column, and they point out different ways to achieve these functions. Creativity should be exercised during this activity. The different possibilities may be listed in text or in sketches. Both methods are exercised in this situation. The available power sources are

- Hydraulic
- Magnetic
- Gravity
- Thermal
- Sound
- Pneumatic
- Solar
- Electric
- Combustion

Several designs can be generated from the morphological chart shown in Figure 8.21.

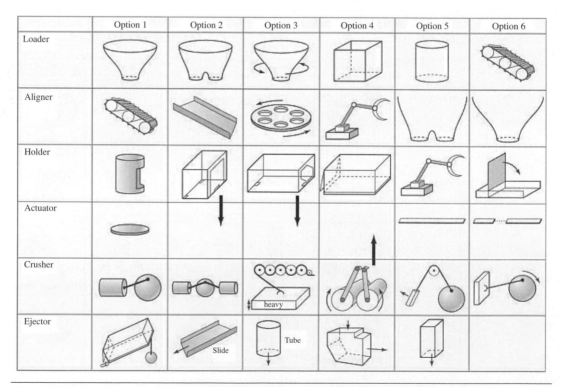

Figure 8.21: Morphological Chart of Automatic Can Crusher (*continues*)

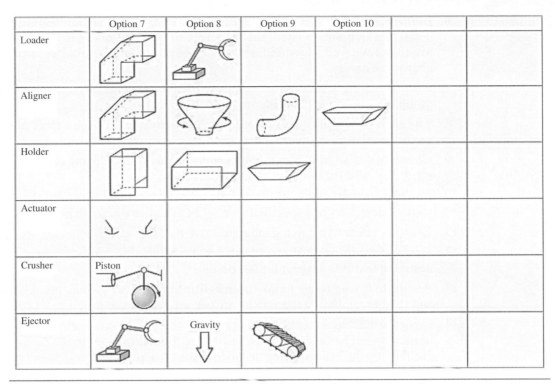

Figure 8.21: Morphological Chart of Automatic Can Crusher (*continued*)

8.8 CHAPTER SUMMARY

1. The objective of conceptual design is to produce these three outputs: (1) A sketch of the overall system describing the scheme; (2) sketches of the important subsystems; and (3) rough dimensions of the proposed product.

2. *Morphological analysis* is a method for identifying and investigating the total set of possible "configurations" contained in a given problem. In design, the functions are the parameters that seek all possible solutions.

3. Steps to use the morphological chart in design include

   a. Extracting all possible functions

   b. Developing concepts for each function

   c. Listing the methods and means for each function or subfunction

   d. Drawing up a chart containing the functions and subfunctions in the left column and the possible solutions for each of the functions and subfunctions in the right column

   e. Identifying feasible combinations

4. The *tree format* for all possible solutions can be used to identify the most viable combinations.

5. Idea-generation techniques can be broadly categorized as intuitive or logical.

6. In *the KJ method*, relevant facts and information are written on individual 5 × 4 cards, which are collated, shuffled, spread out, and read carefully. The cards are then reviewed, classified, and sorted based on idea similarity, affinity, and characteristics.

7. *Brainstorming* aims to produce a large number of ideas through the interaction of a group of people and then filtering for better ideas.

8. *The structured-questions technique* can be used to produce a large number of ideas to alter or change an existing product.

9. *Checklists* as an idea-generation technique consist of enlisting contributions based on attributes, wishful thinking, or demerit listings.

10. *Collaborative sketches (C-sketches)* collect design ideas through the collaborative work of designers in a specified number of passages among them.

11. *Critique* refers to receiving input on current design ideas.

12. *Design by analogy* utilizes available features in nature or in other products to produce desirable features in the intended design.

13. *Attribute listing* refers to taking an existing product or system, breaking it into parts, and then recombining these to identify new forms of the product or system.

14. Complex machines can be thought of as a harmonious integration of function providers (typically simple machines) that deliver the function set out in the function description outlined by the function tree or the function structure.

15. Approaches to concept design include

    a. Partitioning the solution space

    b. Dividing and conquering

    c. Function, behavior, and structure method

    d. Adaptive design

16. Design thinking is the combination of (1) intuitive grasp of the context of the problem, (2) creativity in the generation of insights and solutions, and (3) rationality in analyzing and fitting various solutions to the problem context.

17. Creativity is used in all steps of the systematic design process.

8.9 PROBLEMS

1. Develop a morphological chart for the following statement. If you drive in the state of Florida, you may notice some traffic congestion when crews trim trees on the state roads and freeways. Assume that the Florida Department of Transportation is the client. Usually, branches 6 inches or less in diameter are trimmed. The allowable horizontal distance from the edge of the road to the branches is 6 feet. The material removed from the trees must be collected and removed from the roadside. To reduce the cost of trimming, a maximum of only two workers can be assigned for each machine. The overall cost (which includes equipment and labor, among other things) must be reduced by at least 25% from the present cost. The state claims that the demand for your machines will follow the price reduction (i.e., if you are able to reduce the cost by 40%, the demand will increase by 40%). Allowable working hours depend on the daylight and weather conditions.

2. Compare (a) the systematic design process that leads to a morphological chart with a large number of possible solutions with (b) a design process, in which the problem is identified and then a brainstorming session is started to generate different solutions. Which of these methods would you use. Why? In the event that a certain aspect of the design needs modification, which of the two methods would be easier to apply toward that modification. Why?

3. Apply the structured-question technique to a product such as a coffee machine. List all possible features that you and your team can work on to improve an existing product.

4. Employ the C-sketch technique to generate possible ideas for a new computer mouse.

5. Use the simple machine technique to generate a device to roast peanuts.

6. Compare and contrast group creativity and individual creativity.

7. Discuss the statement that design is a social activity. How would that statement fit in the systematic process?

8. What is the difference between creativity and innovation?

9. What is a morphological chart?

10. Without a creative objective tree, you could not have a creative morphological chart. Discuss.

11. How would you increase your creativity level? Categorize yourself (analytical, organizational, social, intuitive). You can perform the survey in Lab 2.

12. What is the Herrmann model of the brain?

References

[1] French, M. J. *Conceptual Design for Engineers*, Springer, 1990.

[2] Shah, J. J., Kulkarni S. V., and Vargas-Henandez, N. *"Evaluation of Idea Generation Methods for Conceptual Design: Effectiveness Metrics and Design of Experiments,"* Transactions of the ASME, Vol. 122, pp. 378–384, 2000.

[3] Ford, C. M. "Creative Development in Creativity Theory," *Academy of Management Review*, 25(2), pp. 284–285, 2000.

[4] Dym, C. L. and Little, P. *Engineering Design: A Project-Based Introduction*. Hoboken, NJ: John Wiley & Sons, 2003.

[5] Gentner, D. "Analogy." In W. Bechtel and G. Graham (Eds.), *A Companion to Cognitive Science* (pp. 107–113). Oxford, UK: Blackwell, 1998.

[6] Dance, S. P. "Ship-Worm Inspires Brunel," *Mollusc World*, Issue 12, 2006.

[7] Prater E. L. *Basic Machines*. Naval Education and Training Professional Development and Technology Center, 1994.

[8] Gero, J. S., Tham, K. W., and Lee, H. S., "Behavior: A Link between Function and Structure in Design," in Brown, D. C., Waldron, M., and Yoshikawa, H. (Eds), *Intelligent Computer Aided Design IFIP* pp. 193–225, North Holland: B-4, 1992.

[9] Herrmann, N. *The Creative Brain*. Lake Lura, NC: Brain Books, 1990.

[10] Lumsdaine, E. M., Lumsdaine, M., and Shelnutt, J. W. *Creative Problem Solving and Engineering Design*. New York: McGraw-Hill, 1999.

[11] Kemper, J. D. *Engineers and Their Profession*. New York: Oxford University Press, 1990.

[12] Mccabe, M. P. "Influence of Creativity on Academic Performance." *Journal of Creative Behavior*, Vol. 25, p. 2, 1991.

[13] Gardner, H. "Creative Lives and Creative Works: A Synthetic Approach." *In The Nature of Creativity*, J. R. Sternberg (Ed.), pp. 298–321. Cambridge, UK: Cambridge University Press, 1988.

[14] Nickerson, R. S. "Enhancing Creativity." In *The Nature of Creativity*, J. R. Sternberg (Ed.), pp. 392–430. Cambridge UK: Cambridge, University Press, 1988.

In addition the following books/articles/websites were used in preparing this chapter:

AMABILE, T. M. and CONTI, R. "Changes in the Work Environment for Creativity During Downsizing." *Academy of Management Journal*, Vol. 42, pp. 630–640, 1999.

CRANDALL, R. *Break-Out Creativity*. Corte Madera, CA: Select Press, 1998.

CROSS, N. *Engineering Design Methods: Strategies for Product Design*. New York: Wiley, 1994.

DHILLON, B. S. *Engineering Design: A Modern Approach*. Toronto: Irwin, 1995.

DUNNETTE, M. D., CAMPBELL, J. and JAASTAD, K. "The Effect of Group Participation on Brainstorming Effectiveness for Two Industrial Samples." *Journal of Applied Psychology*, Vol. 47, pp. 30–37, 1963.

DYM, C. L. *Engineering Design: A Synthesis of Views*. Cambridge, UK: Cambridge University Press, 1994.

GALLUPE, R. B., BASTIANUTTI, L. M. and COOPER, W. H. "Unblocking BrainStorms." *Journal of Applied Psychology*, Vol. 76, pp. 137–142, 1991.

GOLDENBERG, J. and MAZURSKY, D. "First We Throw Dust in the Air Then We Claim We Can't See: Navigating in the Creativity Storm." *Creativity and Innovation Management*, Vol. 9, pp. 131–143, 2000.

HENNESSEY, B. A. and AMABILE, T. M. "The Conditions of Creativity." In *The Nature of Creativity*, J. R. Sternberg (Ed.), pp. 11–38. Cambridge, UK: Cambridge University Press, 1988.

HILL, H. W. "Group Versus Individual Performance: Are N_2 Heads Better Than One?" *Psychological Bulletin*, Vol. 91, pp. 517–539, 1982.

ISAKSEN, S. G. and TREFFINGER, D. J. *Creative Problem Solving: The Basic Course*. Buffalo, NY: Bearly Limited, 1985.

KOLB, J. "Leadership of Creative Teams." *Journal of Creative Behavior*, Vol. 26, pp. 1–9, 1992.

MCKIM, R. H. *Experiences in Visual Thinking*. Boston, MA: PWS Publishing Co., 1980.

MUMFORD, M. D. and GUSTAFSON, S. B. "Creativity Syndrome: Integration, Application and Innovation." *Psychological Bulletin*, Vol. 103, pp. 27–43, 1988.

PAHL, G. and BEITZ, W. *Engineering Design: A Systematic Approach*. Springer-Verlag, 1996.

ULMAN, D. G. *The Mechanical Design Process*. New York: McGraw-Hill, 1992.

ULRICH, K. T. and EPPINGER, S. D. *Product Design and Development*. New York: McGraw-Hill, 1995.

WEISBERG, R. W. *Creativity: Genius and Other Myths*. New York: Freeman, 1986.

WHITE, D. "Stimulating Innovative Thinking." Research-Technology Management," Vol. 39, pp. 31–35, 1996.

CHAPTER 9

CONCEPT EVALUATION AND SELECTION

"Decision is the spark that ignites action. Until a decision is made, nothing happens."

~Wilferd A. Peterson

A concept is a scheme to achieve the desired behaviors from a structure that meets the requirements and fulfills the originating societal need. At the end of the concept-generation process the design team will have produced five or six schemes or viable concepts. Concept evaluation is the process of evaluating the merits and demerits of each concept to come up with a concept that is the best among those given or even better than all of them. Use concept selection when there are several concepts, and a choice is needed to focus on the most viable one for further development. To carry out a selection effectively, the desired behaviors or goals and predicted behaviors of the conceptual solutions should be known. But the difficulty in concept evaluation is that the design team members have to base concept evaluation on limited knowledge and data. Concepts at this point are still abstract, have limited details, and cannot be measured. The essential question then becomes, "Should we spend time refining and adding further details to all concepts and then measuring their outputs?" or "Would developing a mechanism to point out one concept that will most likely become the quality product prove to be a more efficient option?" It is widely believed that choosing a concept and developing it further is the better option. In order to evaluate the concepts a set of criteria has to be formulated. This chapter describes how to use a set of criteria formulated with objective trees and the specifications to evaluate the given concepts and choose an optimal one.

9.1 OBJECTIVES

By the end of this chapter, you should be able to

1. Use different methods to evaluate the different concepts that were generated in the previous design step.

2. Select a design alternative for further development.

3. Use the *decision matrix* method to choose a design concept using the given criteria and weights.

4. Use Pugh's *concept selection* method to improve and choose a concept from the alternatives given.

9.2 CRITERIA FOR CONCEPT SELECTION

The starting point for product development is the *design brief* as explained in Chapter 4. One of the inputs to produce the design brief is the objective tree. One of the main purposes of the design brief is to explain to the design team what the product is. The findings from the objective tree are recorded *as benefits to be delivered* in the design brief. The main criteria used to satisfy the objectives of the client or company are the benefits to be delivered. On the other hand, the customers rate the products according to what the product will do, and if the customers give the product a low rating, it will be a failure. Thus the degree of satisfying the customer requirements should be the basis for concept selection. An alternative should meet the customer's demands; otherwise it will be dropped in the initial screening. In the QFD-based method of establishing specifications, their relative importance ratings are also calculated. Therefore it is easy to pick and incorporate specifications that have importance ratings above a certain threshold with the benefits to be delivered to form a complete set of criteria for concept selection. If, however, the specifications are written using any of the other methods, the importance rating of the originating customer requirements should be used. Calculation of the weighting factors for each criterion can be made as in the following example.

Example 9.1:

Let the criteria be the cost, appearance, portability, convenience, and durability of a product. Prepare a square matrix with the criteria arranged in the following way:

1. Enter a "–" for the diagonal elements. Take the first row and compare it with all the other elements. If appearance is preferred over cost, enter a "0" for the cell in the cost column. Similarly, if portability is preferred over cost, enter a "0" in the cell. If cost is preferred over convenience, enter a "1" in the cell. If durability is preferred over cost, put a "0" in the cell. The row sum is the score for cost. Enter "½" if both are preferred alike.

2. Repeat the process for the second, Appearance row.

3. Keep on filling the remaining rows, and compute the row sum.

4. Add 1 to all the score values so that there are no 0s in the score.

5. Sum the scores, and divide each score by the total. These are the weights for each individual criterion.

6. For the matrix in Table 9.1, the weights are [0.1, 0.2, 0.2, 0.3, 0.2] and they will add up to 1. The weight is computed by dividing each of the scores by the total of all scores.

Table 9.1: Matrix for Deciding Weighting Factors

Objectives	Cost	Appearance	Portability	Convenience	Durability	Score
Cost	-	0	0	1	0	1
Appearance	1	-	1	0	0	2
Portability	1	0	-	0	1	2
Convenience	0	1	1	-	1	3
Durability	1	1	0	0	-	2

9.3 CONSTRAINTS

Constraints are restrictions on some aspect of the designed object. They eliminate some otherwise viable solutions. They must be reviewed at the proposed component level for each function and at a later stage in the design development since some constraints may be the material type or specific weight or material strength or manufacturability, and so forth.

At this stage in the design process, the two primary constraints to watch for are that components are not violating a function or specifications. For example, the cutting fluid from a machine tool carries small pieces of debris with it that must be removed before they are sent through the pump. Normally they are passed through a filter where the debris pieces are caught and the clean coolant returns to the tank for normal use. The filter element or media is cleaned at regular intervals and eventually replaced. The European Union imposed a constraint banning the use of media. This eliminated the viable media-based cleaning of the cutting fluid. Knowledge of applicable standards and the component performance environment is crucial to eliminate components that do not meet the performance requirements.

9.4 CONCEPT EVALUATION

Evaluation of a design concept implies an assessment of its value made from explicit goals or how well the behaviors of the concept correspond to the desired behaviors. How well the design solution will solve the task can be assessed either relatively to other alternatives or absolutely toward the defined goals.

Evaluation can be carried out in two different ways: (1) selecting the most appropriate solutions and (2) excluding solutions that don't work. To select the most fitting solution, determine or estimate the "value" of the degree of fulfillment of various concepts according to the criteria derived as explained previously. The outcomes of the evaluation depend to a large extent on how well thought out the criteria are in extent and objectivity and on the design team's knowledge and understanding of the respective solutions. The exclusion of inadequate solutions, on the other hand, focuses on the limitations of a solution—that is, their shortcomings or disadvantages.

Concept evaluation is achieved in two stages—preliminary and detailed. In the preliminary stage the concepts are assessed to see whether they can meet the specifications and whether they are technically viable. You can accomplish the preliminary screening of concepts in the following way:

1. The design team directly assesses the concepts with respect to a fixed set of limits derived from the specifications. If a concept doesn't satisfy the specification

requirement, it should be dropped. Although certain functional presentations are on the morphological chart, eliminate them from consideration if they do not meet the specification criteria established previously.

2. The design team uses their experience and engineering sense to eliminate some concepts because the technology is not yet available to facilitate the concept, or, based on their judgment, the product is not feasible. Why use a parachute to stop the wheelchair if another function that is more sound from an engineering standpoint could perform the same function?

9.4.1 Constituents of Concept Evaluation

The product development process thus far can be described as a sequence of three activities: (1) information collection, (2) evaluation, and (3) decision making. The information-collection activity requires three types of information: (1) the goals or benefits to be delivered, (2) important specifications, and (3) synthesized solutions. In the evaluation part of the process, the criteria are formed and their relative importance ratings calculated. In the evaluation process, construct the Pugh's matrix or the decision matrix. In the decision-making process, the better or optimal solution is chosen based on the matrices. Figure 9.1 schematically illustrates the entire process.

INFORMATION GATHERING			CRITERIA FORMATION	EVALUATION	DECISION
Benefits to be delivered	Important specifications	Concepts	Importance of each criterion	Decision matrix Pugh's matrix	Chosen concept chosen concept

Figure 9.1: Concept Evaluation and Selection Process

9.5 PUGH'S CONCEPT EVALUATION METHOD

When considering a collection of conceptual solutions to a design problem, there can be some solutions that are obviously weak. However, these weak solutions may still have a few good features. It is a good practice to eliminate the weak ones and then to apply creative thinking, using the good features to generate some new concepts and eliminate the weak features in the good designs. Pugh introduces a stage-wise improving model which is explained in Figure 9.2.

The entry stage takes all concepts that have to be evaluated. Constructing Pugh's matrix, which is explained in Section 9.5.1, leads to the method of evaluation. After inspecting the Pugh's matrix and rejecting the weak concepts, new concepts that are modifications to eliminate weaknesses and to incorporate the identified good features are added to the chosen lot. A second Pugh's matrix is constructed now, and the weak concepts are again rejected. New concepts are added to the chosen lot and Pugh's matrix is formulated for the third time. There will now be one or two concepts that incorporate all good features; only these will remain.

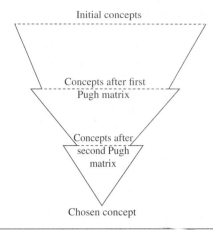

Figure 9.2: Pugh's Stage-Wise Concept Evaluation and Selection

9.5.1 Pugh's Matrix

In Pugh's Matrix, the columns represent the concepts and the rows represent the criteria. Enumerate the concepts in the top row and the criteria in the left-hand column. Perform the Pugh evaluation method using the following steps:

STEP 1. *Choose the comparison criteria*:

If all alternatives fulfill the demands on the same level, then the criteria should be listed in the specification table or with the design criteria. Remember, the design criteria were organized in an ordered fashion before they were used in the house of quality. The order is based on the level of importance according to the design team. The weighting-factor method is based on the design team's assessment of how important the attribute may be to the final product. If the alternatives contain differences in the levels at which demands may be fulfilled, the demands can be entered among the comparison criteria.

STEP 2. *Select the alternatives to be compared*:

From the morphological chart, you will generate different alternatives. Some of these alternatives will be dropped because they do not satisfy the customer demands or are not feasible. The rest of the alternatives are possible candidates. However, by using the initial screening stages, a few alternatives will be left for the final stage. Not all feasible alternatives will be allowed to enter the final stage. Typically *up to six alternatives will be allowed*.

STEP 3. *Generate scores*:

After careful consideration, the design team chooses a concept to become the benchmark or datum against which all other concepts are rated. The datum is generally either

- An industry standard, which can be a commercially available product or an earlier generation of the product.
- An obvious solution to the problem.
- The most favorable (measured according to a vote by the design team) concepts from the alternatives under consideration.
- A combination of subsystems that have been combined to represent the best features of different products.

For each comparison, the concept being evaluated is judged to be better than, about the same as, or worse than the datum. If it is better than the datum, the concept is given a positive [+] score. If it is judged to be about the same as the datum, the concept is given a zero [0] or a letter [S] for similar. If the concept does not meet the criterion as well as the datum does, it is given a negative [−].

After a concept is compared with the datum for each criterion, three scores are generated: the number of plus scores, the number of minus scores, and the number of zero or [S] scores. The scores can be interpreted in a number of ways.

a. If a concept or a group of similar concepts has a high positive total score, it is important to notice what strengths it exhibits. In other words, notice which criteria it meets better than the datum. Likewise, the grouping of negative scores will show which requirements are difficult to meet.

b. If most concepts get the same score on a certain criterion, examine that criterion closely. It may be necessary to develop more knowledge in the area of the criterion to generate better concepts. It may be that the criterion is ambiguous, or it may be interpreted differently by different members of the design team.

c. To learn even more, redo the comparison making the highest scoring concept the new datum. Redo this iteration until the clearly best concept or concepts emerge. *This process becomes a must when the datum is one of the alternatives under consideration.*

After each team member does this procedure, the entire team should compare their individual results. Then conduct a team evaluation.

Consider the example matrix shown in Figure 9.3. In it, eleven concepts and six criteria are entered.

Concept Criteria	1	2	3	4	5	6	7	8	9	10	11
A	+	−	+	−	+	−	D	−	+	+	+
B	+	S	+		−	−		+	−	+	−
C	−	+	−	−	S	S	A	+	S	−	
D	−	+	+	−	S	+		S	−	−	S
E	+	−	+	−	S	−	T	S	+	+	−
F	−	−	S	+	+	+		+	−	+	S
Σ+	3	2	4	1	2	2	U	3	2	4	2
Σ−	3	3	1	4	1	3		1	3	2	2
ΣS	0	1	1	1	3	1	M	2	1	0	2

Source: Pugh, Total Design, Addison-Wesley Pub (Sd) (February 1991), Fig. 4.2, p. 77.

Figure 9.3: Pugh's Matrix (published with permission)

Now follow the steps explained previously to fill the cells in the matrix.

1. Select one design, the one that in the opinion of the design team is the best in the lot. Call this the DATUM (concept 7)

2. Consider the first design. Compare it with the DATUM. For criterion A, if concept 1 is better enter a + in the cell A1. For criterion B, if concept 1 is better enter a + in cell B1. For criterion C if the DATUM is better enter a − in the cell C1. If they are similar enter an S. Finish the process with criteria D, E, and F.

3. Now consider the second design and repeat Step 2.

4. Repeat evaluating all eleven designs in the matrix.

5. Now add all the + in the column, all the − in the column, and all the S in the column.

6. A design with many − is a definite loser. Scrap it, but note any good feature in it.

9.6 DECISION MATRIX

This method utilizes a more numerical approach. Here, the designer rates the conceptual ideas against criteria that are deemed relevant to the task. The conceptual ideas go on the rows, and the design criteria on the columns. The designer decides on importance ratings for each design criteria. For example, let us assume a simple product has only three criteria: *long*, *fast*, and *strong*, each with the following importance weighting:

- *Long*—35%
- *Fast* is very important—50%
- *Strong* is the least important—15%

Remember, all criteria weightings must add up to 100%. In the actual matrix, these would be scaled down to a total of 1.0; therefore, long would be 0.35, fast would be 0.5, and strong would be 0.15. These numbers are referred to as the *weighting factors* or W.F.

For each concept, the designer tries to see how well it achieves each design criteria on a rating from 1 to 10, 10 being the best. These numbers are referred to as the *rating factors* (R.F.).

Example 9.2: Yard Leaf Blower

This example studies several concepts for a yard leaf remover, a devise to remove leaves that have fallen from trees. Here are the design criteria the designer decides on.

Use of standard parts	8%
Safety	12%
Simplicity and maintenance	10%
Durability	10%
Public acceptance	18%
Reliability	20%
Performance	15%
Cost to develop	3%
Cost to buyer	4%

These values are scaled down to add up to 1.0 and are referred to as the W.F. For example, the W.F. for the use of standard parts can be 0.08, and the W.F. for safety can be 0.12. The designer then develops four conceptual alternatives to satisfy the product's requirements following a similar process as described throughout this book. The four concepts are displayed in Figure 9.4. Each concept is then rated against each design criterion on a scale from 1 to 10 (the rating factor), and this is placed in the top-left triangle of each corresponding cell within the decision matrix (Figure 9.5). Therefore, the *rating factor* for the leaf bailer concept is 3 for use of standard parts and is 5 for safety and so on, as shown in Figure 9.4.

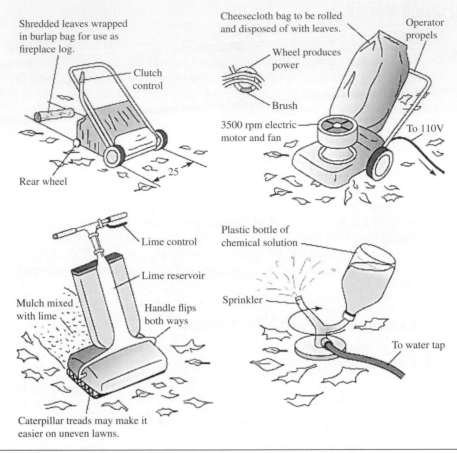

Figure 9.4: Conceptual Sketches of Yard Leaf Collector

Weighting factor / Design criteria	Use of standard parts	Safe	Simplicity and maintenance	Durability	Public acceptance	Reliability	Cost to develop	Cost to buyer	Performance	Sum
Alternatives	0.08	0.12	0.10	0.10	0.18	0.20	0.03	0.04	0.14	1.0
A) Leaf bailer	3 / 0.24	5 / 0.60	2 / 0.20	4 / 0.40	9 / 1.62	6 / 1.20	1 / 0.03	1 / 0.04	3 / 0.45	4.78
B) Vacuum collector	9 / 0.72	10 / 1.20	10 / 1.00	8 / 0.80	6 / 1.08	7 / 1.40	10 / 0.30	10 / 0.40	8 / 1.24	8.14
C) Shredder	5 / 0.40	6 / 0.72	7 / 0.70	7 / 0.70	8 / 1.44	6 / 1.20	3 / 0.09	4 / 0.16	5 / 0.75	6.16
D) Chemical decomposer	8 / 0.64	10 / 1.20	9 / 0.90	8 / 0.80	9 / 1.62	7 / 1.40	2 / 0.06	8 / 0.32	8 / 1.24	8.18

Weighting factor (W.F.) = Measure of relative importance (0 to 1.0∑ = 1.0)

Rating factor (R.F.) = Measured value of alternatives against design criteria (0 to 10)

Figure 9.5: Decision Matrix

The bottom-right triangle of each cell is calculated as W.F. × R.F. and presents the *weighted rating factor* of each conceptual alternative with respect to the individual design criteria. For example, the weighted rating factor for the leaf bailer for use of standard parts would be calculated as

R.W.F. = R.F. × W.F. = 3 × 0.08 = 0.24

For the same concept with respect to safety,

R.W.F. = R.F. × W.F. = 5 × 0.12 = 0.60

Then add the weighted rating factors for each conceptual alternative and compare the final sums for each alternative. The concept with the highest rating is the one that most closely satisfies the set design criteria. In the example given in Figure 9.4, both the chemical decomposer and the vacuum collector have similar top ratings. In this case, common sense is also required to see which design is more suitable. It is also possible to try to incorporate some of the strong points of one design into the other, if this is feasible, in order to make it a better design.

9.7 CONCEPT SELECTION EXAMPLE

Example 9.3: Wheel Barrow to Transport Steel Balls

This example is based on an assignment given to students at Brunel University in London. It is based on designing a wheelbarrow for transporting steel balls that are to be used in a cement mill as grinding media. The students were asked to act as a group of young engineers in the research and development department in a cement factory to produce a conceptual design for a new wheelbarrow. The company would make only six of them and hence encouraged using standard parts. The students were also asked to start their design process by analyzing a garden wheelbarrow. The students followed a four-step strategy.

1. Analysis of a garden wheelbarrow
2. Requirement analysis of a garden wheelbarrow and adaptation to the industrial wheelbarrow
3. Generation of conceptual designs
4. Concept evolution and selection

Function Analysis of the Garden Wheelbarrow

The purpose of this analysis is to make a function tree of the garden wheelbarrow, which could then be used as a guide to the industrial wheelbarrow.

Five functional subsystems were identified. They are (1) receive effort, (2) support the machine and load, (3) balance and act as a brake, (4) supply motion, and (5) contain and carry load. The function tree developed is given in Figure 9.6.

Requirement Analysis

A rigorous analysis of the requirements of both garden and industrial wheelbarrows was carried out, and the requirements and their importance were identified as shown in Table 9.2.

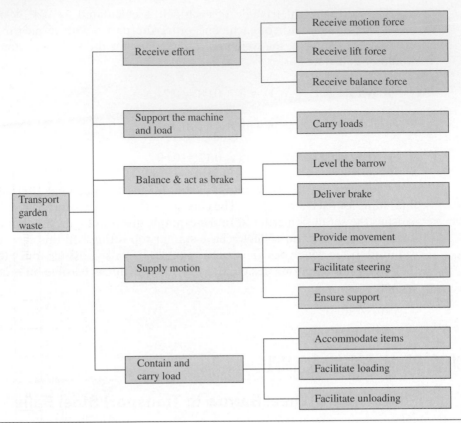

Figure 9.6: Function Tree of a Garden Wheelbarrow.

Table 9.2: Prioritized Requirements of Garden Wheelbarrow

Easy to control	6
Easy to lift	9
Easy to move	6
Easy to load	9
Easy to unload	9
Low manufacturing and operating cost	8
Does not hurt back	6

House of Quality and Weights

The requirements and their importance ratings were used to construct the house of quality chart. Figure 9.7 shows the house of quality. Since there were no identified benefits to be delivered, it was decided to use all the quality functions as the criteria. and the calculation of their weights is included in the house of quality chart.

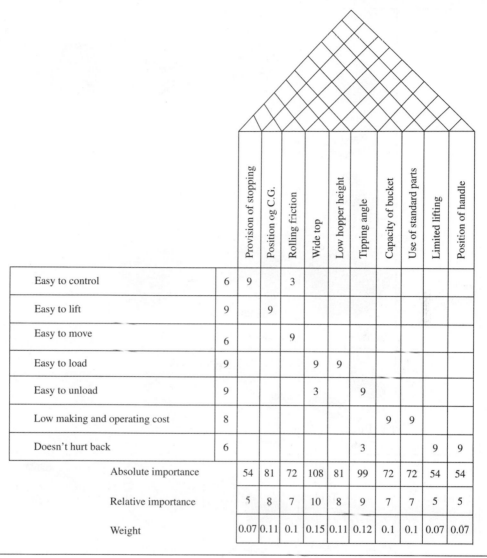

Figure 9.7: House of Quality for the Industrial Wheelbarrow

		Provision of stopping	Position og C.G.	Rolling friction	Wide top	Low hopper height	Tipping angle	Capacity of bucket	Use of standard parts	Limited lifting	Position of handle
Easy to control	6	9		3							
Easy to lift	9		9								
Easy to move	6			9							
Easy to load	9				9	9					
Easy to unload	9				3		9				
Low making and operating cost	8							9	9		
Doesn't hurt back	6						3			9	9
Absolute importance		54	81	72	108	81	99	72	72	54	54
Relative importance		5	8	7	10	8	9	7	7	5	5
Weight		0.07	0.11	0.1	0.15	0.11	0.12	0.1	0.1	0.07	0.07

Morphological Analysis

The morphological analysis was based on the first-level functions or the purpose functions as explained in this book. They are

1. Support
2. Moving
3. Ball holding
4. Control
5. Operator interface
6. Loading and unloading

Figure 9.8 shows the morphological chart for the industrial wheelbarrow.

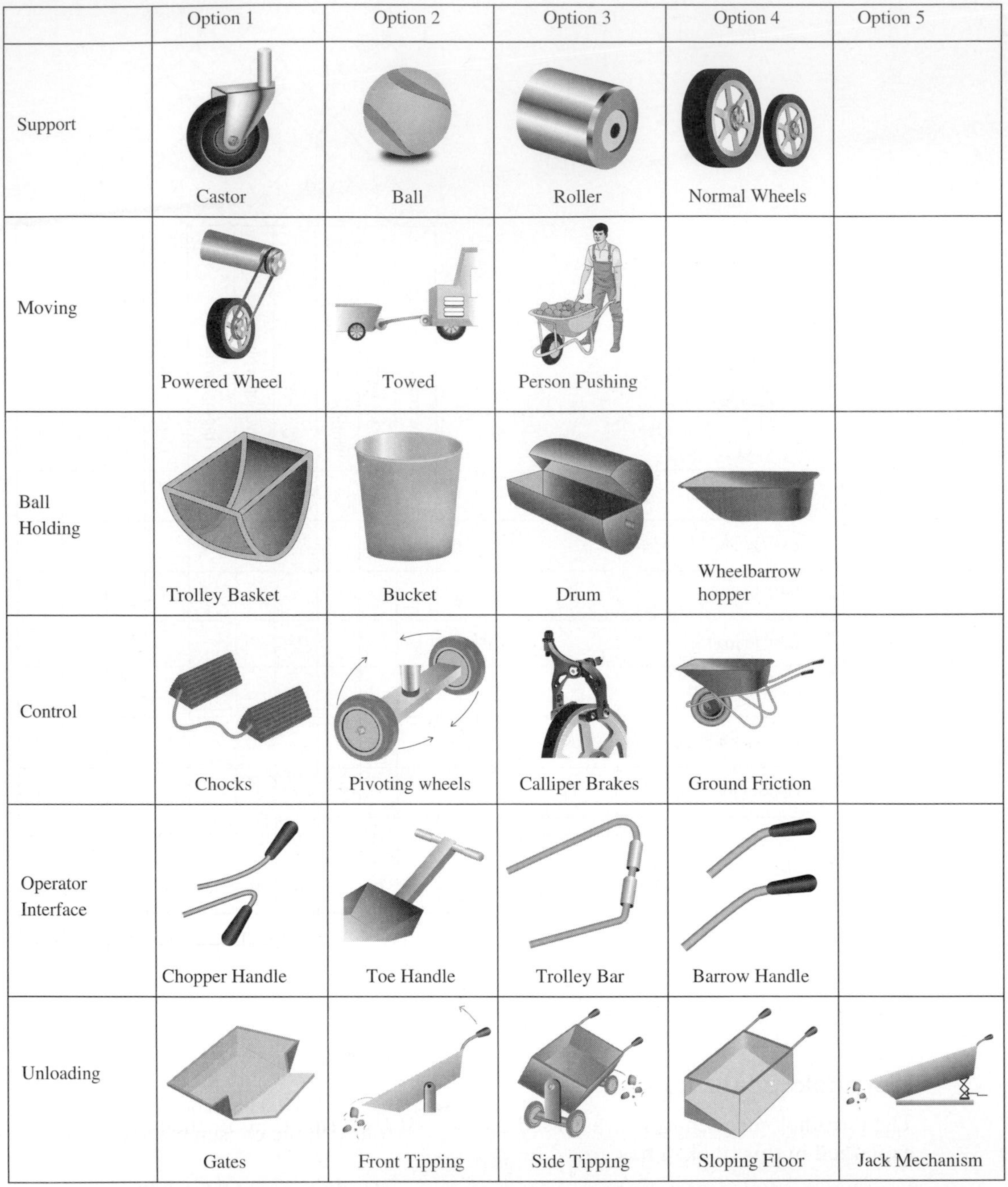

	Option 1	Option 2	Option 3	Option 4	Option 5
Support	Castor	Ball	Roller	Normal Wheels	
Moving	Powered Wheel	Towed	Person Pushing		
Ball Holding	Trolley Basket	Bucket	Drum	Wheelbarrow hopper	
Control	Chocks	Pivoting wheels	Calliper Brakes	Ground Friction	
Operator Interface	Chopper Handle	Toe Handle	Trolley Bar	Barrow Handle	
Unloading	Gates	Front Tipping	Side Tipping	Sloping Floor	Jack Mechanism

Figure 9.8: Morphological Chart for the Industrial Wheelbarrow

Based on the morphological chart, eight preliminary concepts were developed. Figure 9.9 shows them.

Pugh's Concept Selection/Evolution Matrix

In the development of the stage 1 concept evolution matrix, the standard garden wheelbarrow was chosen as the DATUM. Figure 9.9 shows the completed Pugh's matrix. From the matrix concepts 2 and 3 were chosen for further consideration.

Figure 9.9: Developed Conceptual Designs

Second-Level Concept Evolution

In the second attempt, concepts were improved using the lessons learned from the first Pugh's matrix. Three concepts were generated and compared with the chosen wheelbarrows from the first stage. The designs are shown in Figure 9.11.

Pugh's matrix was constructed for the second set with concept 3 from the previous set and the new three, developed using the garden wheelbarrow as the DATUM. Figure 9.10 shows the Pugh's matrix. From this, the chosen design was Concept 9.

Decision Matrix

As an alternative to the second-stage Pugh's matrix, the decision matrix was formed using the 10 quality functions and the weights identified in the house of quality chart as shown in Figure 9.5. Figure 9.12 shows the second stage Pugh's matrix using concept 1 as the Datum. The decision matrix is shown in Figure 9.13.

Again concept 9 was chosen as the final concept.

Criteria	(wheelbarrow)	(hand truck)	(basket cart)	(wagon)	(barrow)	(single wheel barrow)	(roller)	(wagon handle)
Provision of stopping	DATUM	S	–	–	–	S	–	–
Position of C.G.		+	+	–	S	+	–	–
Rolling effort		S	–	–	–	–	–	–
Wide top		–	S	S	s	s	–	S
Low hopper height		S	S	–	–	S	–	S
Tipping effort		–	+	–	S	S	–	–
Capacity of bucket		S	+	+	+	+	–	+
Use of standard parts		S	S	–	–	–	–	–
Limited lifting		+	+	+	+	S	+	+
Position of handle		+	S	–	S	S	–	S
Σ+		3	4	2	2	2	1	2
Σ–		5	2	7	4	2	9	5
ΣS		2	4	1	4	6	0	3

Figure 9.10: Pugh's Concept Evaluation Matrix

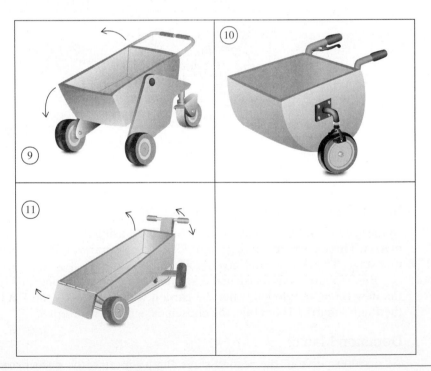

Figure 9.11: Improved Conceptual Designs

Criteria	1	3	9	10	11
Provision of stopping		−	+	−	+
Position of C.G.		+	+	−	S
Rolling effort		−	+	S	+
Wide top		S	S	S	+
Low hopper height		S	−	+	S
Tipping effort		+	+	+	−
Capacity of bucket		+	S	+	+
Use of standard parts	DATUM	S	+	S	S
Limited lifting		+	+	+	+
Position of handle		S	+	S	S
$\Sigma+$		4	7	4	5
$\Sigma-$		2	2	2	1
ΣS		4	1	4	4

Figure 9.12: Second-Stage Pugh's Matrix

	Provision of stopping	Position of C.G.	Rolling effort	Wide top	Low hopper height	Tipping effort	Capacity of bucket	Use of standardized parts	Limited lifting	Position of handle	
	0.07	0.11	0.1	0.15	0.11	0.12	0.1	0.1	0.07	0.07	
Concept 3	6 / 0.42	7 / 0.77	6 / 0.6	6 / 0.9	6 / 0.66	7 / 0.84	7 / 0.7	6 / 0.6	6 / 0.42	6 / 0.42	6.33
Concept 9	7 / 0.49	8 / 0.88	9 / 0.9	6 / 0.9	6 / 0.66	7 / 0.84	7 / 0.7	6 / 0.6	6 / 0.42	6 / 0.42	6.81
Concept 10	6 / 0.42	6 / 0.66	6 / 0.6	6 / 0.9	6 / 0.66	7 / 0.84	7 / 0.7	6 / 0.6	6 / 0.42	6 / 0.42	6.03
Concept 11	6 / 0.35	7 / 0.77	5 / 0.5	7 / 1.05	7 / 0.77	3 / 0.36	6 / 0.6	6 / 0.6	7 / 0.49	5 / 0.35	5.84

Figure 9.13: Decision Matrix for Industrial Wheelbarrow Concepts

9.7.1 Chosen Concept for the Industrial Wheelbarrow

Based on the decision matrix, Figure 9.14 shows the chosen concept.

Figure 9.14: Chosen Concept

CHAPTER SUMMARY

1. The design brief, the objective tree, and the specifications constitute the design criteria that are used to compare alternatives.
2. An alternative should meet the customer demands; otherwise, it will be dropped in the initial screening.
3. Evaluation of a design concept implies an assessment of its value made according to explicit goals.
4. Construct the Pugh's matrix or the decision matrix during the evaluation process.
5. Pugh's method evaluates alternatives against criteria and a datum. It compares the satisfaction of the criteria for each alternative as it compares to a datum.
6. The decision matrix rates the concepts against criteria the designer develops.

PROBLEMS

1. To enhance your engineering sketching ability, consider the paper stapler. Disassemble the stapler to its basic components. Sketch these components, and then sketch an assembled stapler.
2. Why would a designer place items (demos of function) in the morphological chart if he or she knows that they would not be feasible?
3. Define engineering sense. If someone asks you to demonstrate a speed of 10 cm/sec, how would you do so using your engineering sense?
4. Why would you need to evaluate alternatives if your gut feeling is pointing you toward one of these alternatives?

5. Discuss the following: "Using different alternatives for each function allows design engineers to substitute that particular element (in case of a problem) rather than changing the whole design."

6. What difference does it make if you use the absolute or relative scale in the weighting function? Which one would you recommend and why?

7. Why would you evaluate the different alternatives individually rather than as a team?

8. In all of examples presented in this chapter, the design team has the opportunity to establish the evaluation criteria, which thus may be biased by the design team. What would you use as mechanism to establish the evaluation criteria if you are the design manager at Ford Motor Company?

   a. How would you conduct the evaluation?

   b. How did the house of quality contribute to the evaluation mechanism?

9. In the event that the mechanism that satisfies a function or its presentation is found to be difficult to attach to another function(s), what would you do and why?

10. Name situations in which you have to perform the evaluation chart more than once.

11. Use the Pugh method to evaluate this course. Use your physics course as a datum.

References

The following books, articles, and websites were used in preparing this chapter

AMBROSE, S. A. and AMON, C. H. "Systematic Design of a First-Year Mechanical Engineering Course at Carnegie Mellon University." *Journal of Engineering Education*, pp. 173–181, 1997.

BURGHARDT, M. D. *Introduction to Engineering Design and Problem Solving*. New York: McGraw-Hill, 1999.

EEKELS, J. and ROOZNBURG, N. F. M. "A Methodological Comparison of Structures of Scientific Research and Engineering Design. Their Similarities and Differences." *Design Studies*, Vol. 12, No. 4, pp. 197–203, 1991.

PAHL, G. and BEITZ, W. *Engineering Design: A Systematic Approach*. New York: Springer-Verlag, 1996.

PUGH, S. *Total Design*. Reading, MA: Addison-Wesley, 1990.

SUH, N. P. *The Principles of Design*. New York: Oxford University Press, 1990.

ULMAN, D. G. *The Mechanical Design Process*. New York: McGraw-Hill, 1992.

ULRICH, K. T. and Eppinger, S. D. *Product Design and Development*. New York: McGraw-Hill, 1995.

VIDOSIC, J. P. *Elements of Engineering Design*. New York: The Ronald Press Co., 1969.

WALTON, J. *Engineering Design: From Art to Practice*. New York: West Publishing Company, 1991.

PART 5

EMBODIMENT DESIGN

CHAPTER 10

CONCEPT PROTOTYPES

"*Avoid hand-waving syndrome, prototype it.*"

The word *prototype* can have several meanings. The Merriam Webster dictionary defines prototype as an original or first model of something from which other forms are copied or developed. During the development of a product, there are several stages at which physical prototypes are built, particularly when new concepts are developed. They are often simplified versions of the concept, created to explain, test, measure, or prove some aspects of the product. If you are confronted with difficulties while working on a new venture, the common advice is to "get back to the basics." But what are the so-called basics? In the context of product design, the basics are the conditions all parties concerned can understand and agree on without difficulty. Consider how children learn math. At the beginning, the basics are the addition, subtraction, multiplication, and division of numbers. After children have mastered these concepts, they move on to higher-level classes where the four rules of algebra are introduced. The basics move up to this new level when a problem in handling a complex operation is involved. After that has been mastered, trigonometry is introduced. Again the basics move up.

A similar view can be taken of product development, and this requires the movement of basics during the whole journey of product development from an abstract set of needs to the definition of a physically realizable system. Concept prototyping and experimenting with the prototypes are the only means available for the designer to lift the basics up during this journey. This chapter explains prototypes and how to build and experiment with them.

10.1 OBJECTIVES

By the end of this chapter, you should be able to

1. Discuss the different types of presentations of a product.
2. Identify prototype dimensions.
3. Understand the term *design for "X."*

10.2 PROTOTYPES: WHAT ARE THEY?

Houde and Hill [1] define *prototype* as any representation of a design idea, regardless of medium. Four presentation techniques are available for designers.

1. *Mock-up*: This is generally constructed to scale from plastics, wood, cardboard, and other materials to give the designer a feel for the design. It is often used to check the clearance, assembly techniques, manufacturing considerations, and appearance. This the least expensive technique, is relatively easy to produce and can be used as a tool in selling the idea to clients or management. A solid model using the computer-assisted design (CAD) system can often replace the mock-up. Mock-ups are usually referred to as a *proof of concept prototype*.

2. *Model*: The model relates the physical behavior of the system through mathematical similitude. The modeling is usually referred to as a *proof-of-product prototype*. Different types of models are used to predict the behavior of a real system.
   a. A *true* model is an exact geometric reproduction of the real system, built to scale, and satisfying all restrictions imposed by the design parameters.
   b. An *adequate* model is constructed to test specific characteristics of the design and is not intended to yield information concerning the total design.
   c. A *distorted* model purposely violates one or more design conditions. This violation is often required when it is difficult or impossible to satisfy the specific conditions because of time, material, or physical characteristics and when it is felt that reliable information can be obtained through the distortion.

3. The *prototype* is the most expensive technique and the one that produces the greatest amount of useful information. The prototype is a constructed, full-scale, and working physical system. Prototypes can be comprehensive or focused. A comprehensive prototype corresponds to a full-scale, fully operational version of the whole product. An example of a comprehensive prototype is the *beta* prototype, which is given to customers in order to identify any remaining design flaws before committing to production. In contrast, focused prototypes correspond to one or a few of the product elements. Examples of focused prototypes include foam models used to explore different forms of a product.

4. *Virtual prototyping:* Computer-aided design (CAD) and computer-aided engineering (CAE) software suppliers have fulfilled the goals of virtual prototyping through the delivery of 3D feature–based modeling capabilities. Solid modeling has enabled the use of geometry to visualize product models quickly and to detect any gross interference problems. One of the unique features in feature-based modeling software is its ability to automate design changes. It provides integrated capabilities for creating detailed

solid and sheet-metal components, building assemblies, designing welding, and producing fully documented production drawings and photorealistic renderings.

Feature-based modeling is built on combining the commands needed to produce a common feature. For example, a combination of commands to create a hole of prescribed dimensions and relations to keep the parts of the hole maintained is stored in a feature that is given the name *hole*. The hole is an example of feature-based modeling.

Parametric modeling allows the presentation of the dimensions of the parts in terms of parameters. Virtual prototyping can have its biggest payoff in the manufacturing and analysis stages. Using solid models as the base for a finite-element analysis tool would reduce prototyping time and save money. In manufacturing, parametric feature–based models will reduce the time needed to produce a feasible prototype.

For the purpose of this book, *a prototype is a physical model of some kind*. Simulating a design through prototyping can reduce design risk without committing to the time and cost of full production. Questions about a design or specific aspects of a design can be answered concretely by building prototypes of design concepts [2].

10.2.1 Types of Prototypes

There are several different ways to categorize prototypes. Ullman [4] describes four types of prototypes based on their function and stage in product development:

1. *Proof-of-Concept Prototype*: A proof-of-concept prototype is used in the initial stages of the design process to better understand the approach to take in designing a product.
2. *Proof-of-Product Prototype*: A proof-of-product prototype clarifies a design's physical embodiment and production feasibility.
3. *Proof-of-Process Prototype*: A proof-of-process prototype shows that the production methods and materials can result in the desired product.
4. *Proof-of-Production Prototype*: A proof-of-production prototype demonstrates that the complete manufacturing process is effective.

Neumann [3] identifies several ways to categorize prototypes:

1. Prototype domain:
   a. Features—typically describes the number of features of the final product that are present in the prototype.
   b. Functionality—describes the amount of the functionality of the final product that is included in the prototype.
   c. Interaction—describes the degree of similarity in interaction with the final product that is present in the prototype.
   d. Design—describes the degree of similarity between the prototype and the actual product.
2. Low-fidelity and high-fidelity types: Low-fidelity prototypes are often built from available or waste resources to represent design alternatives and typically have limited functionality features and interaction. High-fidelity prototypes are typically built to present the final product with full functionality and interaction.
3. Horizontal and vertical types (based on how a target system is represented): The horizontal prototypes include all features with limited functionality whereas vertical prototypes include limited features with full functionality.

10.2.2 Low-Fidelity and High-Fidelity Prototypes

Low-fidelity prototypes are coarse approximations of the design alternative using common material (paper, cardboard, wood, foam, etc.) to provide a primary presentation of the functional design. It is a fast and cheap method that allows iteration to improve on or alter the design alternative. High-fidelity prototypes, on the other hand, produce a presentation of the design alternative that is almost identical to the production version and are expensive and time consuming. However, they provide detailed information for the designers and stakeholders on how the product will feel and function. As will be shown subsequently in the examples, when you are in doubt about functional connectivity of your design components, it is recommended that you build a low-fidelity prototype (in terms of material) and a high-fidelity prototype in terms of dimensions.

10.2.3 Applications: Why Prototypes?

The conceptual design phase produces outline schemes of the product, and the first uncertainty concerns the viability of the product or whether the planned concept will work. The subsystems that are placed together have to be proportional and should be packable within the allowable space. Interfaces should be placed correctly, and space should be allocated for them.

The design process, as you have experienced thus far, is cognition-intensive and requires massive information gathering to clarify the design requirements and to arrive at a number of possible solutions. A tangible presentation of the selected design concept helps designers and stakeholder cope with design complexity by offloading the cognition into a physical functional structure that can be utilized to reflect on and evaluate the design concept, to communicate with and gather feedback from stakeholders, and to discover unintended advantages or challenges within the functional concept. Prototyping allows a stream of cheap iterations to critique and explore design alternatives and perfect the concept design.

10.3 PROTOTYPE DIMENSIONS

Houde and Hill [1] identify three categories of questions, called *dimensions*, for which appropriately built prototypes can provide answers. They are

1. *Role*: The artifact provides a new functionality for users and plays a new role in their lives. The prototype is intended to answer questions about that role and its features.

2. *Look and feel*: The artifact presents the functionality in a novel way, but its role is known. The prototype is intended to answer questions about the way the new presentation looks and the users' feelings about it.

3. *Implementation*: The artifact presents a new technique for a known product. The prototype should answer questions about the way the new technique is implemented.

In essence the prototypes are built to answer the previously mentioned three types of questions. The three dimensions can be visualized by means of a triangular model. The apexes of the model represent the three dimensions. A prototype that is developed to examine the role of a product will be placed near the *role* apex. A prototype that is intended to explore the look and feel aspects will be placed near the *look-and-feel* apex. In the same fashion a prototype that is aimed at exploring the implementation of technologies will be placed near the *implementation* apex. Figure 10.1 shows the model.

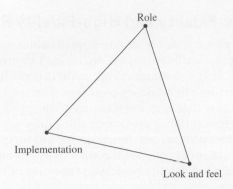

Figure 10.1: Model of the Three Dimensions of a Prototype

Thus the most important question in experimenting with prototypes lies in selecting the focus of the prototype.

10.4 PLANNING A PROTOTYPE STUDY

The prototyping stage is the right time to catch design flaws and improve alternatives. To achieve the right focus and outcome of prototyping, the answers to the following questions must be considered.

1. Who are the people who will use this prototype ? Are they users, sponsors, the design team, or others?
2. What features must be included to facilitate communication?
3. What are the dimensions of the prototype—role, look and feel, or implementation?
4. Is the prototype horizontal or vertical with respect to the dimensional features?
5. Is it a low-fidelity prototype or otherwise?
6. What resources are available in terms of money, time, and manpower?

10.5 EXAMPLES

Example 10.1: Shuttle of an Operating Table

This example is developed from a graduation project at United Arab Emirates University where a group of students designed a surgeon's operating table. The students had visited the nearby hospital a few times to understand the product and then to talk to personnel to identify their requirements. The main function of an operating table is to present the patient in different positions, which requires (1) linear up-and-down motion, (2) turning the head up or down, (3) tilting right and left, and (4) linear horizontal motion. They carried out a design interpretation on one of the tables and established the function tree for it. Their design had the table as an assemblage of four sections: (1) head section, (2) torso section, (3) hip section, and (4) leg section. The table is mounted on a carriage, which slides on the bars fitted to the upper plate.

The full details of the operating table and its parts are shown in Figures 10.2 and 10.3. The column carrying the lower plate section moves up and down in the casing. The lower plate rotates about the middle axle to provide the head-up and down motion. The upper plate is fitted on the lower plate, thus providing the tilt motion. The hip section is fitted to the carriage and is carried by the shafts in the upper plate, which permits the table to move horizontally. The assembly of the column—lower plate, upper plate, and the carriage—is called the shuttle.

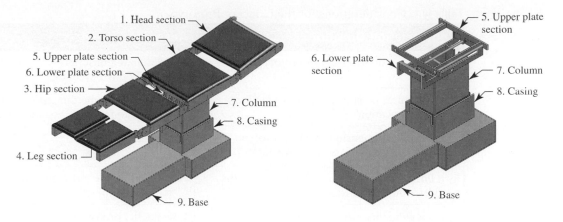

Figure 10.2: Labeled Solid Model of an Operating Table—Part 1

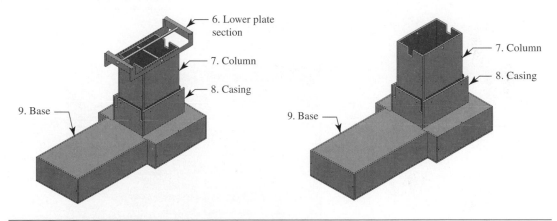

Figure 10.3: Labeled Solid Model of an Operating Table—Part 2

Visiting the hospital, talking to the hospital personnel, and carrying out the design interpretation of an existing operating table gave the students a clear understanding of the requirements, but the concept of providing the motions was not clear to them. At this stage, they planned to build the concept prototype of the shuttle. Table 10.1 shows the considerations.

The prototype was built in an afternoon with the help of a technician in a workshop using scrap timber and an aluminum shaft. The shuttle consists of column A; the axis carrying the rotating frame; the tilting frame fitted onto the rotating frame, capable of tilting about the tilting axis; and the shafts providing for linear motion along the x axis as shown in Figure 10.5. The rectangular column A, which is the main part of the shuttle, moves up and down inside a sleeve. The concept prototype separated each function with a separate structure to provide the function.

10.5.1 Insights Gained from the Prototype

The prototype was made to the dimensions planned for the parts except for the thicknesses of the various parts made from the timber. This gave the means for anchoring the various hydraulic cylinders needed for the product. But the important finding that was made by experimenting with the prototype was identifying the interferences. The grooves shown in Figure 10.4 were made to prevent interference.

Table 10.1: Considerations for the Concept Prototype

Consideration Question	Answers
The people who will use this prototype	The design team, supervisor, coordinators, and possibly other students
What features must be included to facilitate communication?	Possibly to the same dimension as the real product to facilitate visualization
What are the dimensions of the prototype – role, look and feel, or implementation?	Implementation. Height, tilt, rotation, and linear motion
Is it horizontal or vertical with respect to the dimensional features?	A horizontal type, providing insight into the motions of the table.
Is it a low-fidelity prototype or otherwise?	Low-fidelity prototype
What resources are available in terms of money, time, and manpower?	Maximum of $50 with three hours of technician time, a heap of scrap wood, and a few aluminum rods.

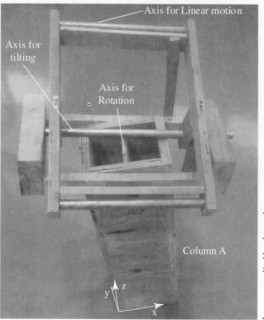

Figure 10.4: Concept Prototype

Example 10.2: Concept Prototype for Laying Out

This example also was developed from a graduation project undertaken by a group of students from the United Arab Emirates University. The aim of the project was to design and build an intermittently rotating portable display using a Geneva mechanism. The product can have up to four screens displaying different announcements, pictures, and the like. It can have a central address system to broadcast audio announcements. The students had no similar items to study and become familiar with the product. They made a conceptual sketch, shown in Figure 10.5. However, the parts have to be visualized in three dimensions, and a conceptual prototype was proposed.

Here again the difficulty was in implementing the dimension of the prototype. The idea was to develop a prototype made to full scale as far as possible so that the rough dimensions for the location of various parts could be determined by taking measurements. The prototype is shown in Figure 10.6.

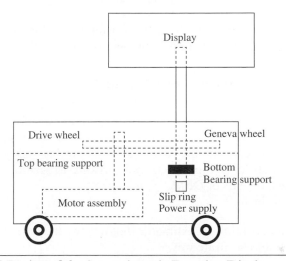

Figure 10.5: Conceptual Design of the Intermittently Rotating Display

Figure 10.6: Concept Prototype of the Rotating Display

10.6 DESIGN FOR "X"

Unlike smaller engineering design projects, in an industrial setting many teams contribute to create new products including nonengineering teams. Several attributes are put into the product to keep up with competitors and to produce reliable, marketable, safer, and less costly products in shorter times. To integrate these attributes through various activities and maintain cost-effective benefits, a process called *concurrent engineering* has been developed. The use of this approach implies systems theory thinking. Synergistic results become the main goal of its processes.

Designing for such different attributes (e.g., manufacturability, assembly, environment, safety, etc.) are usually referred to as a design for "X."

10.6.1 Design for Manufacturing

Design for manufacturing (DFM) is based on minimizing the costs of production, including minimizing the time to market while maintaining a high standard of quality for the product. DFM provides guidance in the selection of materials and processes and generates piece-part and tooling-cost estimates at any stage of product design. You can find several companies on the Internet that sell products and software to facilitate the capability of a company to perform the elements of DFM.

These include

1. An accurate cost estimator that reviews the cost of parts as they are being designed in a fast and accurate way.
2. A concurrent engineering implementation that provides quantitative cost information, which allows the design team to make decisions based on real-time information and to shorten the product development time.
3. Providing supplier negotiations with unbiased details of cost drivers.
4. Competitive benchmarking that compares the designs with competitors' products to determine marketability and target cost.

There are several methods that have been used in DFM for analysis, such as process-driven design, group technology, failure mode and effect analysis, value engineering, and the Taguchi method.

10.6.2 Design for Assembly

Design for assembly (DFA) is the study of the ease of assembling various parts and components into a final product. A lower number of parts and an ease of assembly contribute to reducing the product's overall cost. With DFA, every part must be checked. It must be determined whether it is a necessary part, whether it would be better integrated into other parts, or whether it should be replaced by a similarly functioning part that is simpler and costs less. Integrating both design for manufacturing and design for assembly helps contribute to the competitive success of any given product by matching that product's demands to its manufacturability and assembly capabilities.

10.6.3 Design for Environment

It is widely recognized that the resources consumed in the 20 century far exceed what was available during all the preceding 19 centuries. Recently, a trend of environmentally conscious products, along with the rapid implementation of worldwide environmental legislation, has put the responsibility for the end-of-life disposal of products onto the manufacturer. Manufacturers must now implement rules during the design that enforce *design for environment* (DFE). By designing products up-front for environmental and cost efficiency, manufacturers are gaining an edge on slow-to-react competitors who face the same issues unprepared. DFE is concerned with the disassembly of products at the end of their useful life and reveals the associated cost benefits and environmental impacts of a product's design. This quantitative information can then be used to make informed decisions at the earliest concept stages of the product's design.

The products are expected to have the following characteristics for being environmentally friendly:

1. They are less harmful to the environment over all stages of their life cycle.
2. They offer long life along with robust performance characteristics.
3. They provide for incremental improvement throughout their life.

To achieve these objectives design methods including (1) design for assembly, (2) design for disassembly, (3) design for durability, and (4) design for maintainability were developed. Five major groups of characteristics were identified as the evaluation criteria for the environmental friendliness of products. They are illustrated in Figure 10.7. The five groups are identified as pollution, energy, resources, quality, and recycling. Each of these groups contains factors that contribute to that group.

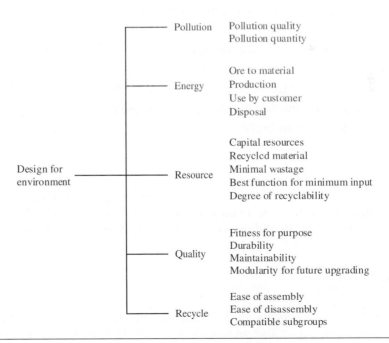

Figure 10.7: Framework for Assessing Environment Friendliness

10.7 SAFETY CONSIDERATIONS

One very important consideration in engineering design maintains that the resulting product should be safe for humans. In recent years, the liability decisions made by the courts have further increased the importance of safety standards. According to the U.S. National Safety Council, over 300,000 American workers were accidentally killed and over 10 million suffered a disabling injury in 2011, which translates into $753 billion in economic loss from accidental injuries. Safety is hardly a new issue; it has been important for thousands of years. There are many safety functions used during the design process, including

1. Developing accident prevention requirements for the basic design.
2. Participating in design reviews.
3. Performing hazard analyses during the product design cycle.

10.7.1 Safety Analysis Techniques

The two techniques used for design stage and reliability analysis—failure modes and effect analysis (FMEA) and fault trees—can also be applied for safety analyses.

Failure Modes and Effect Analysis

This method was originally developed for use in the design and development of flight control systems. The method can also be used to evaluate design at the initial stage from the point of view of safety. Basically, the technique calls for listing the potential failure modes of each part as well as the effects on the parts and on humans. The technique may be broken down into seven steps.

1. Define system boundaries and requirements.
2. List all items.
3. Identify each component and its associated failure modes.
4. Assign an occurrence probability or failure rate to each failure mode.
5. List the effects of each failure mode on relevant items and people.
6. Enter remarks for each possible failure mode.
7. Review and initiate appropriate corrective measures.

Fault Trees

This technique uses various symbols. It starts by identifying an undesirable event (called the top event) and then successively asks, "How did this event occur?" This process continues until the fault events do not require further development. If you know the occurrence data for the basic or primary fault event, you can calculate the occurrence measure for the top event.

10.8 HUMAN FACTORS

Return to Section 4.4. It is essential that the designers consider human factors when they evaluate a prototype and subsequently the product, designing the product to meet the humanly desired behaviors for successful products.

10.9 CHAPTER SUMMARY

1. There are four presentations available for a designer to express the functionality and feel of a product prior to the production phase; these are mock-up, model, prototype, and virtual prototype.

2. For the purpose of this book a *prototype* is a physical model of some kind.

3. There are different types of prototypes: proof of concept, proof of product, proof of process, and proof of production.

4. Low-fidelity prototypes are commonly used by design students due to economical considerations associated with time and financial resources. Using them presents the concept design with limited functionality, features, and interaction; still, they provide good insights into the design concept.

5. Appropriately built prototypes provide answers to role, look and feel, and implementation questions.

6. Before commencing a prototyping study, designers should plan for the study by answering a set of questions to better align expectations with the available resources and intended need.

7. Design for "X" is the design for different attributes such as manufacturability, assembly, or environment and should be considered by the designer at the prototyping and embodiment design stage.

10.10 PROBLEMS

1. What is the difference between sequential and parallel prototyping?

2. What might be possible reasons for prototyping prior to full-scale manufacturing?

3. What factors affect the number of iterations needed for a successful prototype that leads to product manufacturing?

4. Survey your school CAD software and list the capabilities of each program. Some of this software has a very small learning curve, which may allow you to produce a 3D model of your product in a very short time. If you have engineering graphics experience from a previous course, use it to determine whether the software has an analysis module that allows you to feed your model into the analysis module. Check the requirements of the analysis module.

References

[1] Houde S. and Hill C., "What Do Prototypes Prototype?" *Handbook of Human-Computer Interaction* (2nd ed.). Helender M., Landauer, and Prabhu P. (Eds.), Amsterdam: Elsevier Science B.V., 1997.

[2] Yang C. M. and Epstein G .J., "A study of prototypes, design activity, and design outcome," *Design Studies* 26 (2005) 64–669.

[3] Neumann P., *Prototyping*, Report 681, October 2004. Accessed online at http://pages.cpsc.ucalgary.ca/~saul/pmwiki/uploads/Main/topic-neumann.pdf in August 2015.

[4] Ulman, D. G. *The Mechanical Design Process*. New York: McGraw-Hill, 1992.

CHAPTER 11

EMBODIMENT DESIGN

"Machines say
for all our power and weight and size,
We are nothing more than children of your brain!"

~Rudyard Kipling

One dictionary meaning of "embodiment" is given as "the representation or expression of something in a tangible or visible form." It therefore follows that *embodiment design* is the physical expression or representation of the conceptual design. This chapter introduces methods for embodiment design. One of these embodiment design methods is concerned with three major sections: (1) product architecture, (2) configuration design, and (3) parametric design.

1. *Product architecture* divides the overall system into subsystems and modules. Decisions on how physical components can be arranged to carry out the functional duties of the design are made here.

2. *Configuration design* is the process of determining all required features in the parts and how they are arranged relative to each other. It also decides the materials and manufacturing processes.

3. *Parametric design* establishes the exact dimensions and tolerances. It examines the robustness of the product.

11.1 OBJECTIVES

By the end of this chapter, you should be able to

1. Recognize and distinguish among three methods for embodiment design.
2. Identify the main embodiment-determining requirement.
3. Stepwise add components and features to establish the embodiment for the given concept.
4. Determine the layout and dimensions in a stepwise fashion.
5. Explain the embodiment design of a given artifact.

11.2 EMBODIMENT DESIGN

At this stage in the design process, conceptual designs have been evaluated to produce a single concept for further development. Embodiment design is the realization of conceptual design from abstraction to physical product structure. Kesselring [1] was the first to refer to embodiment design as a set of four minimum principles: manufacturing cost, requirements, weight, and losses. Pahl and Beitz [2] defined embodiment design as the part of the design process that controls the design in accordance with engineering and economic criteria. Langeveld [3] defines embodiment design as designing with materials, manufacturing, and geometry to fulfill a new function or an update of an existing function.

The definitions of embodiment design refer to the translation of a product from the conceptual domain to the physical domain by maintaining technical and economic aspects during the transformation process. In this book, we refer to embodiment design stage as the bridge between conceptual design and detail design. In it the functions explored in the conceptual design are translated into physical structure by the appropriate choice of engineering materials and the manufacturing processes. Design for "X," as defined in the previous chapter, occurs more clearly in the embodiment-design stage. It helps designers to focus on constructive design solutions. Engineering analysis and economical considerations are addressed in the subsequent chapters.

There are a number of methods for embodiment design, but in this book we concentrate on the three methods discussed in the following sections.

11.3 DIETER'S EMBODIMENT DESIGN METHOD

Dieter [4] suggests that embodiment design is made up of three major components as listed here:

1. *Product architecture*: This is the arrangement of physical elements to carry out a function This can be done by
   a. Arrangement into modules
   b. Layout (with subassemblies and critical components)

2. *Configuration design*: This includes a preliminary selection of material and manufacturing process, along with
   a. Initial sizing (idealized analysis)
   b. Design process for parts

3. *Parametric design*: This includes
   a. Final sizing (numerical analysis or model test)
   b. Dimensions and tolerances
   c. Making assemblies work

To ensure precise understanding, the following terms are defined:

11.3.1 Product Architecture

The architecture of the product is established by defining the product's basic building blocks. The building blocks are responsible for the functions and their interfaces and are called *chunks*. Each chunk is made up of a collection of components meant to carry out a specified function. The building blocks implement only one or a few intended functions, and the interactions between two building blocks are well defined in modular architecture.

Consider the trailer carrying a cargo with the intended functions: (1) protecting from weather; (2) couple to a vehicle; (3) carry the cargo; (4) suspend the trailer; and (5) transfer the load to the ground. The blocks for these functions can be as shown in Figure 11.1.

With this mapping, the architecture for the trailer can be similar to what is shown in Figure 11.2. There are interfaces between the suspension and the wheel axle, between the suspension and the bed, between the bed and the hood, and between the hinge bar and the bed. Optional interfaces may be necessary to attach the cargo to the bed.

The product architecture design can be realized using the following three steps:

1. **Define the arrangement of functional elements.** In this step, consider the purpose functions or the functions used for morphological analysis.

2. **Map the physical components from the functional elements.** This can be the specifying of any component (i.e., part, standard part, or special part).

3. **Define the specifications of the interfaces among interacting physical components.** This is a major activity. There may be geometrical, thermal, and/or electrical interfaces between any two components. Creating a geometric layout often highlights whether the geometric interfaces between the components are feasible or not. The interface between the hood and the bed in the example calls for the consideration of (1) the dimensions of the contact surfaces between the two components, (2) the positions and the sizes of the bolt holes, and (3) the maximum force the interface is expected to sustain.

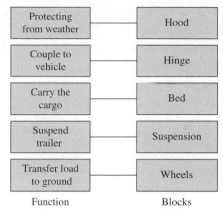

Function	Blocks
Protecting from weather	Hood
Couple to vehicle	Hinge
Carry the cargo	Bed
Suspend trailer	Suspension
Transfer load to ground	Wheels

Based on IIT. Lecture 3: Embodiment Design, http://nptel.ac.in/courses/112101005/downloads/Module_1_Lecture_3_final.pdf

Figure 11.1: Trailer Design

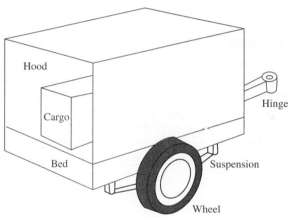

Based on IIT. Lecture 3: Embodiment Design, http://nptel.ac.in/courses/112101005/downloads/Module_1_Lecture_3_final.pdf

Figure 11.2: Architecture of the Trailer

11.3.2 Configuration Design

The shape and the general dimensions of components are realized in configuration design. It provides a preliminary selection of materials and manufacturing processes and modeling or sizing of parts, which strongly depends on the availability of the materials and production techniques that would be used to create the form from the material. Decisions about the design of a component cannot proceed further unless the decisions are made about the material from which the product (or the components) will be made, as well as the manufacturing process that will convert a raw material to a functional part of a component or product. The following steps are given as a guideline:

1. *Determine the spatial constraints* that are related to the product and the subassembly being designed. It is important to consider the constraints pertaining to the human interaction with the product, the product's life cycle, and the constraints related

to providing access for maintenance and repair in addition to the physical spatial constraints.

2. *Create and refine the interfaces and connections* between the components.
3. *Ensure the maintenance of functional independence* in the design of an assembly or of the components. This means that changing of a critical dimension should affect only a single function.
4. *Ensure that wherever possible, redundant parts are identified and eliminated or combined.*
5. *Use standard parts wherever possible.*

Results of Configuration Design

1. List of standard parts
2. List and definition of special-purpose parts
3. List of standard assemblies (could be used as subassemblies of components)
4. Definition of overall assembly and special-purpose subassemblies

11.3.3 Parametric Design

Parametric design is the process of assigning values to the attributes of the various design elements that are found in the configuration design. An attribute of a part whose value is under the control of the designer is called a design variable. This category includes the dimensions or tolerances, materials, shapes, manufacturing processes, and assembly and finishing processes that must be undertaken to create the part. Parametric design aims to assign values for the design variables that will produce the optimal design in terms of performance and manufacturability. By the end of the parametric design, the prototype of the product should be refined and finalized.

 ## 11.4 EMBODIMENT DESIGN METHOD BY PAHL AND BEITZ

Part is a designed object that has no assembly operation in its manufacture.

Standard part has a generic function and is manufactured routinely without regard to a particular product.

Special-purpose part is designed and manufactured to a specific purpose in a specific production line.

Component includes special-purpose parts, standard parts, and standard assemblies and modules.

Assembly is a collection of two or more parts.

Standard assembly or module is an assembly or subassembly that has a generic function and is manufactured routinely—for example, gearboxes or electric motors.

Pahl and Beitz [2] propose a 13-step method for embodiment design. The process from concept to preliminary layout takes 10 steps, and the process from preliminary layout to definitive layout takes 3 steps. The steps suggested here are progressively refining steps and are very general. They help the designer to plan his or her work and as such are very useful. The model is shown in Figure 11.3, and Table 11.1 shows an explanation of the steps stipulated in Figure 11.3.

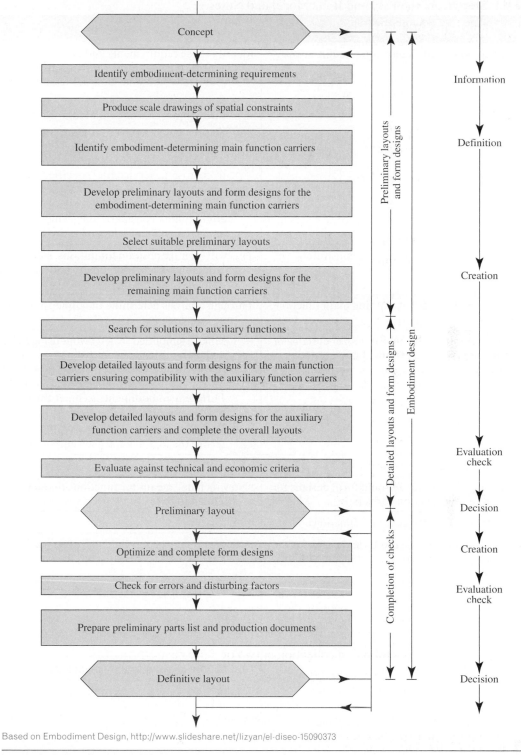

Based on Embodiment Design, http://www.slideshare.net/lizyan/el-diseo-15090373

Figure 11.3: The Pahl and Beitz Model

Table 11.1: Steps from the Pahl and Beitz Model and Notes

Steps from Pahl and Beitz	Notes
Identify embodiment-determining requirements.	Identify those requirements and engineering characteristics that drive physical form. Others can be accommodated through later additions or modifications. This includes 1. Size-determining requirements, such as output, throughput, size of connectors, etc. 2. Arrangement-determining requirements, such as direction of flow, motion, position, etc. 3. Material-determining requirements, such as resistance to corrosion, service life, specified materials, etc. 4. Requirements such as those based on safety, ergonomics, production, assembly, and recycling involve special embodiment considerations, which may affect the size, arrangement, and selection of materials.
Produce scale drawings of spatial constraints.	This will identify position limitations, sides of application, location of the center of gravity and similar mandatory conditions.
Identify embodiment-determining main function carriers.	Through analysis determine the dimensions of the main function carriers.
Develop preliminary layouts and form designs for the embodiment-determining main function carriers.	1. Make a simplified scaled layout using CAD with blocks and blobs instead of components. Then propose different layouts to deliver the same concept, similar to proposing concepts at the conceptual design stage. This may include 2. Different embodiments to meet the same concept. 3. Different ways to give physical form to the same concept. 4. Significantly different ways to find new arrangements and record the logic behind each embodiment alternative.
Select suitable preliminary layouts.	
Develop preliminary layouts and form designs for the remaining function carriers.	"Now include" means for functions that have not been attended to so far.
Search for solutions for auxiliary functions.	
Develop detailed layouts and form designs for the main function carriers, ensuring compatibility with auxiliary function carriers.	Alterations may be required at this stage.
Develop detailed layouts and form designs for the auxiliary function carriers, and complete the overall layouts.	Now everything is included. The overall layout should make sense.
Evaluate against economic and technical criteria.	
Create the preliminary layout.	Now choose a layout from the viable ones.

Based on Pahl and Beitz

 ## 11.5 METHOD BY VDI

Another method for embodiment design has been proposed by the German Engineering Institute (Verein Deutscher Ingenieure). It is a complete methodology for form design given by the form design process model shown in Figure 11.4. It consists of 11 steps.

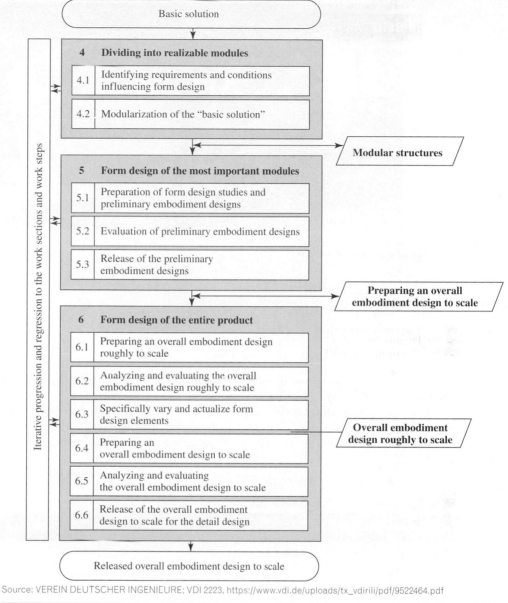

Source: VEREIN DEUTSCHER INGENIEURE; VDI 2223, https://www.vdi.de/uploads/tx_vdirili/pdf/9522464.pdf

Figure 11.4: Form Design Process Model by VDI2223

11.6 EMBODIMENT DESIGN EXAMPLES

Example 11.1: Plug Assembling Fixture

Figure 11.5 shows the components of an electrical plug. These are made in lots of thousands, and they have to be assembled. The assembling is achieved in two stages: (1) keeping the base plate facing up and (2) keeping the base plate facing down. While the base plate is facing up, the assembler carries out the following operations:

1. Insert the grounding pin
2. Inserting the return pin

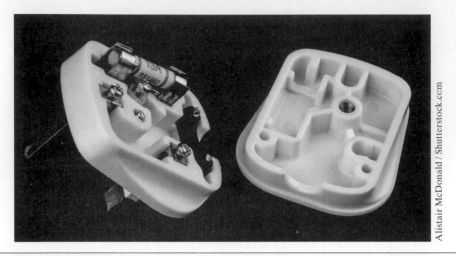

Alistair McDonald / Shutterstock.com

Figure 11.5: Plug Components

3. Inserting the life pin
4. Inserting the fuse support
5. Inserting the grips
6. Inserting the cap

The assembly is then turned upside down, and the holding screw is inserted and tightened to complete the assembly.

The operations are precise, and certain amounts of force have to be used to insert the components. A fixture is needed to hold the base to facilitate the assembling process. Communication with the stakeholders revealed the requirements, which are shown in Table 11.2.

Table 11.2: Prioritized Requirements

Requirements	Importance	Type
The plug base should be held flat.	9	1
The base should be held reasonably firm.	8	3
Free access for component insertion from top is needed.	8	4
There should be free space under the base.	7	1
Should be easy to insert and remove plug base	7	6
The base should be stable in the fixture.	6	3
It should have positive locking for easy use.	7	2
Pleasing appearance	5	7
It should be portable.	6	5
It should be cheap.	5	7
The fixture should be strong and stable.	8	5

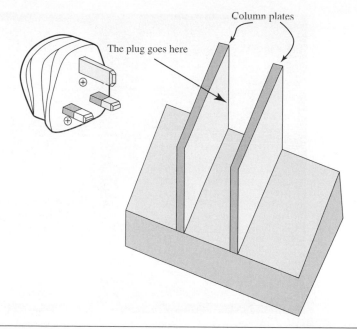

Figure 11.6: Conceptual Design of the Plug Fixture

The scheme that describes the conceptual design consists of two column plates supporting the base by its ears fitted on a base, as shown in Figure 11.6.

The first set of embodiments that determine requirements are that (1) the plug base should be held flat when it is placed in the fixture, and (2) there should be free space under the base. These two determine the height of the column plates; thus, the function carrier is the column plate (marked 1 in Table 11.2 in the type column). The plug will be carried by the ears in the fixture; hence, the groove is the next main function carrier (marked 2 in Table 11.2). The main constraint is that the width of the bearing surface (groove depth) should be greater than the maximum depth of the ear. The associated constraint is the distance between the column plates. The next embodiment that determine requirements are that (1) the base should be held reasonably firm, and (2) the base should be stable in the fixture. These two requirements cause the groove to have an easy-running fit as the function carrier. "Needs free access for component insertion from top" is the next embodiment-determining requirement, and it determines the location of the groove to be at the top. The next embodiment-determining requirement is "should be portable." This makes the requirement call for a base and an interface. The base and the locating grooves on the base become the function carriers for this requirement.

Now that we've determined the main function carriers—the column plates and the base—we can determine solutions for auxiliary functions. "Should be easy to insert the plug base" requires a wider converging entry at the inlet of the groove. The final requirements are "pleasing appearance" and "should be cheap." The first of these is achieved by having smooth rounding of the corner; the latter is achieved through the choice of cheap material (timber). The available thickness of timber was found to be 22 mm, and this was chosen to keep cost at minimum. The completed product is shown in Figure 11.7.

Example 11.2: Reading Assistant

Figure 11.8 shows the reading assistant whose conceptual design was developed to be analogous to a rubber plant. The books should be within reachable height fpr a reader who is seated next to the reading assistant. The distance between the presenters should be greater than

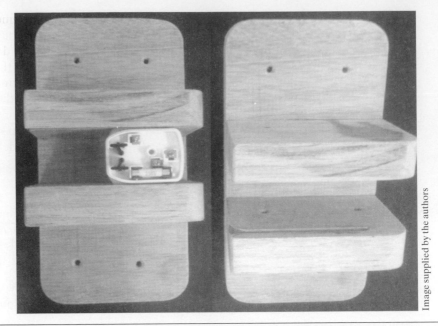

Figure 11.7: The Fixture Showing the Groove and Carrying the Plug Base

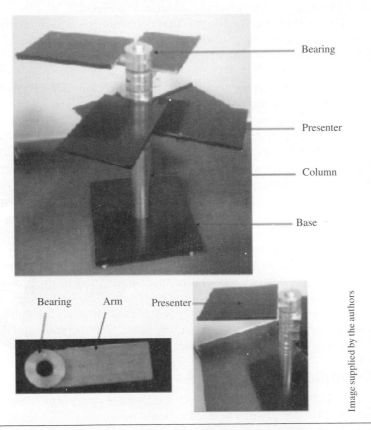

Figure 11.8: Reading Assistant

that of the tallest book. Thus the embodiment-determining requirement is the height of the tallest book. The presenter should support the broadest book, and this determines the size of the planar surface of the presenter. This and the arm determine the distance of the center of gravity from the axis of the column. The wheelbase should be greater than this distance to prevent the reading assistant from toppling. Thus the embodiment-determining requirements are (1) height of the tallest book and (2) width of the widest book. From these the bearing heights and wheelbase are determined. The height of the column is the next embodiment-determining requirement. This and the heights of the bearings determines the step sizes. Thus the embodiment design starts with the determination of the main embodiment-determining requirement and continues with stepwise definition of components and features.

11.7 CHAPTER SUMMARY

1. Embodiment design is the bridging stage between conceptual design and detail design.
2. Embodiment design is the physical presentation of a concept.
3. The Dieter method includes
   a. Product architecture wherein the overall system is divided into subsystems and arranged to carry out functional performances;
   b. Configuration design wherein all features are arranged relative to each other; and
   c. Parametric design that establishes the exact dimensions and tolerances of the product.

11.8 PROBLEMS

1. Consider an operating table shown in Figure 11.9 and the behaviors of the subassembly shown in Table 11.3. Develop a structural connectivity at the subassembly level.

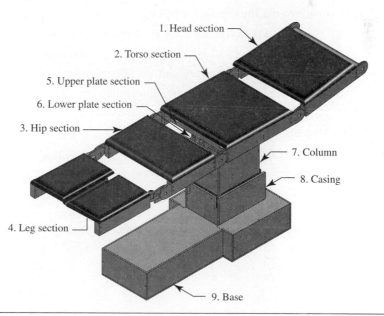

Figure 11.9: Operating Table

Table 11.3: Behaviors at the Subassembly Level

Head section	Takes the push from person and moves the table. Support and position head and transfers the load to the torso.
Torso section	Support the head and hip and transfers the load to upper plate. Permit rotations of head and hip section.
Hip section	Support the leg sections and transfer load from hip and transfer them to torso. Permit rotation of leg sections.
Leg section	Support and position legs and transfer load to hip section.
Upper plate	Support carriage and tilt cylinder. Transmit tilt and inclination torque.
Lower plate	Secure height, allow tilt and inclination rotation to upper plate, and support hydraulic valve and pipes.
Casing	Support lower plate and column.
Cylinder control	Control the valve of the cylinder.
Electrical system	Give the signals to the control system to start working.
Hydraulic system	Provide motions.
Inclination cylinder	Allow inclination motion.
Tilt cylinder	Allow tilting motion.
Height cylinder	Allow vertical motion.
Base	Support column, stabilize the entire structure, transmit load to the ground, and translate in X and Y DOF.

2. Consider an adhesive tape dispenser.

   a. Perform a redesign exercise as explained in Section 6.9.

   b. For the new selected design, develop product components and assembly.

   c. Generate a behavior table at the subassembly level.

   d. Compose a part list and its materials.

References

[1] Kesselring, F., *Technische Kompositionslehre*, Berlin: Springer Verlag, 1954.

[2] Pahl, G., and Beitz, W. *Engineering Design: A Systematic Approach*. New York: Springer-Verlag, 1996.

[3] Langeved, L. "Product Design with Embodiment Design as a New Perspective," in *Industrial Design—New Frontiers*, Prof. Denis Coelho (Ed.), InTech. ISBN: 978-953-307-622-5, 2011.

[4] Dieter, G. *Engineering Design*. New York: McGraw-Hill, 1983.

DETAIL DESIGN

CHAPTER 12

DETAIL DESIGN

"*The details are not the details. They make the design.***"**

~Charles Eames

Part 5 completed the presentation of the definition and layout of all constituent parts that form the product. The overall dimensions of parts, subassemblies, and assemblies have been established. Interfaces have been developed, and the entire product has been packaged within the acceptable envelope. In Part 6 the design process is completed with the definition of a product that will perform as expected and as stipulated by the specifications as the output. This requires two activities: (1) the production of the complete definition of the product and subassemblies as amended or modified by analyses, and (2) the proof of the product defined by all necessary analyses. These two processes are intermixed and are referred to as *detail design* or *detailed design*. For clarity, this book will call the complete design with all details included as the detail design and any analyses needed to support and prove the design as detailed design. This chapter describes the completion and definition of the subassemblies, assemblies, and the product.

12.1 OBJECTIVES

By the end of this chapter, you should be able to

1. Plan and extract the elements of an exploded view of an assembly and bill of quantities and place them in a labeled drawing.
2. Prepare individual production drawings of the constituent parts.
3. Combine these elements and prepare a set of production drawings.

12.2 PRODUCTION DRAWINGS

The set of production drawings for a product should contain all the details needed to manufacture it. This has two parts: (1) the geometric details or modeling, and (2) the material choice and manufacturing tolerance details called the geometric dimensioning and tolerancing or GDT for short. The details associated with part (2) form a specialized task, one that is very much dependent on the sector and processes involved. There are several standards associated with this process, but they are beyond the scope of this book. However, a summary of the standards is given in Section 12.4. Section 12.2.1 describes the geometric modeling part.

12.2.1 Geometric Modeling

The objective of the geometric modeling process is to define the product so that the user can understand the geometry of the product completely. To achieve this, two kinds of drawings are produced:

1. The labeled assembly drawing with the bill of quantities
2. The dimensioned drawings of the individual components

Labeled Assembly Drawing: This drawing shows an assembly or exploded assembly with all components numbered (labeled). The main purpose of this drawing is to enable the reader of the drawing to understand the specified component. Just on top of the legend section, the bill of quantities for these components provides the labeled number of the component along with the number of pieces required. This will enable the reader to estimate cost, time to manufacture, process planning, and so forth. As a standard practice, no dimensions are labeled in an assembly drawing. Typical examples of an assembly drawing is given in Figure 12.1 and Figure 12.2.

Dimensioned Drawings of Components: This drawing provides the number of projections and section drawings necessary to provide complete understanding. The legend section uses the same name as that given in the assembly drawing. It uses the sheet number for labeling the component in the assembly drawing. It is mandatory that a drawing sheet should contain only one component. This will enable the management of the workshop to give the drawing sheet to a technician to produce it. Typical examples of the component drawings are shown in Figures 12.3 to 12.4. These drawings can be produced using contemporary software packages (e.g., Solidworks and CATIA).

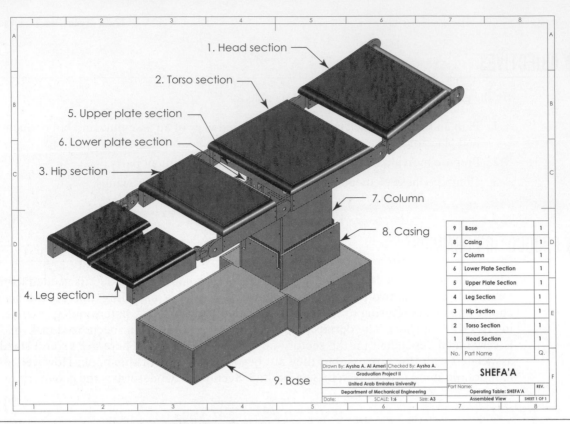

Figure 12.1: Assembly Drawing of an Operating Table

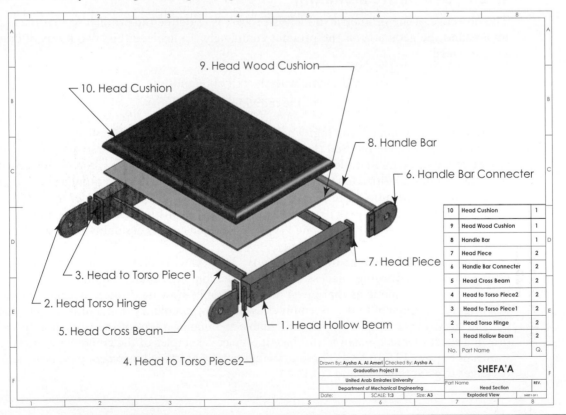

Figure 12.2: Assembly Drawing of the Headpiece

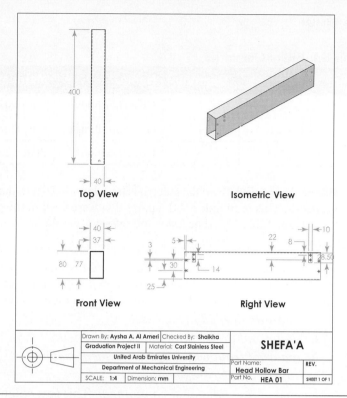

Figure 12.3: Drawing of the Head Hollow Bar

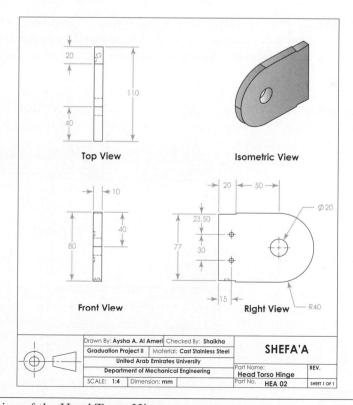

Figure 12.4: Drawing of the Head Torso Hinge

Table 12.1: Bill of Quantities

Item	Part	Quantity	Name	Material	Source
1	G-9042-1	1	Governor body	Cast aluminum	Lowe's
2	G-9138-3	1	Governor flange	Cast aluminum	Lowe's
9	X-1784	4	Governor bolt	Plated steel	Fred's Fine Foundry

Bill of Quantities: Once the materials have been selected, they should be represented in a *bill of quantities*. This is an index of the parts that were used in the product. A typical bill of quantities is shown in Table 12.1. The following information should be included in a bill of materials:

1. *The item number:* This is a key to the components on the assembly drawing.
2. *The part number:* This is a number used throughout the purchasing, manufacturing, and assembly system to identify the component. The item number is a specific index to the assembly drawing; the part number is an index to the company system.
3. *The quantity needed in the assembly.*
4. *The name and description of the component.*
5. *The material from which the component is made.*
6. *The source of the component.*
7. *The cost of the individual component:* This part will be kept for the design team.

12.3 ARRANGEMENT OF DRAWINGS

ISO 5457 of 1999 specifies the size and layout of preprinted sheets for technical drawings in any field of engineering, including those produced by computer. The original drawing should be made on the smallest sheet that allows for the necessary clarity and resolution.

12.3.1 Layouts

Figure 12.5 shows the generic dimensions of layouts of sheets from sizes A_0 to A_3. A standard sheet is size $a_1 \times b_1$. Inside this sheet a rectangle of size $a_2 \times b_2$ is drawn. The area contained

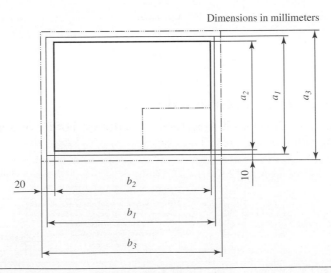

Dimensions in millimeters

Figure 12.5: Generic Layout of Sheets from Sizes $\mathbf{A}_0$ to $\mathbf{A}_3$

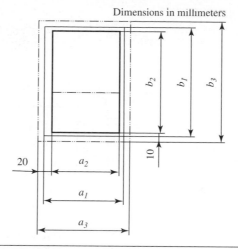

Figure 12.6: Generic Layout of A_4 Sheets

Table 12.2: Sizes of Trimmed and Untrimmed Sheets

Designation	Figure	Trimmed Sheet		Drawing Area		Untrimmed Sheet	
		a_1	b_1	a_2	b_2	a_3	b_3
A_0	Figure 1	841	1189	821	1159	880	1230
A_1	Figure 1	594	841	574	811	625	880
A_2	Figure 1	420	594	400	564	450	625
A_3	Figure 1	297	420	277	390	330	450
A_4	Figure 2	210	297	180	277	240	330

by this rectangle is the drawing area. It is worth noting that the rectangle is 20 mm away from the left end of the paper. This facilitates the filing of the drawing produced. Figure 12.6 shows the generic layout dimensions of a sheet of size A_4.

Table 12.2 gives the sizes of trimmed and untrimmed sheets as well as the drawing space in the ISO series.

Figure 12.7 shows the corner of a typical untrimmed sheet.

12.3.2 Grid Reference Border

In modern drawings it is customary to draw a border of 3-mm width inside the border that shows the drawing area. The border is marked with the centers in all four sides so that the center point of the drawing area can easily be located. Figure 12.8 shows the grid reference system, and Figure 12.9 shows a complete untrimmed A_3 sheet.

12.3.3 Folding Methods

There are two methods for folding the drawings. In this book, only the method suitable for filing will be discussed. Figure 12.10 illustrates the folding of an A1 sheet. It is folded in two stages, first lengthwise [Figure 12.10 (b)] and second crosswise [Figure 12.10(c)]. The title block will be at the top of the folded drawing.

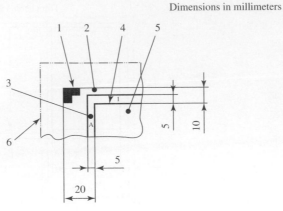

Dimensions in millimeters

Key
1 Trimming mark
2 Trimmed format
3 Grid reference border
4 Frame of drawing space
5 Drawing space
6 Untrimmed format

Figure 12.7: Corner of an Untrimmed Sheet

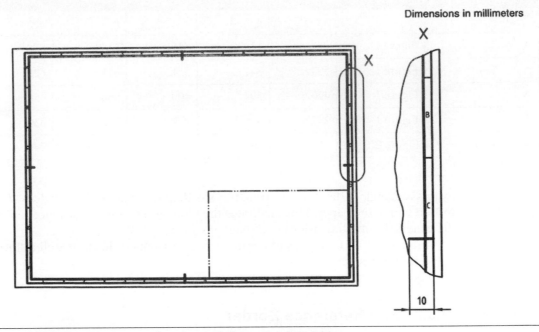

Dimensions in millimeters

Figure 12.8: Grid Reference System and Centering Markings

Table 12.3 shows the dimensions for folding various sizes of drawing sheets by this method.

12.3.4 Lines, Lettering and Scales

All visible outlines and edges should be 0.7 mm thick. Dimension lines, projection lines, hatching, leader lines, center lines, and all other lines should be 0.35 mm thick. It is recommended that spacing between parallel lines should not be less than 0.7 mm.

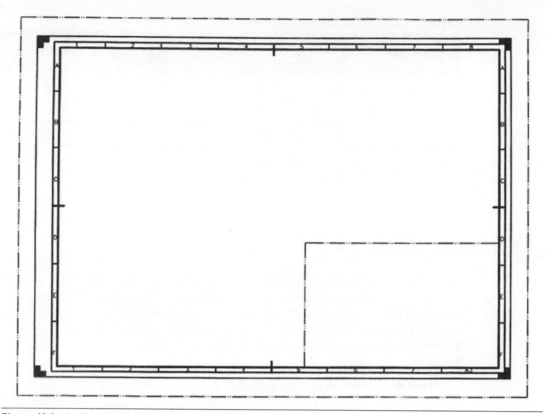

Figure 12.9: A Complete Untrimmed A3 Sheet

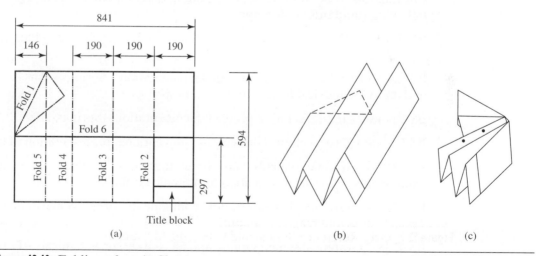

Figure 12.10: Folding of an A_1 Sheet

Lettering is very important in engineering drawings. The recommended sizes are as follows:

Main title	8 mm
Subtitles	5 mm
Notes, dimensions, figures, etc.	3 to 5 mm
Drawing number	10 mm

Table 12.3: Folding of Drawing Sheets

Paper Size	Horizontal Fold Dimensions from Left	Vertical Fold Dimensions from Bottom
A_0	$130+109+190\times5$	$297\times2+247$
A_1	$146+125+190\times3$	$297+297$
A_2	$116+96\times3 +190$	297
A_3	$125+105+190$	297

12.4 TECHNICAL PRODUCT SPECIFICATION STANDARDS

ISO's Technical Product Specification has undergone intensive work, and there are several standards that apply to various aspects of geometric modeling and drawing preparation. British Standard BS8888 implements the ISO system for technical product specification. Essentially it tells the user which standard to look for with any specific aspect of technical product specification. As an example consider item number 14, Representation of Components. It reads as follows:

14 Representation of components

14.1 General

Conventions used for the representation of components shall conform to the following standards, as appropriate.

BS EN ISO 2162-1 Technical product documentation—Springs—Part 1: Simplified representation

BS EN ISO 2162-2 Technical product documentation—Springs—Part 2: Presentation of data for cylindrical helical compression springs

BS EN ISO 2162-3 Technical product documentation—Springs—Part 3: Vocabulary

BS EN ISO 2203 Technical drawings—Conventional representation of gears

BS 2917-1 Graphic symbols and circuit diagrams for fluid power systems and components—Part 1: Specification for graphic symbols

BS 3238-1 Graphical symbols for components of servomechanisms—Part 1: Transductors and magnetic amplifiers

BS 3238-2 Graphical symbols for components of servomechanisms—Part 2: General servo-mechanisms

BS EN ISO 5845-1 Technical drawings—Simplified representation of the assembly of parts with fasteners—Part 1: General principles

BS EN ISO 6410-1 Technical drawings—Screw threads and threaded parts—Part 1: General conventions

BS EN ISO 6410-2 Technical drawings—Screw threads and threaded parts—Part 2: Screw thread inserts

BS EN ISO 6410-3 Technical drawings—Screw threads and threaded parts—Part 3: Simplified representation

BS EN ISO 6412-1 Technical drawings—Simplified representation of pipelines—General rules and orthogonal representation

BS EN ISO 6412-2 Technical drawings—Simplified representation of pipelines—Isometric projection

BS EN ISO 6412-3 Technical drawings—Simplified representation of pipelines—Terminal features of ventilation and drainage systems

BS EN ISO 8826-1 Technical drawings—Roller bearings—Part 1: General simplified representation

BS EN ISO 8826-2 Technical drawings—Roller bearings—Part 2: Detailed simplified representation

BS EN ISO 9222-1 Technical drawings—Seals for dynamic application—Part 1: General simplified representation

BS EN ISO 9222-2 Technical drawings—Seals for dynamic application—Part 2: Detailed simplified representation

12.5 CHAPTER SUMMARY

1. The objective of the geometric modeling is to define the product in engineering drawing.
2. Assembly drawing for all of the device or parts of the device includes the bill of quantities.
3. A detailed bill of quantities includes the item number, part number, quantity needed and name and description of the component, material of the component, source and cost of the component.
4. A dimensioned drawing for all individual components is typically used for the manufacturing of the component.
5. ISO 5457 specifies the size and layout for sheets to print technical drawings.
6. The two methods to fold drawings are, lengthwise and crosswise.
7. Visible outlines should be 0.7 mm thick. Dimension lines, projection lines, hatching, leader lines, centerlines, and all other lines should be 0.35 mm thick.
8. Technical product specifications are as per the ISO standards.

12.6 PROBLEMS

1. Explain the difference between a working drawing and a picture drawing of an object.
2. How are hidden or invisible edges of an object specified in a working drawing?
3. Explain why one view is not sufficient in a complete working drawing.
4. Make a complete drawing of a fork wrench for a ½″ nut, in the form of a complete working drawing.
5. Draw a 1-inch hexagonal bolt and nut—U.S. Standard Thread.

LAB 7: Geometric Dimensioning and Tolerancing

The misperception that CAD models contain all of the information needed to produce the product has led to the importance of dimensioning and tolerancing. For years, the value of geometric dimensioning and tolerancing (GD&T) has been highly regarded because of its ability to provide a clear and concise way to communicate tolerance requirements for parts and assemblies when needed.[1]

One of the most important features of design work is the selection of manufacturing tolerance so that a designed part fulfills its function and can be fabricated with a minimum of effort. It is important to realize the capabilities of the machines that will be used in the manufacturing as well as the time and effort required to maintain a tolerance.

It must be emphasized that for the GD&T to be effectively used, the functional requirements of the product must be well formulated. Vague requirements may result in a product that will not function as intended. Similarly, if tolerances are not well detailed, this may also cause a part not to function as intended or make it too expensive. Figure L7.1 shows two samples of outputs for a cup in which (a) the dimensioning and tolerancing standards are not used, and (b) detailed tolerance requirements are stated. It is clear that the output in part (a) results in a nonfunctional part, whereas in part (b) it results in a functional part. The translations of the tolerancing symbols in part (b) are as follows.

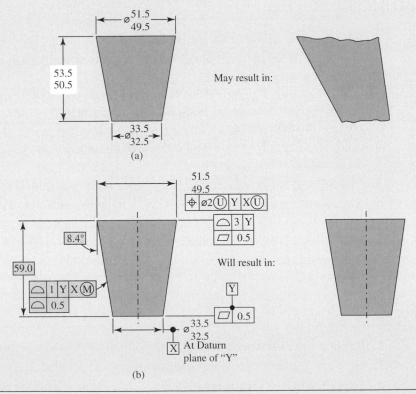

Figure L7.1: Comparison between Absence of Proper Tolerancing and Proper Tolerancing.

---

[1] Murphy, M. presentation at the NSF workshop at Central Michigan University, 1999.

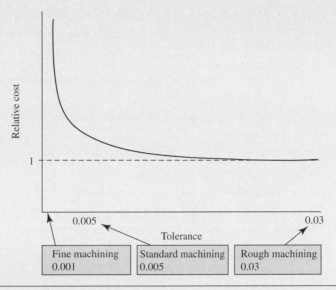

Figure L7.2: Tolerance versus Cost

1. The top must be in line with the bottom within 2 mm.
2. The cup must not rock; the bottom must be flat within 0.5 mm.
3. The cup must be transparent.
4. The cup must have a smooth surface within –0.4 μm.
5. The volume is 0.5 to 0.6 L.
6. The side angle is approximately 20 degrees plus 1 mm uniform zone.

Designers have a tendency to incorporate specifications that are more rigid than necessary when developing a product. Figure L7.2 shows the relation between the relative cost and the tolerance. As the tolerance requirement becomes smaller in size, the cost becomes higher. Because of the high cost of tight tolerance, designers should look closely at their designs and adjust functions or components such that the tolerance will not add cost to the product. For example, if a part such as a bearing calls for a tight tolerance (0.005) when installed inside a casing and may function with (0.03) tolerance when placed outside the casing, designers should change the design to fit the bearing outside the casing.

Procedure

This lab introduces you to geometric dimensioning and tolerancing.[2] The GD&T Trainer demo can be obtained from Effective Training, Inc, and is based on the ASME Y14.5M code.

Follow these instructions:

1. Launch the GT demo. There will be a few slides to explain the system requirements and what the program is all about.
2. Click on the course index.
3. Click on Concepts. Remember, this is just a demo version. About three modules will work in this demo.

---

[2] The information is based on the ASME Y14.5M-1994 standard. The basis of this lab is obtained from The Effective Training Inc. demo.

4. Click on lesson number 3: Modifiers & Symbols. This module will introduce you to the geometric characteristics and symbols. The text that is marked red is clickable and will give you more information.

5. Proceed in this lesson. On page 10 there is a problem that you need to solve. Type that characteristic table as a Word document. You can hand draw the symbols if you can't use Word to reproduce them.

6. Go through the lesson and answer all the questions. Report all your answers in a Word document.

7. Once done with concepts, go back to the course index and click on Forms.

8. After that, choose flatness.

9. Go through this tutorial and report your answers.

10. Once you have finished with flatness, go back and take the quiz located on the page for flatness. Report the quiz questions and answers.

Exit the demo and start the Qr_demo. From the subject index, choose position RFS/MMC and report your findings.

CHAPTER 13

DETAILED DESIGN:
ENGINEERING ANALYSIS

"Common sense is calculation applied to life."

~Henri-Frédéric Amiel

Chapter 12 showed how the geometry of the product is defined. Contemporary computer-aided engineering (CAD) practice is used to create the complete geometric model, (often called the master model) and then to subject this model to various analyses and optimizations as the design team deems appropriate for the product. These analyses and optimizations constitute the detailed design. The detailed-design stage is the final step in the engineering design process before manufacturing and production can be commenced. Most engineering-degree courses are placed within the detailed design–stage framework. During this stage (commonly referred to as analysis and simulation) the designer selects the appropriate materials for each part and calculates accurately the dimensions and tolerances of the product. This process may include calculating variables such as static or dynamic loads, stresses, forces, temperature, pressures, fluid dynamics, electric current, resistance, chemical reactions, and so forth where necessary. At this stage, the designer will also apply a suitable design factor to the design to ensure that the minimum requirements for non-failure of the product are well within the design limits. Obviously, all of these variables are subject areas in their own right and are beyond the scope of this book; however, we will present an introduction to the first phases of detailed design.

13.1 OBJECTIVES

By the end of this chapter, you should be able to

1. Understand the detailed design stage.
2. Identify and select engineering materials that suit a product.
3. Use techniques introduced in this chapter to evaluate and analyze design cost.

13.2 PRELIMINARY ANALYSIS

The first and the most common analyses are concerned with the effects of forces on the product to ensure that the product is safe and will not fail. An object can be subjected to three kinds of external forces:

1. *Surface forces*: forces that are continuously distributed over the surface, such as hydrostatic pressure.
2. *Body forces*: forces that are continuously distributed over the volume, such as gravitational force.
3. *Point forces*: forces that concentrate at one point, such as the tip load in a cantilever beam.

These forces are schematically shown in Figure 13.1.

The drawing of a body, which shows all forces and moments acting on a body, is called the free-body diagram. The forces create strains in the body, and they in turn create stresses in the body. When these stresses exceed material-specific values, they become plastically deformed, and the product loses its structural integrity. Thus, there are two important parameters to watch for:

1. The loading creates stress in the body.
2. The material properties decide the level of the stress the material can safely withstand.

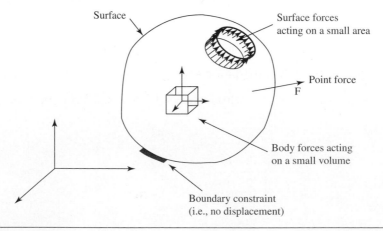

Figure 13.1: Types of Forces

13.2.1 Stress on the Body

A long piece of material can be used in different applications. When it supports transverse loading, it is called a *beam*. When it carries an axial load, it is called a *column*. When it transmits a torque, it is called a *shaft*. The theories of load-deformation behavior are also different under the different structural applications. The loading can be constant or time varying. The time-varying load will induce a different behavior from the material. Furthermore, the uniformity of the load, the suddenness of the variations, and the effect of the environment all affect the behavior of the components under stress.

If the component is simple, the stress fields will be uniform, and simple theoretical calculations called *hand-calculations* can predict the stress field. But complex structures and loading will require computer-based finite-element analysis.

13.3 INTRODUCTION TO FINITE-ELEMENT ANALYSIS

The field variables encountered in solid mechanics are stresses and displacements. Field variables in other areas of study include velocities in fluid mechanics, electric and magnetic potentials in electrical engineering, and temperatures in heat-flow problems.

In a continuum (a thing that has a continuous structure) these unknowns are infinite values at the infinite number of points. The finite-element method reduces this number of unknowns to a finite number by describing the variables at chosen points called the *nodes*. A portion of the continuum is defined by the connection of these nodes, and the region is called an *element*. The field variables at points within the element are found by interpolation, using the values at the nodes and interpolating a function called the **shape function**.

As an example, consider the tension flat shown in Figure 13.2. Due to the application of the force *P*, the flat undergoes strain, and this in turn induces stresses. They are not the same all through the flat because of the variation in the geometry. In order to study the *field-variables stresses* and the *strains* or *displacements*, the finite-element method can be employed. Because of its geometrical symmetry, investigation of the quarter ACDB in Figure 13.2 is sufficient to predict the field variables all through the tension flat.

Consider the division of the portion ACDB as shown in Figure 13.3. This process of division is called discretization. The continuum has an infinite number of points, and the field

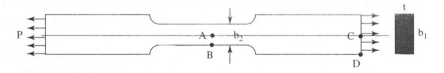

Figure 13.2: A Typical Tension Flat

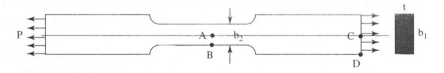

Figure 13.3: Discretization of the Component into Nodes and Elements

Figure 13.4: A Typical Element

variables take distinct values at these points. The 48 points in this division represent these infinite points. These 48 points (called the nodes) have displacements and stresses that must be calculated by some means, and stresses in all other points are calculated by interpolation using those values.

Consider the element 5 shown in Figure 13.4. The four nodes 6, 7, 10, and 11 mark its boundaries. A shape function is used to interpolate the field variables inside the element as a function of the values of the field variables in these four nodes 6, 7, 10, and 11. This shape function reflects the geometric and material characteristics of the element.

Thus, the finite-element method can be described as a method whereby (1) the field variables at the infinite number of points in a continuum are represented by a chosen finite number of points called nodes; (2) their values are then calculated using some material-science principles and other governing relationships; and (3) finally, the values of the field variables at all intermediate points are calculated by interpolation using the shape functions.

13.3.1 Finite-Element Analysis Methods

Several different finite-element analysis methods can be performed on the structure, and the stress variations, displacements, and strain-related properties can be obtained.

1. Real-world engineering problems can be subjected to several types of engineering analyses, including

2. *Linear static analysis*: This is the most common type of analysis. It produces the stress and deflection results of a part or structure. The inputs into the analysis can be forces, pressures, enforced displacements, or temperature differences. This type of analysis is valid for materials that are in their linear range. Also required for this analysis to be valid are boundary conditions that do not change and loads that are applied gradually.

3. *Nonlinear static analysis*: Nonlinear static analysis can be categorized into one of the three common types: (1) material nonlinearity (nonlinear elastic, elastic-plastic, and perfectly plastic), (2) geometric nonlinearity (large displacement and rotation), and (3) boundary nonlinearity (contact).

4. *Buckling analysis*: Buckling analysis calculates the critical load and the mode shape of a structure with compressive loads. Long slender parts and thin-walled parts that have compressive loads applied to them are typical examples of problems that require a buckling analysis.

5. *Modal analysis*: Modal analysis determines the natural frequencies and the mode shapes of a structure or part for a given constraint set. A modal analysis is important for parts that experiences vibrations from a motor, the road, the wind, or from being dropped or struck with a hammer. Knowing the natural frequencies is an important first step in determining whether a dynamic analysis is needed.

6. *Dynamic analysis*: A dynamic or vibration analysis is performed when the input load is time dependent. There are three types of dynamic analyses: (1) frequency, (2) transient, and (3) random response.

7. *Heat-transfer analysis*: A heat-transfer analysis is performed to determine the magnitude and direction of heat flow through a body. In addition, the temperature and temperature gradients are important causes of stress in a body. These analyses are governed by different sets of differential equations.

13.3.2 General Analysis Procedure

The finite-element method was first developed in 1943 to solve for vibrational systems. Later it was utilized to solve for complex engineering problems. In the finite-element method, the structure is discretized into elements, whereby the continuous structure is changed to a mesh of elements. The elements are chosen based on the dimensionality of the engineering problem sought (1-D, 2-D, or 3-D). The mesh contains the material and engineering properties that define how the structure will react to an external stimulus. The points at which the different elements are joined together are called nodes. The nodes are the locations at which the unknown variables are computed based on a mathematical model that describes the physical behavior of the structure. The finite-element method can be thought of as a technique that approximates solutions for the physical continuous structure at discrete locations of the structure. The discretization process allows for the approximation of the solution and often converts the differential equation into a set of algebraic equations that can be solved at the discrete points. The solution for the remaining portion of the continuous structure can be obtained using interpolation schemes. The details theory of the finite-element method is not the subject of this book, but ample resources are widely available.

Due to the complexity of the mathematical models and the engineering structure geometries, the problems are usually analyzed using computer simulation. There is a number of available software solutions that employ the finite-element method to analyze the physical structures using available mathematical models. Most CAD software is now equipped with analytical routines that help designers to evaluate structural and other engineering physical quantities. The example in the following sections provide a sample of analysis conducted using the finite-element method.

Example 13.1: Analysis of a Mechanical Vegetable-Harvesting Machine

In this example, a design team at Florida State University was asked to design and perform structural analysis on a mechanical vegetable collection and packaging system composed mainly of wood. The vegetable harvesting and packaging system must be capable of harvesting the vegetables without damaging a substantial amount of the crop. In addition, the design must not rely on man-made power for its operation, since it will be used in the field. During the design process, the design team developed the function analysis shown in Figure 13.5. A morphological chart for the design is shown in Figure 13.6. Three concepts were generated, which are shown in Figure 13.7.

In the first concept, a horse pulls the yoke, moving the vehicle forward. The front wheels turn a belt, which in turn provides torque on the blade shaft. Blades and sifters sort debris from the vegetables, which are then slung into the carriage compartment where an operator packages the product.

The principle on which the second concept mechanism operates is that the vegetables are harvested very close to the soil surface. By employing a plow, the vegetables will be ripped

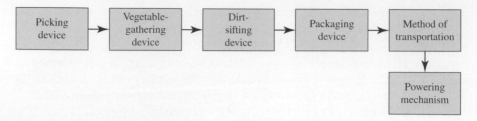

Figure 13.5: Function Analysis for Vegetable-Harvesting Machine

	Option 1	Option 2	Option 3	Option 4
Vegetable-picking device		Triangular plow	Tubular grabber	Mechanical picker
Vegetable-placing device	Conveyor belt	Rake	Rotating mover	Force from vegetable accumulation
Dirt-sifting device	Square mesh	Water from well	Slits in plow or carrier	
Packaging device				
Method of transportation		Track system	Sled	
Power source	Hand pushed	Horse drawn	Wind blown	Pedal driven

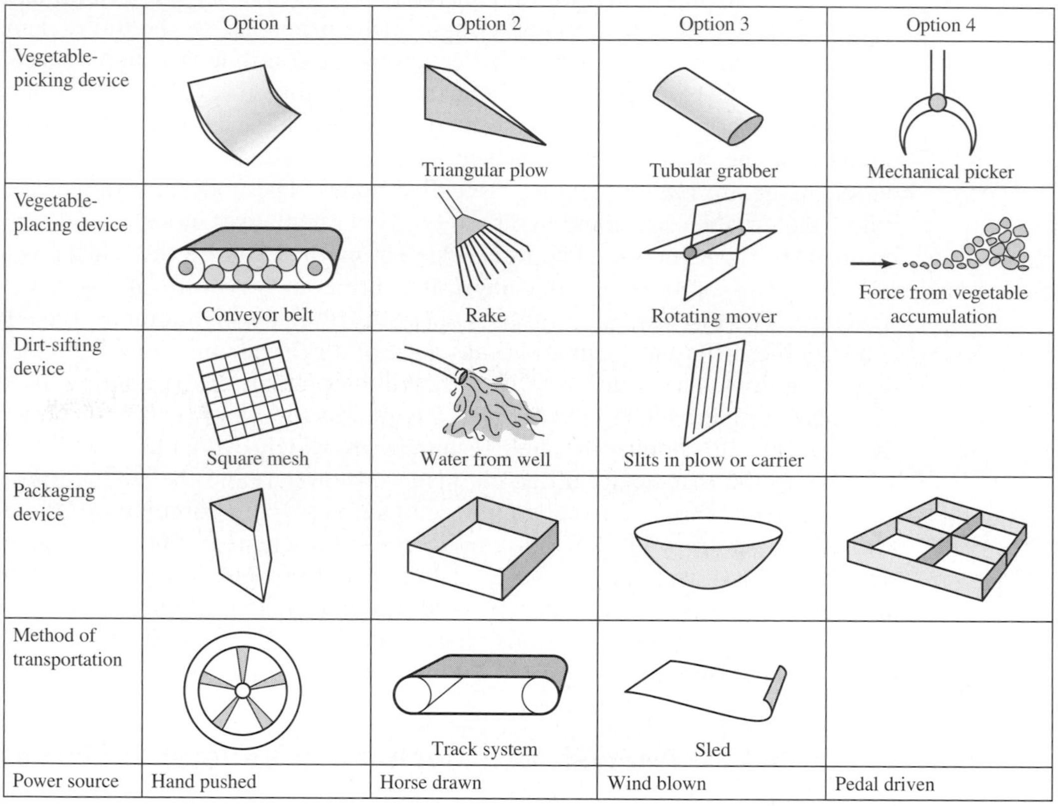

Figure 13.6: Morphological Chart for Vegetable-Harvesting Machine

free from the roots and will travel up the plow to the conveyor system. The only piece of metal on the apparatus is found at the edge of the plow. This edge is reinforced with steel since it will be in direct contact with the soil. The conveyor system is composed of two axles and a conveyor belt; the belt is composed of hemp (the strongest natural plant fiber). The conveyor system sorts out the vegetables from the dirt and small stones, serving more or less as a sifter. The vegetables will stay on the conveyor system, and the soil will fall through the sorter. The rear axle drives the conveyor system via a three-to-one gear ratio from the rear axle to the driving axle of the conveyor system. This allows the conveyor to travel at three times the speed of the tangential velocity of the rear axle. Once at the top of the conveyor system, the vegetables are pushed onto a rear-loading table, complete with guides. The rear axle is attached rigidly

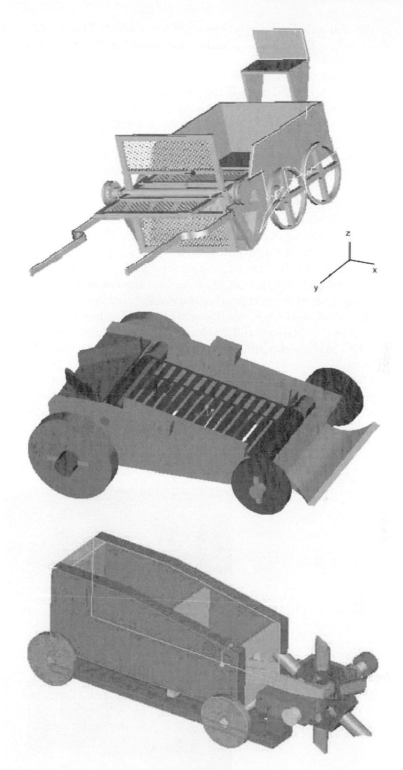

Figure 13.7: Concepts for Vegetable-Harvesting Machine

to the rear axle by a single wooden pin per wheel. The wheel has a carved-out section where this pin lies. There is also a wedge that holds the wheel firmly against the pin to ensure a rigid connection.

The conveyor is driven by means of a friction belt located between the rear axle gears. The belt is arranged to allow the rotation induced upon the belt by the rear axle to be reversed so that the conveyor can travel in the correct direction—toward the back of the machine. The chassis is designed using a single piece of wood. Openings are included in the bottom of the chassis to allow soil or rocks to fall back to the earth. The main portion of the plow is made from the chassis, but it is replaceable if severely damaged. The attachments where the horses are fastened are also made from the chassis. All of the wooden pieces are composed of red oak.

In the third concept, a horse pulls the device in the *y* direction. The front wheels turn, making the collecting wheel rotate. The tubes cut the vegetables, dragging them against the cart. Once the vegetables are against the cart, they are guided inside the pipes to the upper position. There is a hole at the bottom of each pipe where the vegetables exit. At this point, there is a slide that carries the vegetables to the back of the cart.

All three concepts were evaluated for their structural integrity. As an example, the structural analysis of the rear shaft of the second concept is presented here. This analysis was done using Pro/MECHANICA software. Figure 13.8 shows the loads on the rear shaft, and Figure 13.9 shows the stresses and displacement due to these loads.

The load results illustrated in Figure 13.9 show that there would be no failure due to loading.

This type of analysis should be used when different designs are evaluated. Designers should produce evaluation tables based on a part's performance under analysis. Designers can then change a part if the part fails to fulfill its function.

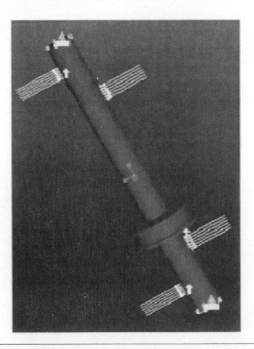

Figure 13.8: Loads on the Rear Shaft

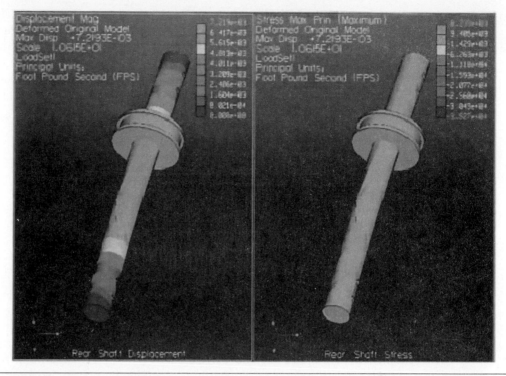

Figure 13.9: The Stress and Displacement on the Rear Shaft

13.4 MATERIAL SELECTION

The design, even of the simplest element, requires the selection of a suitable material and a decision regarding the methods of manufacturing to be used in producing the element. These two factors are closely related, and the choice will affect the product's shape, appearance, cost, and so on. It may also determine the difference between a commercial success and a commercial failure. As the design becomes more complex and involves more elements, the selection of suitable materials and methods of production becomes more difficult. The design engineer must be sufficiently familiar with the characteristics and properties of the materials and the way in which they can be shaped to ensure that his or her decisions are well made.

13.4.1 Material Classifications and Properties

The various types of materials that may be used in a product design are classified as follows:

1. *Metals:* This important materials classification can be further divided into ferrous and nonferrous alloys. Ferrous alloys are based on iron; nonferrous alloys are based on materials other than iron, such as copper, tin, aluminum, and lead.

2. *Ceramics and glass:* These are the result of the combination of metallic and nonmetallic elements. They are good insulators, brittle, thermally stable, and more wear resistant compared with metals. They are also harder and lower in thermal expansion than most metals.

3. *Woods and organics:* These are obtained from trees and plants. Their major advantage is that they are a renewable resource. However, some of their drawbacks include that they absorb water, require special treatment to prevent rotting, and are more flammable.

4. *Polymers or plastics:* These materials change viscosity with variation in temperature; therefore, they are easy to mold into a given shape. There are several benefits of polymers. They
   - Are good insulators.
   - Are resistant to chemicals and water.
   - Have a smooth surface finish.
   - Are available in many colors and do not need to be painted.

   There are several drawbacks to polymers, however, in that they

   - Are low in strength.
   - Deteriorate in ultraviolet light.
   - Have excessive creep at all temperatures.

The properties of materials can be divided into six categories: mechanical, thermal, physical, chemical, electrical, and fabrication. *Mechanical properties* include fatigue, strength, wear, hardness, and plasticity. *Thermal properties* include absorptivity, fire resistance, and expansion. *Physical properties* include permeability, viscosity, crystal structure, and porosity. *Chemical properties* include corrosion, oxidation, and hydraulic permeability. Four important *electrical properties* are hysteresis, conductivity, coercive force, and the dielectric constant. There are several *fabrication properties*, including weldability, castability, and heat treatability. A partial list of selected material properties is provided in Table 13.1

13.4.2 Material Selection Process

Several systematic approaches have been developed for selecting materials; one of these is given here.

STEP 1. *Perform a material requirements analysis:*
Determine the environmental and service conditions under which the product will have to operate.

STEP 2. *List the suitable materials:*
Filter through the available materials and pick several suitable candidates.

STEP 3. *Choose the most suitable material:*
Analyze the material listed, noting factors such as cost, performance, availability, and fabrication ability. Then select the most suitable material.

STEP 4. *Obtain the necessary test data:*
Determine the important properties of the selected material experimentally, under real-world operational conditions.

STEP 5. *Product specification fulfillment:*
Ensure that the selected material satisfies the stated specifications.

STEP 6. *Cost:*
This important factor plays a dominant role in the marketing of the end product.

Table 13.1: Material Properties

Material	Density Kg/m³	v Poisson's ratio	E MPa	σ_y MPa	σ_{ut} MPa	Melting Temperature °C	Thermal Conductivity W/mK
Pure Metals							
Beryllium	1827		3.033	3.792	6.205	1282	218
Copper	8858	0.33	1.172	0.689	2.206	1082	400
Lead	11349	0.43	0.138	0.138	0.172	327	35
Nickel	8858		2.069	1.379	4.826	1440	91
Tungsten	19 376		3.447		20.68	3367	170
Aluminum	2768	0.33	0.689	0.241	0.758	649	235
Alloys							
Aluminum							
2024-T4	2768	0.33	0.731	3.034	4.137	579	121
Brass	8581	0.35	1.034	4.136	5.102	932	109
Cast iron (25T)	7197	0.2	0.896	1.655	8.274	1177	55
Steel: 0.2% C							
Hot Rolled	7833	0.27	2.068	2.758	4.826	1516	89
Cold Rolled	7833	0.27	2.068	4.482	5.516	1516	89
Stainless Steel	7916	0.3	1.999	6.895	9.653	1413	21
Type 302 C. R.							
Ceramics							
Crystalline Glass	2491	0.25	0.862	1.379		1249	1.3
Fused Silica Glass	2214	0.17	0.724			1582	1.1
Plastics							
Cellulose Acetate	1301	0.4	0.017	0.345	1.379		0.17
Nylon	1135	0.4	0.028	0.552	0.896		0.25
Epoxy	1107		0.045	0.483	2.068		0.35

STEP 7. *Material availability:*
The selected material must be available at a reasonable cost.

STEP 8. *Material joining approach:*
Under real-world conditions, it may be impractical to produce a design element that uses a single piece of material. This may call for manufacturing the components with different pieces of material joined together to form a single unit.

STEP 9. *Fabrication:*
Engineering products usually require some level of fabrication, and different fabrication techniques are available. Factors affecting the fabrication method selected include time constraints, material type, product application, cost, and quantity to be manufactured.

STEP 10. *Technical issues:*
Technical factors are mainly concerned with the material's mechanical properties. Examples include strength compared with anticipated load, safety factors, temperature variation, and potential loading changes.

13.4.3 Primary Manufacturing Methods

The primary manufacturing methods that are used to convert a material into the basic shape required are as follows:

1. *Casting:* Casting is a widely used first step in the manufacturing process. During casting, an item takes on its initial usable shape. In the casting process, a solid is melted down and heated to a desirable temperature. The molten material is then poured into a mold made in the required shape. The cast items may range in size and weight from a fraction of an inch to several yards. Typical examples include a zipper's individual teeth and the stern frames of ships.

2. *Forging:* Forging, which is among the most important methods of manufacturing items of high-performance uses, involves changing the shape of the piece of material by exerting force on that piece. The methods of applying pressure include the mechanical press, the hydraulic press, and the drop hammer. Products such as crankshafts, wrenches, and connecting rods are the result of forging. The material to be forged may be hot or cold.

3. *Machining:* Machining involves removing unwanted substances from a block of material according to given specifications—such as size, shape, and finish—to produce a final product. There are many machining processes, such as milling, boring, grinding, and drilling.

4. *Welding:* Welding is a versatile production process that is used for permanently joining two materials through coalescence, which involves a combination of pressure and surface conditions.

13.4.4 Material Properties: An Introduction

Density

Density is one of the most important material properties because it determines the weight of the component. The range of material densities is based on the atomic mass and the volume the material occupies. Metals are dense because they consist of heavy and closely packed atoms. Polymers are formed from light atoms (such as carbons and hydrogen) and are loosely packed.

Melting Point

The melting point of a material is directly proportional to the bond energy of its atoms. Melting point becomes an important factor when the operating conditions of the material are in high-temperature environments, as in internal combustion engines and boilers. In design, the general rule is that the environment temperature should be at least 30% lower than the melting point of the material (($TTTm < 0.3$)). Designers should also keep in mind that when a material is subjected to high temperature and undergoes stress, creep occurs (slow change of dimensions over time).

Coefficient of Linear Thermal Expansion

The coefficient of linear thermal expansion is generally defined as

$$\alpha = \frac{dL}{LdT}$$

where

L is the linear dimension of the object.

dL is the change in the linear dimension of the object.

dT is the change in temperature that causes the change in length.

Thermal Conductivity

The thermal conductivity constant is a property of the material by which it can be determine whether the material is a good heat conductor or insulator. In metals, electrons are the carriers of heat. A relation between electrical conductivity and heat conductivity is established as

$$\frac{k}{s} = 2.45 x 10^{-8} T \ [W - \Omega K^2]$$

where

k is the thermal conductivity.

s is the electrical conductivity.

T is the temperature in Kelvin.

Strength of Material

The strength of the material determines the amount of load a material can sustain before it breaks. In general, the design criterion is based on the yield stress of the material. The strength of the material is measured experimentally using a tensile-testing machine. In a tensile-testing machine, a specimen is pulled with a certain load, and the elongation associated with the load is measured. The load is converted to stress, and the elongation is converted to strain. There are two general representations of the data: the stress–strain diagram and the true-stress–true-strain diagram. In the stress–strain diagram, the stress is measured by

$$\sigma = \frac{\text{load}}{\text{initial area}} = \frac{p}{A_o}$$

and the strain by

$$\varepsilon = \frac{\text{change in length}}{\text{initial length}} = \frac{\Delta L}{L_0}$$

In the true-stress–true-strain diagram, the change in area after each loading is recognized, and the true stress and true strains are obtained by

$$\text{stress} = \frac{\text{load}}{\text{instantaneous area}} = \frac{P}{A_{\text{inst}}}$$

$$E = \int_{L_0}^{L} \frac{dL}{L} = \ln \frac{L}{L_0}$$

The instantaneous area can be obtained by conservation of volume as

$$A_0 L_0 = AL = \text{constant}$$

Given the diagram, the modulus of elasticity can be obtained by finding the slope of the straight line before the yield stress. Designers apply a safety factor based on the application and strength of the material.

Ductility

Ductility determines the capacity of the material to deform before it breaks. It is an important property that can be used to determine whether a material can even be formed.

A material that cannot sustain deformation is called a brittle material. Ductility is measured as either percent elongation, $\%e$, in the material or percent reduction in area, $\%AR$:

$$e = \frac{L_f - L_0}{L_0} \times 100$$

or

$$AR = \frac{A_0 - A_f}{A_0} \times 100$$

Fatigue Properties

When the component is subjected to cyclic loading, knowing the fatigue properties is essential. In this case, the failure will be introduced at a stress lower than the yield stress. A correlation that has been used to estimate the fatigue limit is

$$\text{Fatigue limit} = 0.5 \, S_{\text{uts}}$$

where S_{uts} is the ultimate tensile strength.

Impact Properties

The impact property of a material is its resistance to fracture under sudden impact.

Hardness

Hardness is the measure of the material's resistance to indentation. Hardness can be measured in three tests: (1) Brinell (uses balls to indent the surface), (2) Vickers (uses pyramids to indent the surface), and (3) Rockwell (measures the depth of indentation). The Brinell hardness number (BHN) can be obtained as

$$\text{BHN} = \frac{2P}{D \left[D - \sqrt{D^2 - d^2} \right]}$$

where P is the load, D is the diameter of the indenter, and d is the measure of the diameter of indentation. The BHN is used to estimate the ultimate tensile strength

$$\left(SP_{uts} = \frac{P_{max}}{A_0} \right)$$

using

$$S_{uts} = 3.45 \ BHN$$

in MPa

$$S_{uts} = 0.5 \ BHN$$

in Ksi, where K is kips.

13.5 COST ANALYSIS

Although it is important to generate a rough cost estimate as early as possible in the design process and later to refine the cost estimate as soon as the embodiment phase is reached, it is only during the detailed design phase that a detailed cost analysis can be made. This is because it is only at the detailed design stage that the appropriate materials, manufacturing processes, accurate dimensions, and tolerancing are identified. As the design is refined, the cost estimate is also refined until the final product is produced.

Cost evaluation includes, among other things, estimating the cost of building a plant or the cost of installing a process within a plant to produce a product or a line of products. It also involves estimating the cost of manufacturing a part based on a particular sequence of manufacturing steps.

In design, it becomes important to find techniques that help reduce cost. Some of the cost-reduction techniques that can be used are as follows:

1. Introduce new processes.
2. Take advantage of knowledge gained as you gain more experience.
3. Allow for standardization of parts, materials, and methods.
4. Employ a steady production rate, when feasible.
5. Utilize all production capacity.
6. Have product-specific factories or production lines.
7. Improve methods and processes to eliminate rework, reduce the work in process, and reduce inventory.

It is important to avoid factors that tend to increase the costs of the design process. Factors that increase costs are as follows:

1. Incomplete product design specifications
2. Redesign due to failures
3. Supplier delinquency
4. Management and/or personnel changes
5. Relocation of facilities

6. Unmet deadlines

7. A product that has become too complex

8. Technologies and/or processes that are not developed as well as they were thought to be

9. Inadequate customer involvement

10. Disregarding needs

11. Failing to subject a new product idea to competitive evaluation

12. Not demonstrating that a new design can function properly under realistic conditions

13. Designing without considering manufacturing processes

14. Not continually improving and optimizing the product

15. Relying on inspection (last minute test)

13.5.1 Costs Classifications

Each company or organization develops its own bookkeeping methods. In this section, we present the various classification divisions of costs. Cost estimation within a particular industrial or governmental organization follows a highly specialized and standardized procedure, particular to the organization.

1. *Nonrecurring or recurring:* There are two broad categories of costs:
   a. *Nonrecurring costs:* These are one-time costs, which are usually called capital costs, such as plant building and manufacturing equipment.
   b. *Recurring costs:* These costs are direct functions of the manufacturing operations.

2. *Fixed or variable costs:* Fixed costs are independent of the rate of production of goods; variable costs change with the production rate.
   a. Fixed costs include
      i. Investment costs
         - Depreciation on capital investment
         - Interest on capital investment
         - Property taxes
         - Insurance
      ii. Overhead costs include
         - Technical service (engineering)
         - Nontechnical service (office personnel, security, etc.)
         - General supplies
         - Equipment rental
      iii. Management expenses
         - Share of corporate executive staff
         - Legal staff
         - Share of corporate research and development staff
      iv. Selling expenses
         - Sales force
         - Delivery and warehouse costs
         - Technical service staff
   b. Variable costs:
      i. Materials
      ii. Direct labor
      iii. Maintenance cost
      iv. Power and utilities

     v. Quality-control staff
    vi. Royalty payments
   vii. Packaging and storage cost
  viii. Scrap losses and spoilage

3. *Direct or Indirect:* A cost is called a direct cost when it can be directly assigned to a particular cost center, product line, or part. Indirect costs cannot be directly assigned to a product but must be spread over the entire factory.

  a. Direct costs include
     i. *Material:* This Includes the costs of all materials that are purchased for a product, including the expense of waste caused by scrap and spoilage.
    ii. *Purchased parts:* These are components that are purchased from vendors and not fabricated in-house.
   iii. *Labor costs:* These include wages paid to and benefits awarded to the workforce that is needed to manufacture and assemble the product. This includes the employees' salaries as well as all fringe benefits, including medical insurance, retirement funds, and vacation times.
   iv. *Tooling costs:* These include all fixtures, molds, and other parts specifically manufactured or purchased for manufacture of the product.
  b. Indirect costs include
     i. *Overhead:* This category comprises cost for administration, engineering, cleaning, utilities, leases of buildings, insurance, equipment, and various other costs that are incurred day to day.
    ii. *Selling expenses:* These are the costs for marketing advertisements.

13.5.2 Cost-Estimating Methods

The cost-estimating procedure depends on the source of the components in the product. There are three possible options for obtaining the components:

1. Purchasing finished components from a vendor
2. Having a vendor produce components designed in-house
3. Manufacturing components in-house

There are strong incentives to buy existing components from vendors. Cost is only one factor that determines whether a product or a component of a product will be made in-house or bought from an outside supplier.

If the quantity to be purchased is large enough, most vendors will work with the product design and modify existing components to meet the requirements of the new product. If existing components or modified components are not available off the shelf, they must be produced. In this case, a decision must be made regarding whether they should be produced by a vendor or made in-house. This make-or-buy decision is based on the cost of the component involved, as well as the capitalization of equipment and the investment in manufacturing personnel. Table 13.2 lists factors effecting the make-or-buy decision.

When an element is manufactured, the methods used to develop cost evaluation fall into three categories:

1. *Methods engineering (industrial engineering approach):* The separate elements of work are identified in detail and summed into total cost per part. Machined components are manufactured by removing portions of the material that are not wanted.

Table 13.2: The Make-or-Buy Decision

Reason to Make	Reason to Buy
Cheaper to make	Cheaper to buy
Company has experience making it	Production facilities are unavailable
Idle production capacity available	Avoiding fluctuating or seasonal demand
Compatible and fits in production line	Inexperience with the making process
Part is proprietary	Existence and availability of suppliers
Wish to avoid dependency on supplier	Maintain existing supplier
Part fragility requiring high packing	Higher reliability and quality
Transportation costs are high	

Thus, the costs for machining are primarily dependent on the cost and shape of the stock material, the amount and shape of the material that needs to be removed, and how accurately it must be removed. An example for the production of a simple fitting from a steel forging is shown in Table 13.3.

2. *Analogy:* Future costs of a project or design are based on past costs of a similar project or design, with due allowance for cost escalation and size difference. This method requires a backlog of experience or published cost data.

3. *Statistical approach:* Techniques such as regression analysis are used to establish relations between system costs and the initial parameters of the system: weight, speed, power, and so forth. For example, the cost of developing a turbofan aircraft might be given by

$$C = 0.13937 \times 10.7435 \times 20.0775$$

where C is in millions of dollars, $\times 1$ is the maximum engine thrust in pounds, and $\times 2$ is the number of engines produced.

Table 13.3: Sample Production/Operation Cost Table of a Simple Fitting from a Steel Forging

Operations	Material	Labor	Overhead	Total
Steel forging	37			37
Set-up on milling machine		0.2	0.8	1
Mill edges		0.65	2.6	3.25
Set-up on drill press		0.35	1.56	1.91
Drill 8 holes		0.9	4.05	4.95
Clean and paint		0.3	0.9	1.2
Total	37	2.4	9.91	49.31

13.5.3 Labor Costs

The most accurate method of determining labor costs is to set up a staffing table to find the actual number of people required to run the process line. Be sure to account for standby personnel. The cost of operating labor includes fringe benefits. As an order of magnitude estimate of labor costs,

$$\frac{\text{Operating person hours}}{\text{Tons of product}} = K \frac{\text{Numbers of process steps}}{\left(\dfrac{\text{tons}}{\text{day}}\right) 0.76}$$

where

  $K = 23$ for a batch operation.

  $K = 17$ for average labor requirements.

  $K = 10$ for an automated process.

13.5.4 Product Pricing

Pricing can be defined as the technique used for arriving at a selling price for a product or service that will interest the customer while simultaneously returning the greatest profit. Different methods are available to assist in arriving at the optimum price for a product. One of these methods is known as the break-even chart.

13.5.5 Break-Even Chart

The break-even chart shown in Figure 13.10 is designed to graphically show the profits and losses based on the selling price and manufacturing costs of the product. The manufacturing costs of a product include

1. Material (Credit should be accounted for recycled scrap and by-products.)
2. Purchased parts.
3. Labor

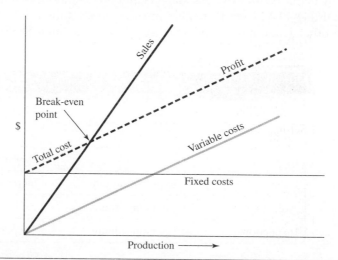

Figure 13.10: Break-Even Chart

4. Tooling
5. Overhead

Materials, purchased parts, and labor are variable costs that depend on the number of goods produced while tooling. Overhead is a fixed cost, regardless of the number of goods produced. The fact that the variable costs depend on the rate or volume of production whereas fixed costs do not leads to the idea of a break-even point. Determination of the production lot size needed to exceed the break-even point and produce a profit is an important consideration. To draw a break-even point chart, use the following steps:

STEP 1. *Determine the variable costs per product.*
Variable costs depend on the number of products and include
  a. Material
  b. Purchased parts
  c. Labor

STEP 2. *Determine the fixed cost.*
The fixed cost does not depend on the volume or the number of products.

STEP 3. *Draw, on the chart, the total cost as a function of the number of products.*
The total cost = fixed cost + variable cost.

STEP 4. *Determine a sale price of the product.*
Draw the sales as a function of the number of products.

RESULT: *The intersection of the sales line with the total price is the break-even point.*

13.5.6 Linear Programming

Linear programming is a tool that is used to represent a situation in which an optimum goal is sought, such as to maximize profit and minimize cost. Mathematically, it involves finding a solution to a system of simultaneous linear equations and linear inequalities, which are optimized in linear form. Consider the following illustrative example for a demonstration of this technique.

A manufacturer of desk staplers and other stationery items produces two stapler models (see Table 13.4). One is manual in operation, retails at $12.50, and returns $2.00 to profit. The other is automatic and sells for $31.25, returning $5.00 to profit. The manufacturer is

Table 13.4: Stapler Models

	Manual	Automatic
Selling price	31.25	12.5
Profit	$2	$5
Constraints	0	Power supply 200 per day
Labor	18 person-minutes	54 person-minutes
Adjustment	3 person-minutes	5.4 person-minutes
Inspection	1 person-minute	1.5 person-minutes

concerned with how to utilize his production facilities to maximize profits. One constraint placed on the maximizing of profits can be written as follows:

$$P = 2Q_M + 5Q_A$$

where P is profit function, Q_M is the quantity of manual staplers, and Q_A is the quantity of automatic staplers.

Assume that the manufacturing operation is such that labor skills and plant facilities are completely interchangeable; thus, varying quantities of manual and automatic staplers may be produced on the same day. Also, assume that the present demand is such that the company can sell as many as it can produce. There are certain production restraints. Most of the major components are prepared on separate assembly lines prior to moving to the final assembly area. There is a sufficient inventory of components for both staplers to satisfy daily production, with the exception of solenoids for the automatic model, which has a daily production capacity of 200. This constraint may be written as

$$Q_A \leq 200$$

Among other duties of employees, fabrication and assembly requires 18 person-minutes for each manual stapler and 54 person-minutes for each automatic stapler. The plant employs 45 people for this task, on an 8-hour shift. Since there is only one shift, the capacity is $8 \times 45 \times 60$, or 21,600 person-minutes per day. This may be defined mathematically by

$$18Q_M + 54_{Q_A} \leq 21,600$$

The final adjustment on the manual stapler requires 3 person-minutes on the assembly line, whereas the automatic model requires 5.4 person-minutes. This phase requires 6 people, and the daily capacity is $8 \times 6 \times 60$ or 2880 person-minutes. This constraint is expressed as

$$3Q_M + 5.4Q_A \leq 2880$$

Final inspection for workability requires 1 person-minute for manual model and 1.5 person-minutes for the automatic model. There are 4 half-time inspectors assigned to this task; the inspection capacity is $8 \times 2 \times 60$ or 960 person-minutes. The inspection constraint is

$$Q_M + 1.5Q_A \leq 960$$

The constraint equations are as follows:

$$P = 2Q_M + 5Q_A$$
$$Q_A \leq 200$$
$$18Q_M + 54Q_A \leq 21,600$$
$$3Q_M + 5.4Q_A \leq 2880$$
$$Q_M + 1.5Q_A \leq 960$$

Plotting these equations yields Figures 13.11 and 13.12.

Linear programming is used to find the isoprofit line (profit is the same along the line) that is farthest away yet still contains at least one point that is technically feasible.

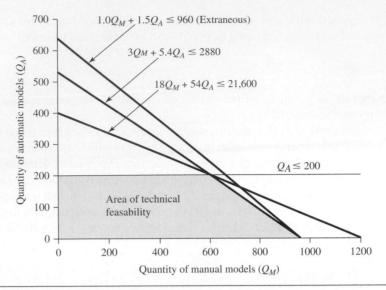

Figure 13.11: Constraints Equations without Profit

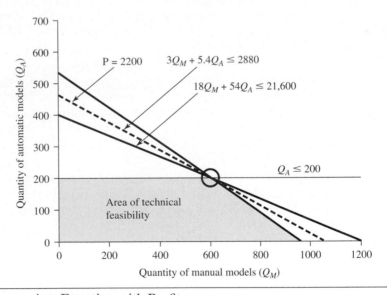

Figure 13.12: Constraints Equation with Profit

A combination of break-even and linear programming approaches can be used to set a price for a product. The linear programming approach can be used to determine the optimum number of goods; that number may then be used on the break-even chart to determine the selling price. The actual setting of a price resides at the heart of business practice, but describing it briefly is not easy. A key issue in establishing a price involves the reaction of the competition and the customer. When competition is fierce, the sales force may push hard for price reductions, claiming that they can make up the lost revenues by increasing the volume.

Special offers, such as promotional discounts, premiums, extras, trade-ins, volume discounts, and trade-in allowances, are often used.

13.6 OUTCOMES OF ANALYSIS

Typically, you should perform an optimization for the concept based on the evaluation criteria and structural and functional analysis. In general, a mathematic model is developed to describe the different parameters influencing design. For example, a model to describe the cost, based on the material choice, a model to describe the structural integrity, a model to describe the geometrical contribution to cost and structural integrity, and so on. These parameters are then optimized to find the alternative that will provide the best performance with the lowest cost. The optimization techniques are mathematical algorithms that have an active area of research.

Many researchers are studying the building of a real-time design engine where the alternatives are modeled using solid modelers. An analysis engine then would evaluate the alternative for its performance criteria if a change is needed based on the analysis. Then a parameter or group of parameters will be changed, and the analysis will be run one more time. This procedure continues until an optimized solution has been obtained.

13.7 COMPLETE DESIGN PROJECT DOCUMENTATION

Documenting the design is a sensitive issue, and companies do not like the design details to go out to the public. However, in a world where mobility of workers has become a norm, the design details should be recorded in sufficient detail so that new personnel can take over and continue from designers who leave the organization. Under these circumstances the following documents can be treated as mandatory:

1. Design brief
2. Customer statements and interpreted and prioritized requirements
3. Requirements metric (quality function) matrix
4. Competitive benchmarking in terms of the metrics
5. House of quality chart
6. Specifications with importance ratings
7. Criteria for concept selection and their importance ratings
8. Function diagram
9. Concepts for consideration
10. Pugh's matrix and decision matrix
11. Chosen conceptual design as a sketch and scheme in words
12. Embodiment design sequence and the calculations involved
13. Complete set of production drawings
14. Chosen analyses
15. Complete details of the analyses

13.8 CHAPTER SUMMARY

1. External forces are classified as surface or body forces.

2. Structural integrity can be determined based on testing if the material can sustain the applicable stresses.

3. Due to the complexity of the engineering models and geometry, computer-assisted methods are used. Among the many methods that solve the mathematical models is the finite element method (FEM). In the FEM, the body is discretized into elements. The mathematical model is solved at the nodes. The solution for the remainder of the body is obtained through an interpolation scheme.

4. Most solid modeling software has analysis packages that solve for structural, heat, fluid, and other physical configurations.

5. Suitable selection of engineering material is essential for the successful completion of the product design.

6. Engineering materials are typically classified as metal, plastic, organic, and ceramics.

7. The systematic material selection process includes
   a. Requirement analysis
   b. A list of suitable materials
   c. Selection of the most suitable material
   d. Obtaining the test data
   e. Specification fulfillment
   f. Cost analysis
   g. Material availability
   h. Joining approaches
   i. Fabrication
   j. Technical and safety issues

8. Primary manufacturing methods include casting, forging, machining, and welding.

9. Cost estimating is a continuous process during all design phases

10. Cost can be classified as either recurring or nonrecurring cost, fixed or variable cost, or direct or indirect cost.

11. Cost is not the only factor that affects the make-or-buy decision.

12. A break-even chart is used to graphically show the profits and losses based on the selling price and manufacturing cost.

13. The outcome of analysis is a model that describe the influence of the different parameters on design.

14. A complete project documentation presents all elements of the design process.

13.9 PROBLEMS

The objective of this exercise is to use computer programming to write a program that can find the cost of different alternatives for your design project. Manufacturing cost is an important factor in evaluating the different alternatives. Some alternatives are eliminated because of their high manufacturing cost.

Outline

In engineering design projects, students are expected to build a device by the end of a limited time period. With a limited amount of money, students need to figure out the manufacturing costs of different designs before they actually start building these devices. A manufacturing cost is a key determinant of the economic success of a product. When computing the manufacturing cost, one can divide the cost into two broad categories: (1) fixed costs and (2) variable costs. Fixed costs are incurred regardless of the number of products that are manufactured, whereas variable costs depend on the amount of goods that are produced. In student design projects, students are not expected to buy major tools or rent warehouses for production; hence the fixed costs and variable costs in this case may use the following components:

1. Fixed costs:
   a. *Tooling cost:* Tooling costs can include buying or renting equipment. Renting depends on the charge per unit time, regardless of the number of units produced. Buying in this case will be for this project only, and it will be assumed that it is paid for in full (that is, no interest or depreciation).
   b. *Setup cost:* This is the cost required to prepare equipment for a production, and it is fixed regardless of the number of units produced.
2. Variable costs
   a. *Material cost:* See Figure 13.13.
   b. *Processing cost:* This varies with the type of manufacturing equipment used and includes charges for both machine time and labor.

Programming Requirements

The program should be readable; all items from the terminal are sorted into "ready" and "need-manufacturing" categories. The information should be fed by the user. Ready items need prices from the vendor, while items that need manufacturing require the following calculation:

1. User inputs the following information:
   a. Mass of the material (see Table 13.5)
   b. Total number of units of this type
   c. Setup cost
   d. Tooling cost
   e. Processing cost
2. The computer calculates the manufacturing cost for each element. The output should be in the format shown in Table 13.5.
3. Name three cost classifications, and state the differences among them.
4. What are the three cost estimate methods, and what are the differences among them? Which method is more appropriate for your design project and why?
5. What is the break-even chart?
6. Using the break-even method, find break-even points if
   a. The fixed cost is $12,000.
   b. The variable cost is $17.95 per unit.
   c. The sale price is $29.50 per unit.
7. What would be the break-even cost in question 6 if the fixed cost is the same but the variable cost becomes $29.50 and the selling price becomes $37.50?

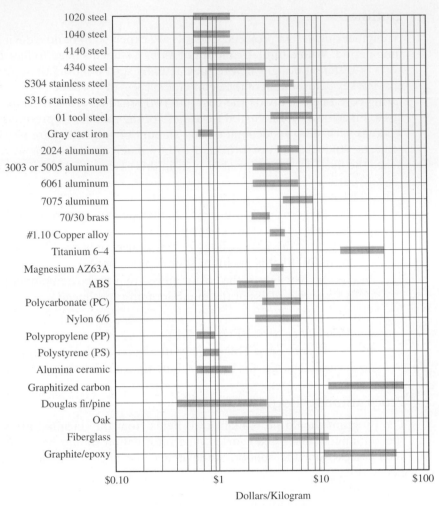

Based on David G. Ullman, The Mechanical Design Process, McGraw-Hill, New York, 1992.

Figure 13.13: Material Cost. Range of costs for common engineering materials; price ranges shown correspond to various grades and forms of each material, purchased in bulk quantities (1994 prices).

8. The following data are for a company that produces washers and dryers.
    * Dryers retail for $198 and contribute $15 to profit.
    * Washers retail for $499.95 and contribute $45 for profit.
    * The washer blade is limited in production capacity to 50 blades, while all other components have no limits.
    * Chassis assembly requires 6 person-hours for each dryer set and 18 person-hours for each washer. The plant employs 225 workers for an 8-hour shift to perform chassis assembly operations.
    * A dryer requires 1 person-hour on the assembly line, a washer requires 1.6 person-hours. There are 30 people on a single 8-hour shift assigned to assembly.
    * Final inspection requires 0.5 person-hour for a dryer and 2.0 person-hours for a washer. The plant employs 20 full-time inspectors and one part-time employee for 2 hours per day.

a. Write a linear programming model.
b. What is the optimum number of dryers and washers?
c. Calculate the maximum profit.
d. The linear-programming and break-even approaches are used to find the selling price for both the dryer and the washer, based on $10,000 fixed cost for both dryer and washer with a variable cost of $160 for the dryer and $330 for the washer. Determine how many days will be required until the company starts making a profit for the washers and the dryers. Use $198 and $500 as selling prices for the dryer and the washer, respectively.

Material Selection

1. In two opposing columns, list five uses of metals that are not appropriate for plastics and five uses of plastics that are not appropriate for metals.
2. From the table of mechanical properties of selected materials (see Table 13.5), list the differences between pure metals and metal alloys.
3. Based on the table, compare between metals, alloys, ceramics, and plastics in terms of weight, strength, stiffness, and thermal expansion.
4. What is the difference between forging, casting, and welding?
5. What is meant by each of the following terms?
   a. Elastic limit
   b. Yield point
   c. Ultimate strength
   d. Stiffness
6. What is the major requirement of a material if minimum deflection is required? Give an example.
7. What is meant by ductility? Suggest a method to measure ductility.

Table 13.5: Material Properties

Material	Density lb/in³	v Poisson's Ratio	E 10⁴ psi	σ_y 10³ psi	σ_{ult} 10³ psi	Melting Temperature °C	Thermal Conductivity lb/sec°F
Pure Metals							
Berylium,	0.066		44	55	90	2340	19.9
Copper,	0.32	0.33	17	10	32	1980	51
Lead	0.41	0.43	2	2	2.5	621	4.5
Nickel	0.32		30	20	70	2625	7.9
Tungsten,	0.70		50	3.5	300	6092	26.2
Aluminum	0.1	0.33	10		11	1200	29.0
Alloys, Aluminum 2024-T4	0.1	0.33	10.6	44	60	1075	15.8
Brass	0.31	0.35	15	60	74	1710	14.9

Continued on next page.

Material	Density lb/in³	v Poisson's Ratio	E 10⁴ psi	σ_y 10³ psi	σ_{ult} 10³ psi	Melting Temperature °C	Thermal Conductivity lb/sec°F
Cast Iron (25T) Steel: 0.2% C	0.26	0.2	13	24	120	2150	5.8
Hot Rolled	0.283	0.27	30	40	70	2760	6.5
Cold Rolled	0.283	0.27	30	65	80	2760	6.5
Stainless Steel							
Type 302 C.R.	0.286	0.3	29	100	140	2575	1.9
Ceramics							
Crystalline Glass	0.09	0.25	12.5	20		2280	0.24
Fused Silica Glass	0.08	0.17	10.5			2880	0.17
Plastics							
Cellulose Acetate	0.047	0.4	0.25	5	20		0.032
Nylon	0.041	0.4	0.41	8	13		0.03
Epoxy	0.04		0.65	7	30		0.10

LAB 8: Material Selection Tutorial[1]

This lab describes an example of selecting optimal materials for a car key. Figure L8.1 shows a schematic diagram of a car-key structure. As seen from Figure L8.1, a car key has two main components: the grip and the shaft. The grip should be soft and cool to the touch, ensuring the driver's comfort. On the other hand, the shaft should not break while twisted. At the same time, the shaft should be a good conductor of electricity so that it helps close an electrical circuit when turning on the ignition.

GRIP The indices for the material selection of a grip are explained in the following text.

Figure L8.1: A Schematic Diagram of a Key

[1]AMM Sharif Ullah, Dr. Engineering, Department of Mechanical Engineering, Kitami Institute of Technology, Japan.

A material feels soft (to the touch) if its hardness is low and it is flexible. As such, an index of softness can be formulated as

$$S = C(HE) \tag{1}$$

In Equation (1), H is the hardness, E is the modulus of elasticity, C is a constant, and S is an index. To obtain a soft material, S has to be minimized. The hardness of material and its strength are directly correlated as

$$H = C^{'\sigma}y \tag{2}$$

In Equation (2), (σ_y is the yield strength of a material. Rearrangement of Equations (1) and (2) yields:

$$S = CC^{'}(\sigma yE) \tag{3}$$

This implies a material index M_S that has to be maximized to get a soft material, as

$$Ms = \frac{1}{\sigma yE} \tag{4}$$

What about the coolness? Suppose that Q amount of heat has been applied to V volume of material. This will raise the temperature of the material. This implies that

$$Q = mCp\Delta T = \rho VCp\Delta T \tag{5}$$

In Equation (5), m is the mass, Cp is the heat capacity, ρ is the density, and T is the rise of temperature. Rearranging the variables in Equation (5) yields:

$$\Delta\Delta T = \left(\frac{Q}{V}\right)\frac{1}{\rho Cp} \tag{6}$$

To achieve coolness, ΔT has to be minimized. This implies a material index M_C that has to be maximized:

$$M_C = \rho Cp \tag{7}$$

To obtain an optimal material for the grip, the following formulation can be used:

$$Moptimal = wsM^{'s + W}cM^{'}c \tag{8}$$

where

$$W_S + W_C = 1$$

$$M^{'}S = \frac{Ms - \min(Ms)}{\max(Ms) - \min(Ms)}$$

$$M^{'}C = \frac{MC - \min(MC)}{\max(MC) - \min(MC)}$$

Figure L8.2 shows an MS Excel™-based system for selecting optimal materials for the grip of a car key. As seen from Figure L8.2, among five materials (aluminum alloy, silica glass, PET, PVC, and natural rubber), PVC is the optimal material, given that equal importance has been placed on softness and coolness. The solution might change if the weights w_S and w_C are redefined.

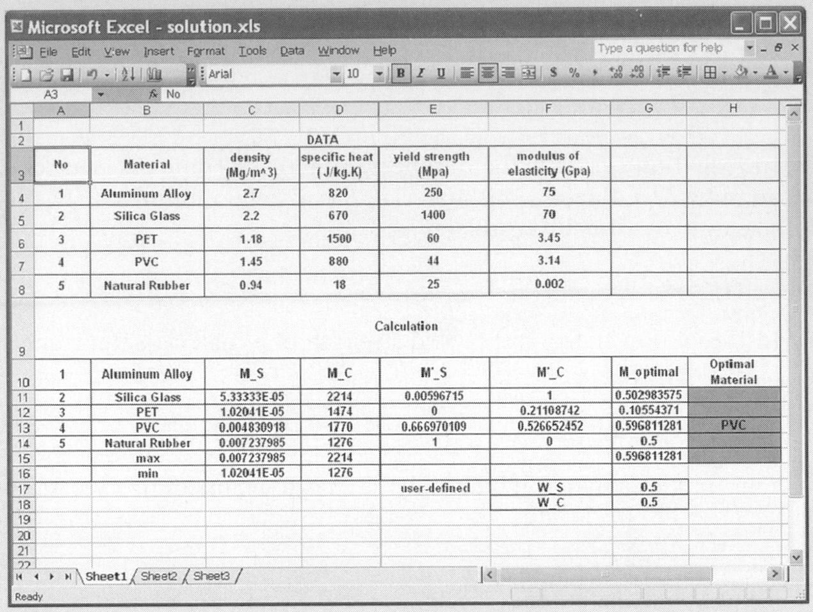

DATA

No	Material	density (Mg/m^3)	specific heat (J/kg.K)	yield strength (Mpa)	modulus of elasticity (Gpa)		
1	Aluminum Alloy	2.7	820	250	75		
2	Silica Glass	2.2	670	1400	70		
3	PET	1.18	1500	60	3.45		
4	PVC	1.45	880	44	3.14		
5	Natural Rubber	0.94	18	25	0.002		

Calculation

1	Aluminum Alloy	M_S	M_C	M'_S	M'_C	M_optimal	Optimal Material
2	Silica Glass	5.33333E-05	2214	0.00596715	1	0.502983575	
3	PET	1.02041E-05	1474	0	0.21108742	0.10554371	
4	PVC	0.004830918	1770	0.666970109	0.526652452	0.596811281	PVC
5	Natural Rubber	0.007237985	1276	1	0	0.5	
	max	0.007237985	2214			0.596811281	
	min	1.02041E-05	1276				
				user-defined	W_S	0.5	
					W_C	0.5	

Source: Microsoft

Figure L8.2: Material Selection for the Grip of a Car Key

GROUP TASK Use the aforementioned procedure and determine the optimal material for the shaft of the same car key. Refer to the first paragraph of this Lab for the shaft's desired attributes.

LAB 9: Use of Pro/MECHANICA® for Structural Analysis

Purpose

Once the materials have been selected and the initial dimensions and tolerances specified, it is advisable to perform a detailed analysis of the product. One way to do this is to perform structural analysis using the finite-element method. This Lab will introduce you to the use of Pro/MECHANICA® as an analysis tool for structural components. Pro/MECHANICA is a tool available for use in conjunction with the Pro/ENGINEER® software. Pro/ENGINEER® (Pro/E) is a solid-modeling software that also can be used to perform virtual prototyping.

Supported Beam

A 2-D simply supported beam by definition has a pin support at one end and a roller support at the other end. The essential feature of a pin support is that it restrains the beam from translating both vertically and horizontally but does not prevent rotation.

As a result, a pin joint is capable of developing a reactive force for both vertical and horizontal components, but there will be no moment reaction. A roller support prevents translation in the vertical direction but not in the horizontal direction. The roller also allows free rotation in the same manner as the standard pin joint.

Point and curve constraints on solid elements can cause singular stresses. The following procedure illustrates techniques for modeling a solid, simply supported beam using rigid connections.

- Generate the part using Pro/E. The part is a square beam as shown in Figure L9.1
- Launch MECHANICA Application>MECHANICA. The default units are in (in.-lbm-sec). For now, assume the default units.
- Select **Structure** from the MECHANICA menu.
- Create datum planes at the ends of the beam. Select **insert Datum Plane**. From the menu shown in Figure L9.2, select **Parallel>Plane**. Select one of the end faces. Repeat for the other face.
- Add rigid connections on each end surface. Click on **Model>Idealizations>Rigid Connection>Create**, and pick one of the end surfaces. Repeat at the other end of the beam.
- Add constraints to each of the end planes. Click on **New Edge/curve constraints**, and pick one of the end surfaces. Fix all translations as shown in Figure L9.3. Repeat for the other edge; this time leave the Z-Translation free.

Figure L9.1: Square Beam

Figure L9.2: Menu Manager

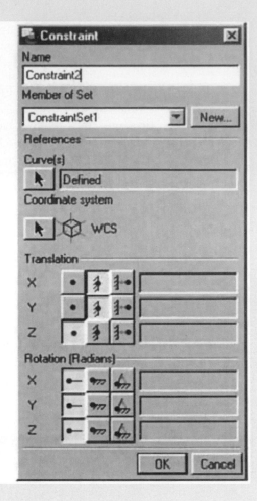

Source: Pro/MECHANICA

Figure L9.3: Constraints

- Apply Loads (see Figure L9.4). Click on **Model>Loads>New**.
- Apply Material Properties **Model>Material**.
- Check Model for errors **Model>Check Model**.
- Select the analysis type. Click on **Model>Analysis>Static>New>Close**.
- Create **Design Study Model>Design Study**.
- Run the analysis. Check the setting. Make sure you know where the files will be saved.
- Check the results. Create a results window. Check for the displacement (Figure L9.5) and stresses (Figure L9.6).

1. Use Pro/MECHANICA to solve for the stress displacement for a shaft of diameter = 2.0 and length = 10.00. A pin on one end and a roller on the other end support the shaft. Solve for the following loads:
   a. Material is steel, and the load is uniformly distributed and equal to 10,000.
   b. Material is Al and the load is uniformly distributed and equal to 10,000.
   c. The load is concentrated at the middle and is equal to 100,000 (for both materials).
   d. The load is concentrated at a point 1/3 from the pin joint and is equal to 100,000 (for both materials).

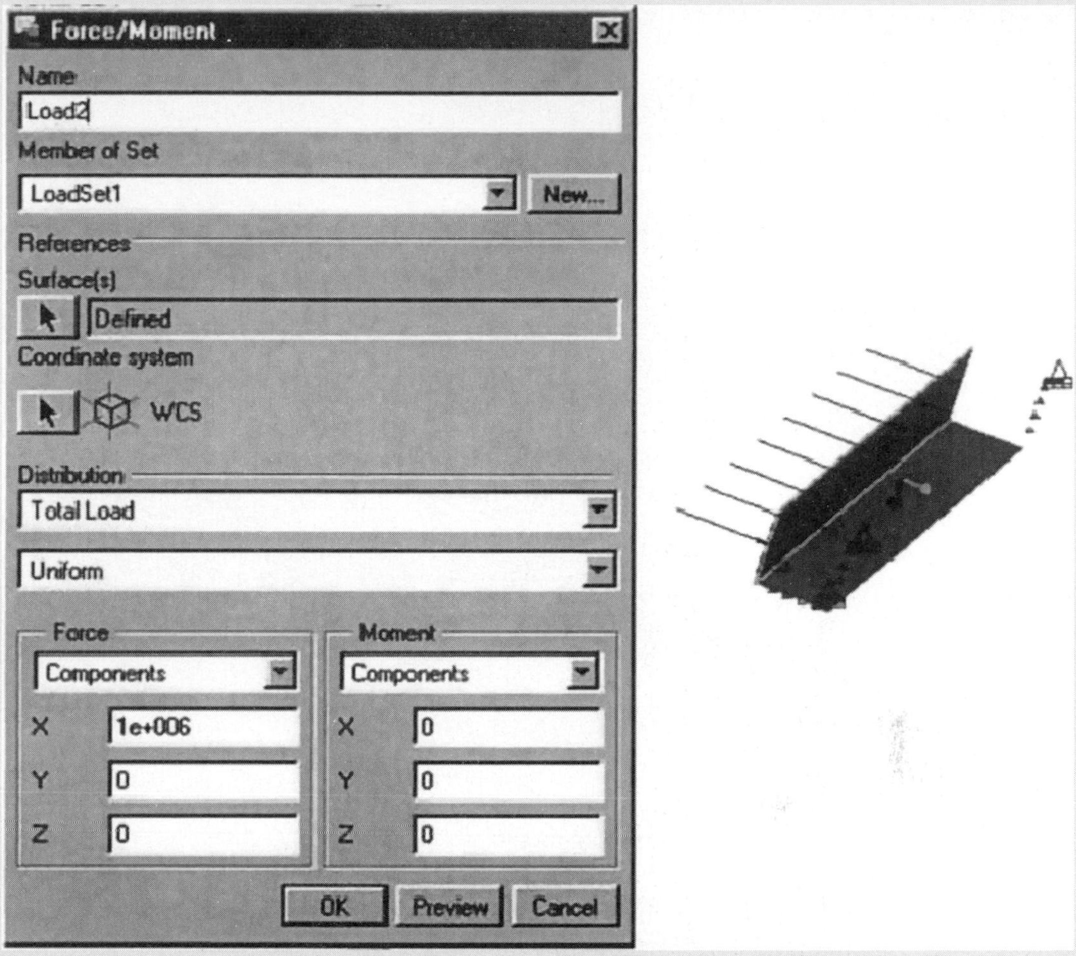

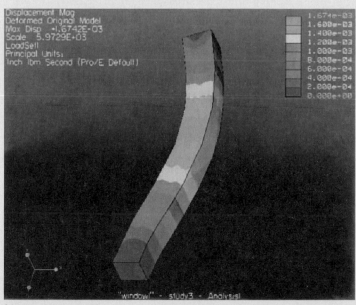

Source: Pro/MECHANICA

Figure L9.4: Load on Beam

Source: Pro/MECHANICA

Figure L9.5: Displacement

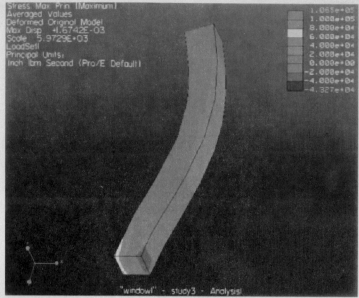

Source: Pro/MECHANICA

Figure L9.6: Stresses

    e. The load is concentrated at a point 2/3 from the pin joint and is equal to 100,000 (for both materials).

    f. Compare the results.

2. Using the beam in the example, change the dimensions of the beam to (10 × 2 × 100). Apply a uniform load of 10,000. (Material is bronze.)

    a. Find the displacement and stresses under this condition.

    b. Assume the analysis is 2-D (use the model type option with plane stress). Find the displacement and stresses; compare with part (a).

References

DHILLON, B. S. *Engineering Design: A Modern Approach.* Toronto: Irwin, 1995.

DIETER, G. *Engineering Design.* New York: McGraw-Hill, 1983.

DYM, C. L. *Engineering Design: A Synthesis of Views.* Cambridge, UK: Cambridge University Press, 1994.

HILL, P. H. *The Science of Engineering Design.* New York: McGraw-Hill, 1983.

PUGH, S. *Total Design.* Reading, MA: Addison-Wesley, 1990.

RADCLIFFE, D. F. and LEE, T. Y. "Design Methods Used by Undergraduate Students." *Design Studies,* Vol. 10, No. 4, pp. 199–207, 1989.

RAY, M. S. *Elements of Engineering Design.* Englewood Cliffs, NJ: Prentice Hall, 1985.

SUH, N. P. *The Principles of Design.* New York: Oxford University Press, 1990.

ULLMAN, D. G. *The Mechanical Design Process.* New York: McGraw-Hill, 1992

ULRICH, K. T. and EPPINGER, S. D. *Product Design and Development.* New York: McGraw-Hill, 1995.

VIDOSIC, J. P. *Elements of Engineering Design.* New York: The Ronald Press Co., 1969.

WALTON, J. *Engineering Design: From Art to Practice.* New York: West Publishing Company, 1991.

CLOSURE

CHAPTER 14

CASE STUDY

By now, you should have experimented with all stages of the design process in a segmented fashion. We have discussed each of the stages along with various design methods. In this chapter we will show all of the design stages from requirements to detailed design. We have selected the Painter's Pal single-stage scissor as the design example.

The process starts with the design brief. Remember that the design brief states what the product would be, the benefits to be delivered, who the stakeholders are, and its position in the market. A team of students, just like you, started work in earnest once they received the design brief. Establishing the target specification (what the product would do) is the product of the *requirements* stage in the design model. The function structure and the specifications mark the beginning of the *product concept* stage. The design team evaluates the conceptual solutions and further develops the best solution. The team then considers the available types of raw materials and their capabilities to form the *embodiment design* stage. In the *detailed design* stage, all parts are modeled and assemblies are constructed using computer-assisted design (CAD) software. As part of the detailed design stage, the power train, which includes the gearbox and the power screw, is designed using scientific engineering principles.

OBJECTIVES

The primary objective of this chapter is to demonstrate the four stages of design using one integrated example. The purpose is not to re-explain the process and methods; rather it is to apply the process to a selected example. Students who have skipped ahead in the study of the process are strongly advised to read the chapters associated with each of the design stages.

14.2 STAGE 1: THE REQUIREMENTS

14.2.1 Design Brief

The design process starts with a *design brief*, an instruction from senior management. The design brief for the professional painters' pal is shown in Table 14.1. Although there are no

Table 14.1: Design Brief

Design Brief of Professional Painter's Pal, an Access Platform	
Drafted by	Simple Simon Company Ltd. (made up for the purpose of the example)
Product Description	A portable, compact, power-driven and very sturdy access platform that enables a professional painter to reach the ceiling of rooms in domestic houses for painting. It should have facilities to hold the paint tray and brushes of various types and sizes at an appropriate height for the painter to easily access while working. It should be capable of being transported in the domestic elevators from floor to floor and to pass through standard domestic doors.
Product Concept	A single scissor-type power-driven mechanism carrying a platform of size about $0.64\,m^2$ and designed to comply with the EN280 standard as needed.
Benefits to Be Delivered	Access to paint heights of up to the ceiling of the rooms of domestic houses with accessories Ability to locate at different intermediate positions Handy control of the up and down power-driven motion Easily movable from place to place Compact for transportation in cars Firmly lockable in positions Withstand and remaining stable under loads up to 100 N side force by hand
Positioning and Target Price	Middle-of-the-range 400 kg payload and price of $800 with attractive appearance
Target Market	DIY enthusiasts, professional painters, women painters, handy workers
Assumptions and Constraints	Sold through the current outlets EN280 will be followed Small enough to be moved around
Stakeholders	Various painters in target market Dealers and retailers

Continued on next page.

Design Brief of Professional Painter's Pal, an Access Platform	
Possible Features and Attributes	Name of the product: Professional Painter's Pal Caption reads "BORN TO ELEVATE" Nonsticky surfaces for easy cleaning
Possible Area for Innovation	Compact packaging Aesthetically pleasing Safety provisions

standard lists of details that must be stated in the design brief, there is a general understanding as to the typical items that should be stated in it. The Instructor in this case represents the senior management or the owner of the project who was encouraged to include product description, product concept, the benefits to be delivered, positioning and target price, target market, assumptions and constraints, stakeholders, possible features and attributes, and possible areas for innovation. This information enables the design team to understand the kinds of design features that must be incorporated in the product.

14.2.2 Selection Criteria

Various concepts would be proposed during the concept generation process; some of which would be viable and some would not. The viable concepts are taken further for evaluation, and one or two will be selected for further development. This selection has to be made based on some criteria, which must be drawn from the design brief. While it is fundamentally important that the product meet customer requirements, the selection has to be based on the intended benefits, positioning in the market, and similar factors. The criteria should be based on responses to the designed product. Table 14.2 lists the criteria drawn from the design brief.

Table 14.2: Selection Criteria and Importance Ratings

Criterion	Importance %
Access height	8%
Auxiliary space for accessories	10%
Size for elevators	7%
Size for doors	8%
Access to controls	18%
Stability for side force and moment	18%
Ease of mobility	10%
Pleasing appearance	8%
Price	5%
Compactness	8%

14.2.3 Customer Verbatim

From the design brief, the list of stakeholders can be compiled. These are the people whose requirements for the product have to be elicited. The process is to record the requirements as told by the stakeholder (verbatim); then the design team translates the verbatim into a list of needs. Table 14.3 shows the customer verbatim established for the Professional Painter's Pal. Students are required to produce an objective tree here.

14.2.4 Customer Needs

Customer needs are the translations and elaborations of the customer verbatim established by the team. The needs are then assigned their relative importance ratings. The design team that established the verbatim and its translations can assign the importance ratings, or that process can be referred back to the customers. The needs for the Professional Painter's Pal are given in Table 14.4, and Table 14.5 gives the needs and their relative importance.

Table 14.3: Customer Verbatim

Customer Verbatim
I want it to be solid enough so that I feel safe while working on it.
I don't want to feel the vibrations while it's rising.
It should be easy to step on.
I don't want to slip on the surface.
I want it to be easy to move.
I want it to be safe.
I want to be able to control the up and down motion with a button.
I want it to not get damaged easily.
It should not get affected by the weather (corrosion).
Foldable and easy to use
Mobile and easy to move
Speed of rise is constant and comfortable.
There should be a rail at least on one side.
The surface should be rough to stop slipping.
There should be a small guard at the edge to warn the user if he/she gets close to the edge.
There should be a brake on the tires used to move it.
The inner components should be covered.
It should look nice.
Some parts should be easy to disassemble to make it easy to move without it being heavy.
The structure should be stable with no tilting.
An emergency stop button

Table 14.4: Customer Verbatim and the Originating Needs

Verbatim	Needs
I want it to be solid enough so that I feel safe while working on it.	It should be rigid laterally.
I don't want to feel the vibrations while it's rising.	It should be rigid longitudinally.
It should be easy to step onto.	It should have a minimum stowed height.
I don't want to slip on the surface.	The surface should be rough.
I want it to be easy to move.	It should have wheels, and its size shouldn't be excessive.
I want it to be safe.	It should have an emergency button, a rail, a wheel brake, and a handle.
I want to be able to control the up and down motion with a button.	It should have a height control.
It should not get damaged easily.	The material should be strong; steel should be used.
It should not get affected by the weather (corrosion).	It should be painted.
It should be foldable and easy to use.	It should have an on/off button, a button to change the height, and an emergency button.
It should be mobile and easy to move.	It should have wheels.
The speed of rising should be constant and comfortable.	Acceleration should be zero.
There should be a rail on at least one side.	It should have a rail with a handle placed where the user can find it while painting.
The surface should be rough to prevent slipping.	Use a rough steel texture.
There should be a small guard at the edge to warn users if they get close to the edge.	There should be a partition around the edges.
There should be a brake on the tires used to move it.	For safety, each wheel should have a brake to keep it from moving.
The inner components should be covered.	The motor and the gearbox should be covered to prevent children or others from touching it.
It should look nice.	It should be painted with an attractive color.
Some parts should be easy to disassemble so it is easy to move without being too heavy.	The rail should be easy to disassemble.
The structure should be stable with no tilting.	It should be rigid.
It should have an emergency stop button.	Design the product for safety.

14.2.5 Needs and Metrics

A metric is a precise and measurable characteristic that represents a need. Achieving a required value or range of values for this characteristic means that the need from which this metric originated has been satisfied. The metrics for the needs of the Professional Painter's Pal are established by developing the need metric matrix shown in Table 14.6.

Table 14.5: Needs and Their Relative Importance

Needs	Importance
It should be rigid laterally.	5
It should be rigid longitudinally.	5
It should be designed for easy storage.	4
The surface should be rough.	3
It should have wheels, and its size shouldn't be excessive.	3
It should have an emergency button, a rail, a wheel brake, and a handle.	4
It should have a height control.	4
The material should be strong; steel should be used.	3
It should be painted,	2
It should have an on/off button, a button to change the height, and an emergency button.	4
Acceleration should be zero.	3
It should have a rail with a handle placed where the user can find it while painting.	2
There should be a partition round the edges.	3
Each wheel should have a brake to prevent it from moving. It should be designed for safety.	4
The motor and the gear box should be covered to prevent children or others from touching it.	3
It should be painted with an attractive color.	2
The rail should be easy to disassemble.	2
It should be rigid.	5

14.2.6 Competitive Benchmarking

Competitive benchmarking is the process of comparing a competitor's products in the market using the metrics established in Table 14.6. The benchmarking for the Professional Painter's Pal is given in Table 14.7.

14.2.7 Target Specifications

The target specifications are values for the metrics as shown in Table 14.6, which, if satisfied, would be acceptable to the customers. These are established considering the needs and the benchmarking results. For the Professional Painter's Pal the target specifications are given in Table 14.8.

Table 14.6: Need-Metric Matrix

Metric	It should be rigid laterally.	It should be rigid longitudinally.	It should have a minimum stowed height.	The surface should be rough.	It should have wheels.	Wheels should be lockable.	The volume shouldn't be excessive.	Weight should be low.	It should easily pass through doors.	It should fit in a domestic elevator (1.5m × 1.5m).	It should have an emergency button.	Rails should be easy to dismantle.	It should have a height control.	Its material should be strong.	It should be painted.	It should be slow and comfortable when rising.	It should have a paint-handling table.	The motor and gearbox should be covered.	It should be stable.
Permissible T/moment Tilt angle, θ degree)																			X
Safety list 4																		X	
Work table (mm²)																	X		
Average rising velocity (mm/s)																X			
List															X				
Overall deflection (mm)														X					
Push button switch													X						
Safety list 3												X							
Safety list 2											X								
Weight (Kg)								X											
Length (mm)							X			X									
Width (mm)							X		X										
Safety list 1						X													
Wheel size (radiuses, mm)					X														
List				X															
Stowed height (mm)			X																
longitudinal deflection (mm)		X																	
Lateral deflection (mm)/100N	X																		

Table 14.7: Competitive Benchmarking

	Presto XL 36-20	HymoMX5 – 8/8	Autoquip 36S15	Stationary scissor lift table SJGI-1.7
Travel distance	36 in	800 mm	36 in	
Raised height	43 in	880 mm	42.5 in	2160 mm
Stowed height	7 in	80 mm	6.5 in	460 mm
Capacity	2000 lb	500 kg	1500 lb	1000 kg
Width of base	24 in	600 mm	24 in	
Length of base	48 in	1350 mm	48 in	
Width of platform	24 in	600 mm	24 in	2000 mm
Length of platform	48 in	1350 mm	48 in	2500 mm
Lifting time	17 sec	19 sec	28 sec	30 sec
Motor power	1 HP	0.48 kw	0.5 HP	2.2 kw
Power supply— Single phase	115 V			
Weight	690 lb	175 kg	580 lb	1200 kg
Price	$2574			

Table 14.8: Target Specifications

Metric No.	Need Nos.	Metric	Importance	Unit
1	1	Lateral deflection (mm)/100N	5	5 mm/100 N
2	2	Longitudinal deflection (mm)	5	10 mm
3	3	Stowed height (mm)	4	400 mm
4	4	List	3	Rough surface (List)
5	5	Wheel height (radius, mm)	3	150 mm
6	6	Safety list 1	4	Brake wheels (List)
7	7, 9	Width (mm)	4	600 mm
8	7, 10	Length (mm)	3	1425 mm
9	8	Weight (kg)	2	500 kg
10	11	Safety list 2	4	Emergency button (list)
11	12	Safety list 3	3	Rails (list)
12	13	Push-button switch	2	Push-button control (list)
13	14	Overall deflection (mm)	3	20 mm

Continued on next page.

Metric No.	Need Nos.	Metric	Importance	Unit
14	15	List	4	Painted chassis (list)
15	16	Average rising velocity (mm/s)	3	50 mm/s
16	17	Work table	2	650 × 150 mm
17	18	Safety list 4	2	Gear box covered (list)
18	19	Permissible T/moment Tilt angle, θ degrees	5	2 degrees

14.3 PRODUCT CONCEPT

The product concept defines (1) the functions the product performs and (2) the specifications of the product.

14.3.1 Functional Modeling

The scissor lift can be divided into five functional sub-groups as shown in the following list:

1. Supply and remove power at required points—*Electrical System*.
2. Power is available through the motor as rotary power with low torque and high speed. This must be converted into rotary power at low speed and high torque. Alternatively, this can be converted to pressure energy if a hydraulic system is used. *Torque speed conversion system*.
3. The rotary power and motion has to be converted to linear or translating power. If it is pressure energy, it must be used for linear translating power—*Rotary to linear power conversion*.
4. The mechanism should provide lifting and lowering motions—*Scissor mechanism*.
5. The frame provides support when the unit is moved from one site to another or is stored—*Basic frame and housing*.

Each part of the of the unit incorporates several functions; thus, these five listed items can be identified as the purpose functions. Table 14.9 shows the purpose functions and the corresponding action functions.

Table 14.9: Purpose Functions and Action Functions

Purpose Functions	Action Functions
Distribute and control power	Receive supply from mains
	Distribute supply to hand control
	Distribute supply to platform control
	Cut power when maximum height is reached
	Cut power when minimum height is reached
	Maintain power supply by pressing the control switch
	Install extra power socket

Purpose Functions	Action Functions
Convert low torque and high speed to high torque and low speed	Receive high speed power
	Reduce the speed to a medium value
	Reduce the speed further to a low value
	Support the high-speed receiver
	Support the medium-speed runner
	Support the low-speed runner
	House the components
	Interface with motor shaft
	Interface with the motion converter
Convert rotary to linear motion	Receive rotary motion from the torque magnifier
	Drive the nut and accessories linearly
	Pull/push the cross beam
	Move the cross beam and rotate the scissor arms
Lift, lower, and hold the platform by rotating the arms of the mechanism	Bring the feet of the arms closer together
	Move the feet of the arms far apart
	Keep the feet of the fixed and free arms level
	Restrict movement only to the vertical direction
	Hold the arms in position
Support and house the entire machine and permit movement from site to site	Establish mobility
	Support and locate the fixed ends of the arm
	Support and guide the moving ends of the arms
	Ensure stability of the machine
	Ensure stability of the machine, payload, and lateral force
	House the drive, torque converter, and motion converter

14.3.2 Scissor Lift as a Connected Mechanism

Figure 14.1 shows the mechanism for the lift. It is used to compute the base and height for the system.

The stowed height is the minimum height of the lift, and the maximum height is the fully extended height of the platform. The difference is the height the lift gives as the lifting height.

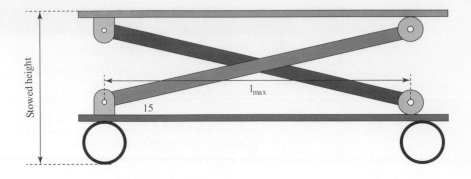

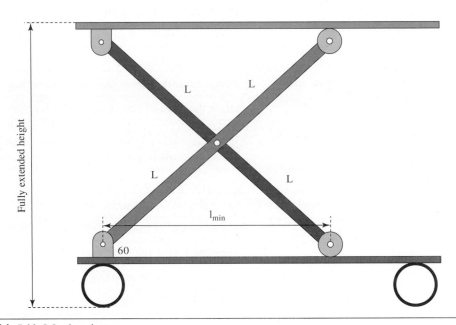

Figure 14.1: Lift Mechanism

If 2L is the length of the scissor arm, and the angles are 15° and 60° from the lifting height.

$$= 2L(\sin 60 - \sin 15) = 1.21L$$

Similarly, if the angles are 10° and 75° and the length is 1250 mm, then the lifting height is

$$= 1250(\sin 75 - \sin 10) = 990 \text{ mm}$$

Minimum scissor separation $= 1250 \times \cos 75 = 323.5 \text{ mm}$

Length	Minimum θ	Maximum θ	Minimum Separation	Maximum separation	Lift
1250	15	60	625	1207.4	759
1250	10	75	323.5 mm	1231	990 mm
1150	10	60	625	1231	865.5
1150	10	75	323.5	1231	865.5

Image supplied by the authors

Figure 14.2: Concept Prototype

14.3.3 Concept Prototype

Figure 14.2 shows the concept prototype built with timber from a scrap yard. It consists of a base frame, a platform frame, two pairs of scissors, and six pins. It mimics a full-scale lift and provides an opportunity to have a better understanding of the product.

The concept prototype process can help you learn the following lessons:

- Understanding the whole mechanism
- Developing new ways of thinking
- Opening a discussion with the design team to come up with more ideas
- Improving a concept to a better one
- Estimating how much time and effort you'll need
- Evaluating the physical principles that govern the behavior of the prototype
- Evaluating the technologies necessary to realize the final product
- Evaluating the basic architecture and the manufacturing process

14.3.4 Special specifications:

Some of the specifications are developed after the concept prototype, such as the rails and handles shown in Figure 14.3.

14.4 SOLUTION CONCEPT

The starting point for developing solution concepts is the functional model developed during the product concept stage. The objective is to provide the purpose functions and the associated action functions, thus enabling the formulation of structural solutions. However, the action functions are many and cannot be considered at the conceptual stage due to a large number of solution concepts that result from combinatorial expansion and because a substantial portion of them would not be viable. The students working on this project chose to form a morphological chart.

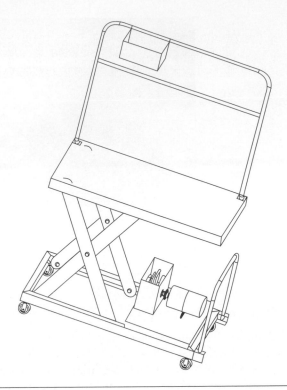

Figure 14.3: Rails and Handles of the Scissor Lift

14.4.1 Morphological Chart

The purpose functions in the function structure are enumerated in the first column, where each row is occupied by a single-purpose function. The electrical system was split into two sections for easy handling. Means were then enumerated in the succeeding columns. Means for the first purpose function occupied the second row and were designated as A_1 and A_2. In a similar fashion, means for the second purpose function were designated as B_1 and B_2. The completed morphological chart is given in Table 14.10.

14.4.2 Concept Selection

Theoretically the morphological chart provided $2 \times 2 \times 4 \times 2 \times 2 \times 2 = 128$ combinations, yielding 128 conceptual solutions. A substantial portion of them are not viable. From a close examination the following solutions, including $A_1B_1C_1D_1E_1F_1$, $A_1B_1C_1D_1E_2F_1$, and several others (such as $A_1B_1C_1D_1E_1F_2$) are identified as viable.

The selection criteria have to come from three sources: (1) the design brief, (2) customer needs, and (3) the design team (on technical ease). At this point all conceptual solutions considered were feasible. From the design brief 10 items were identified for selection (see Table 14.11). Of these only (1) auxiliary space for accessories, (2) compactness, and (3) pleasing appearance could be used at this stage. The customer needs can be represented by feeling safe and sturdy. The considerations from the design side can be (1) number of components, (2) ease of manufacture, and (3) ease of packaging. The selection criteria chosen and their relative importance are shown in Table 14.11.

Table 14.10: Morphological Chart

Purpose Function	Means 1	Means 2	Means 3	Means 4
Electrical System	Motor On/Off & Direction Control	A1 Push Button Switch	A2 Joystick Switch	
	Max, Min Control	B1 Limit Switch	B2 Sensor Switch	
Speed Torque Energy Conversion	C1	C2 Motor and Pump	C3 Two-Stage Chain Drive	C4 Belt Drive
Linear Motion System	D1 Screw and Nut	D2 Hydraulic Cylinder		
Scissor Mechanics	E_1 Center Pivot	E2 Off-Center Pivot		
Frame	F1 Channel Iron	F2 Angle Iron		

A decision matrix was formed, and the viable concepts chosen were evaluated. The decision matrix is shown in Table 14.12:

Concept 1 ($A_1B_1C_1D_1E_1F_1$) shows the highest total, so we can select concept 1 as a final choice.

Table 14.11: Selection Criteria and Importance Rating

Criterion	Rating
Auxiliary space for accessories	12%
Compactness	8%
Pleasing appearance	8%
Feeling safe and sturdy	18%
Number of components	18%
Ease of manufacture	18%
Ease of packaging (all components)	18%

Table 14.12: Decision Matrix

	Auxiliary space for accessories	Compactness	Pleasing appearance	Feel safe for rigidity	Number of components	Ease of Manufacture	Ease of packaging all components	Total
Weight	0.12	0.08	0.08	0.18	0.18	0.18	0.18	1.0
$A_1B_1C_1D_1E_1F_1$	4	8	4	9	9	8	7	7.38
	0.48	0.64	0.32	1.62	1.62	1.44	1.26	
$A_1B_1C_1D_1E_1F_2$	4	8	4	6	9	8	7	6.3
	0.48	0.64	0.32	1.08	1.62	1.44	1.26	
$A_2B_1C_1D_1E_1F_1$	3	7	3	9	9	6	7	6.74
	0.36	0.56	0.24	1.62	1.62	1.08	1.26	
$A_1B_1C_1D_1E_2F_1$	4	6	4	5	9	8	7	6.5
	0.48	0.48	0.32	0.9	1.62	1.44	1.26	
$A_1B_2C_1D_1E_1F_1$	4	8	6	9	9	6	7	7.18
	0.48	0.64	0.48	1.62	1.62	1.08	1.26	

14.4.3 Description of the Chosen Concept

A scissor lift consists of a moving platform that can be used to lift a physical load or people up and down. Linked folding supports in a crisscross "X" pattern resembling a scissor are used to achieve this vertical motion. Consider Figure 14.4.

The members AB and CD are connected by a pin-joint at E. The end C of the member CD is fixed, and the member CD can rotate only about C. When CD rotates, the end D will move along the x and y directions. The movement in the y direction is used to move the payload up and down. When the member CD is rotating about C, the member AB rotates about A, and the end B moves toward D. This permits the end A to move up and toward C. The pin at E acts as the interface to balance the structure. The actuator in Figure 14.5 is the power screw that moves B toward C, and this in turn lifts the ends A and D and thus the load.

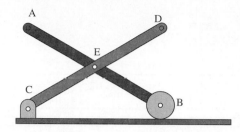

Figure 14.4: Folding Support

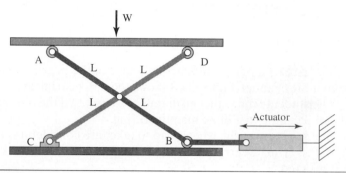

Figure 14.5: Power Screw

14.5 EMBODIMENT DESIGN

Embodiment design represents the first attempt to define the structural components of the product that act together to provide the required functions. Engineering principles, available material forms and dimensions, and manufacturing technologies are combined to form the embodiment design.

The concept assembly is shown in Figure 14.6. Based on the rough calculations that follow and the observation from the prototype, steel was selected as a material for the device. Other material was also selected as shown subsequently.

The embodiment for the frame starts with the choice of the dimensions from the available channel sections. Figure 14.7 shows the sections available (parallel flange channel 75 × 40 mm).

Figure 14.6: Concept Embodiment

Description	Weight per Meter	External Surface Area per Meter	Depth of Section	Flange Width	Flange Thickness	Web Thickness	Depth Between Flanges
	kg/m	m²/m	mm	mm	mm	mm	mm
PARALLEL FLANGE CHANNEL 75 X 40mm	5.92	0.296	75	40	6.1	3.8	62.8
PARALLEL FLANGE CHANNEL 100 X 50mm	8.33	0.385	100	50	6.7	4.2	86.6
PARALLEL FLANGE CHANNEL 125 X 65mm	11.9	0.494	125	65	7.5	4.7	110
PARALLEL FLANGE CHANNEL 150 X 75mm	17.7	0.579	150	75	9.5	6	131
PARALLEL FLANGE CHANNEL 180 X 75mm	20.9	0.638	180	75	11	6	158
PARALLEL FLANGE CHANNEL 200 X 75mm	22.9	0.678	200	75	12	6	176
PARALLEL FLANGE CHANNEL 230 X 75mm	25.1	0.737	230	75	12	6.5	206
PARALLEL FLANGE CHANNEL 250 X 90mm	35.5	0.834	250	90	15	8	220
PARALLEL FLANGE CHANNEL 300 X 90mm	40.1	0.932	300	90	16	8	268
PARALLEL FLANGE CHANNEL 380 X 100mm	55.2	1.13	380	100	17.5	10	345

Figure 14.7: Material Sections

There should be a simultaneous choice of the bearing that will act as the wheel that runs through the channel. Figure 14.8 shows the ball bearing available (305 K ball bearing).

The students chose the pin diameter 25 mm. All the pins in the product were standardized to be of 25 mm to facilitate manufacturing.

The selection of the wheels had to take into consideration what was available on the market and their load-carrying ability. A set of wheels with locking capability and 150 mm tall were selected.

Numerical calculations were made (shown in the detailed design) for the base and scissor arms given the lift size of 1500 mm and door widths of 700 mm. The material selected was steel. The base was kept 75 mm longer than the platform to accommodate the handle. From the available flat bars, a section of 10 mm × 100 mm was chosen.

The power train consists of the motor, the gearbox, and the power screw. The arrangement is shown in Figure 14.9.

Based on the component selection a part tree was developed, as shown in Table 14.13

DIMENSIONS – TOLERANCES

Bearing Number	Bore d		tolerance +0.000 mm +0.0000" to minus		Outside Diameter D		tolerance +0.000 mm +0.0000" to minus		Width C		tolerance +0.000 mm +0.0000" to minus		Fillet Radius[1]		Wt.		Static Load Rating C₀		Extended Dynamic Load Rating Ce[2]	
	mm	in.	mm	in.	mm	in.	mm	in.	mm	in.	mm	in.	mm	in.	kg	lbs.	N	lbs.	N	lbs.
300K	10	0.3937	0.008	0.0003	35	1.3780	0.011	0.00043	11	0.433	0.12	0.005	0.6	0.024	0.054	0.12	3460	780	9200	2080
301K	12	0.4724	0.008	0.0003	37	1.4567	0.011	0.00043	12	0.472	0.12	0.005	1.0	0.039	0.064	0.14	3620	815	9400	2120
302K	15	0.5906	0.008	0.0003	42	1.6535	0.011	0.00043	13	0.512	0.12	0.005	1.0	0.039	0.082	0.18	5240	1180	13300	3000
303K	17	0.6693	0.008	0.0003	47	1.8504	0.011	0.00043	14	0.551	0.12	0.005	1.0	0.039	0.109	0.24	6550	1480	15300	3450
304K	20	0.7874	0.010	0.0004	52	2.0472	0.013	0.0005	15	0.591	0.12	0.005	1.0	0.039	0.141	0.31	7800	1760	17900	4050
305K	25	0.9843	0.010	0.0004	62	2.4409	0.013	0.0005	17	0.669	0.12	0.005	1.0	0.039	0.236	0.52	12200	2750	26600	6000
306K	30	1.1811	0.010	0.0004	72	2.8346	0.013	0.0005	19	0.748	0.12	0.005	1.0	0.039	0.354	0.78	15600	3550	33900	7650
307K	35	1.3780	0.012	0.00047	80	3.1496	0.013	0.0005	21	0.827	0.12	0.005	1.5	0.059	0.472	1.04	18400	4150	37700	8500
308K	40	1.5748	0.012	0.00047	90	3.5433	0.015	0.0006	23	0.906	0.12	0.005	1.5	0.059	0.644	1.42	25900	5850	50800	11400
309K	45	1.7717	0.012	0.00047	100	3.9370	0.015	0.0006	25	0.984	0.12	0.005	1.5	0.059	0.862	1.90	31500	7100	59500	13400

Figure 14.8: Available Ball Bearings

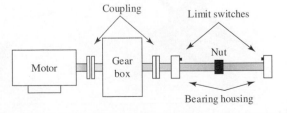

Figure 14.9: Power Train System

Table 14.13: Preliminary Parts Tree

Base	Frame	Front long arm
		Rear long arm
		Left cross arm
		Right cross arm
		Front bush plate
		Rear bush plate
		Wheel assembly
	Motor and Gearbox Mount	Left motor cross beam
		Right motor cross beam
		Base plate, nuts, and bolts
	Power Screw End Support	End support plate
		End support bearing
	Scissor Mechanism	Arms and end pins—4 Nos.
		Center pins
		Bearings—4 Nos.
Gearbox	Motor Assembly	Motor
		Power connections
	Motor Gearbox Coupling	Motor side flange
		Gearbox side flange
		Rubber sheet packing, nuts, bolts
	Gearbox	Pinion 1
		Gear 1
		Pinion 2
		Gear 2
		Input shaft
		Intermediate shaft
		Output shaft
		outer box
		Input partition plate
		Output partition plate
		Transparent top cover

Continued on next page.

Table 14.13: Preliminary Parts Tree (Continued)

Power Screw	Power Screw Gearbox Coupling	Left flange
		Right flange
		Rubber sheet packing, nuts, and bolts
	Power Screw	Threaded shaft
		Nut
		Nut mounting arm
Platform	Frame	Front long arm
		Rear long arm
		Left cross arm
		Right cross arm
		Front bush plate
		Rear bush plate
	Guards	Guard rails
		Mounting bushes
	Auxiliaries	Table for placing brushes and paint
		Extra power socket mount
Electrical and Control	Control	Limit switch for maximum
		Limit switch for minimum
		Press-on switches
		Emergency stop
	Power	Cabling
		Auxiliary power points
		Connection to switches

14.6 DETAIL DESIGN

Figures 14.10 through 14.15 show the details of the product design. Figure 14.10 shows the frame drawing. Figure 14.11 shows the lift arm and its bill of quantities. Figure 14.12 shows the arm drawing. Figure 14.13 shows the gearbox drawings. Figure 14.14 shows the motor drawings. Figure 14.15 shows the shaft drawings. Figure 14.14 shows the motor drawing, and Figure 14.15 shows the shaft drawings.

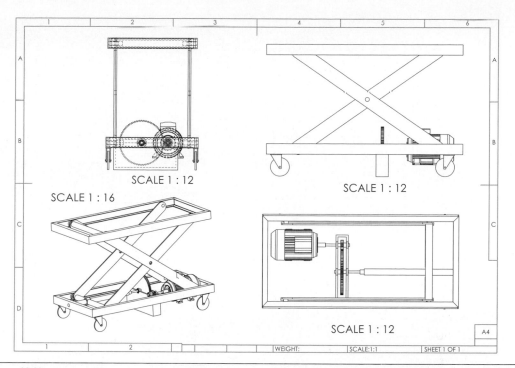

Figure 14.10: Frame Drawing

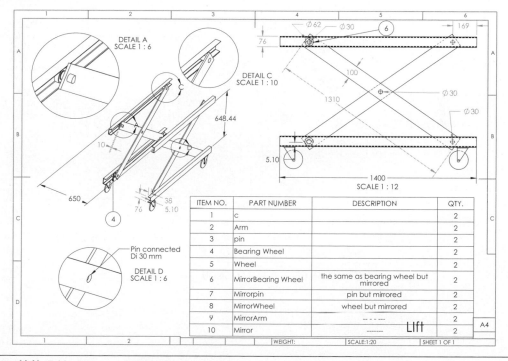

ITEM NO.	PART NUMBER	DESCRIPTION	QTY.
1	c		2
2	Arm		2
3	pin		2
4	Bearing Wheel		2
5	Wheel		2
6	MirrorBearing Wheel	the same as bearing wheel but mirrored	2
7	Mirrorpin	pin but mirrored	2
8	MirrorWheel	wheel but mirrored	2
9	MirrorArm	- - - - -	2
10	Mirror	--------	2

Figure 14.11: Lift Arm and Bill of Quantities

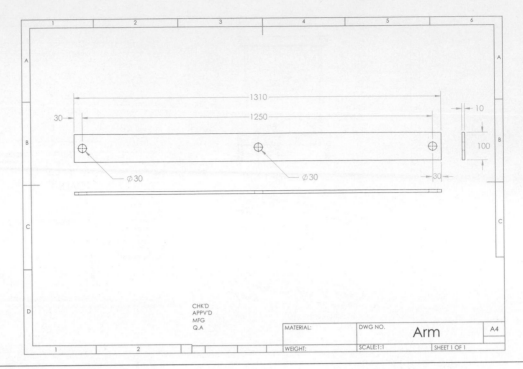

Figure 14.12: Arm Drawing

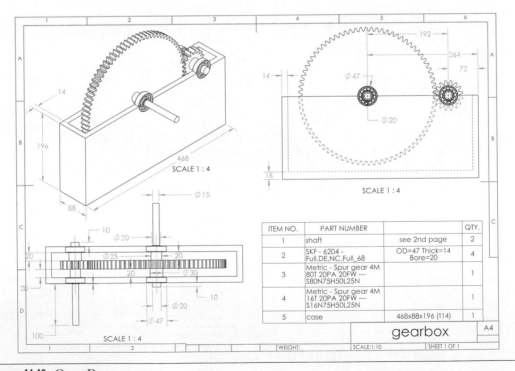

Figure 14.13: Gear Box

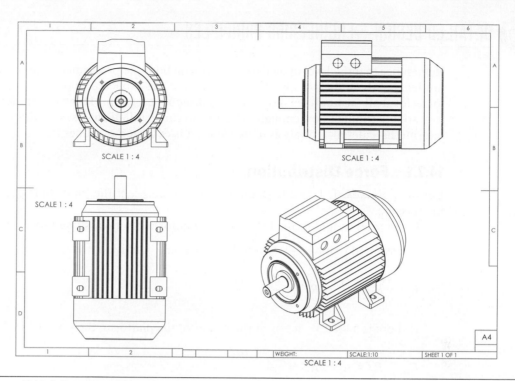

Figure 14.14: Motor Drawings

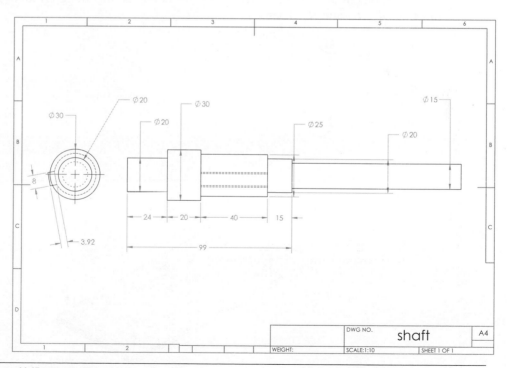

Figure 14.15: Shaft Drawing

14.7 DETAILED DESIGN—ENGINEERING PRINCIPLES

Remember that the level of analysis depends on both the problem and the student's knowledge. Students early in their study of engineering are not expected to possess sufficient knowledge to perform all the analyses shown in the subsequent sections. However, students who are about to graduate should know enough to be able to run detailed analyses. In this section we show an intermediate detailed analysis to gauge what is expected of students.

14.7.1 Force Distribution

Let the points H, I, J, and K (Figure 14.16) to represent the joints that transmit the loads to the scissors.

If the total weight of the lift is B_y, the platform can be assumed to have $\dfrac{B_y}{2}$, and the base can be assumed to have $\dfrac{B_y}{2}$. When this is the only condition, the reactions will be as follows:

$$R_H = R_I = R_J = R_K = \frac{B_y}{2}$$

If there is a load W acting at the center of the platform, the reactions will be

$$R_H = R_I = R_J = R_K = \frac{B_y + W}{2}$$

Similarly, if a moment acts on it as shown in Figure 14.17 this will induce reactions.

The moment M_Z can be divided into two halves acting along IH and JK, inducing reactions R_H, R_I, R_J, and R_K. Then

$$R_I = -R_H = \frac{M_Z}{2a}$$

Similarly

$$R_J = -R_K = \frac{M_Z}{2a}$$

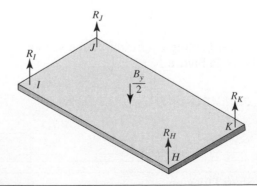

Figure 14.16: Load Distribution

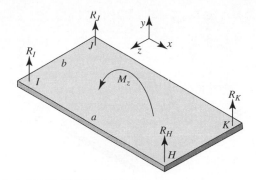

Figure 14.17: Load and Moment

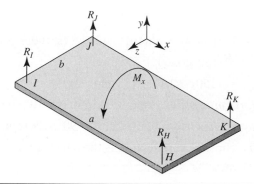

Figure 14.18: Force and Moment

In a similar fashion, consider the moment M_X as shown in Figure 14.18. Then

$$R_I = -R_J = \frac{M_Z}{2b}$$

Similarly,

$$R_H = -R_K = \frac{M_Z}{2b}$$

A force acting on one point can be moved to another point with the addition of a moment; thus, the general loading condition of the scissor lift can be represented by the forces and moments shown in Figure 14.19.

14.7.2 Load and Stability Analysis for the Frame

1. Find L, h, l_{max}, l_{min}:
   a. $1000 = 2L * \sin(60)$
   b. $2L = 1150\,\text{mm}$
   c. $L = 577\,\text{mm}$

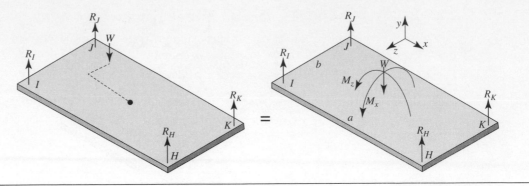

Figure 14.19: Force and Moment Diagram

d. $h = 2L\,[\sin 60 - \sin 10] = 1150 * (\sin 60 - \sin 10) = 796\,\text{mm}$

e. $I_{min} = 1150\,\cos 60 = 575\,\text{mm}$

f. $I_{max} = 1150\,\cos 10 = 1133\,\text{mm}$

g. *Horizontal travel distance* $= 1150\,[\cos 10 - \cos 60] = 558\,\text{mm}$

h. *Stowed height* $= [1150\,\sin 10] + 90 = 290\,\text{mm}$

i. The chosen cross section is the parallel flange as shown in Figure 5.2.1

2. Reactions:

AB

$\Sigma M_B = 0$

$(\dfrac{-w}{2} * \cos 60 * 900) - F_x \sin 60 * 325 - F_y \cos 60 * 325 = 0$

$\Sigma F_x = 0 - R_{x2} \rightarrow\ + F_x = 0 \rightarrow R_{x2} = F_x$

$\Sigma F_y = 0 \rightarrow\ + F_y - \dfrac{w}{2} + R_{y1} = 0 \rightarrow F_y = -R_{y2} + \dfrac{w}{2}$

DC

$\Sigma M_C = 0$

$(\dfrac{-w}{2} * \cos 60 * 900) - F_x \sin 60 * 325 - F_y \cos 60 * 325 = 0$

$\Sigma F_x = 0\ R_{x1} \rightarrow -F_x = 0 \rightarrow R_{x1} = F_x$

$\Sigma F_y = 0 \rightarrow -F_y - \dfrac{w}{2} + R_{y1} = 0 \rightarrow F_y = R_{y1} - \dfrac{w}{2}$

$F_x = R_{x1} = R_{x2} = \dfrac{18w}{13\tan\varnothing} = \dfrac{18 * 130 * 9.81}{13\tan 60} = 1019.5\ \text{N}$

$R_{y1} = R_{y2} = \dfrac{w}{2} = \dfrac{130 * 9.81}{2} = 638\ \text{N}$

14.7.3 Power and Rotational Speed Requirement

When the lift is ascending, the total potential and kinetic energy of the lift and load increase. If constant speed is assumed, kinetic energy remains constant.

The weight of the lift B_y can be assumed to be half on the platform and half on the base frame. Then the energy gained by the system $= \left(H_y + \dfrac{B_y}{2} \right) dh$. Work done by the screw $= F \times dl$. This in unit time can be given as

$$F = \left(H_y + \frac{B_y}{2} \right) \frac{dh}{dl}$$

But $h = 2L \sin \theta$ and hence $\dfrac{dh}{d\theta} = 2L \cos \theta$. Similarly, $l = 2L \cos \theta$ and hence $\dfrac{dl}{d\theta} = -2L \sin \theta$.

Therefore, $\dfrac{dh}{dl} = \dfrac{2L \cos \theta}{2L \sin \theta} = -\cot \theta$; therefore $F = -\left(H_y + \dfrac{B_y}{2} \right) \cot \theta$.

The force is at maximum when the angle is 10°. In addition, $F \times \dfrac{dl}{dt} = \left(H_x + \dfrac{B_y}{2} \right) \dfrac{dh}{dt}$. The ascending speed of the lift $\dfrac{dh}{dt}$ will be given as something like 5 cm per second. Power required by the motor $= \left(H_y + \dfrac{B_y}{2} \right) \dfrac{dh}{dt}$. From this $\dfrac{dl}{dt}$ can be calculated, and this is the distance traveled in a unit time. Hence, $\dfrac{dl}{dt} = \text{Lead} \times \text{RPS}$ of the screw

$$\frac{dh}{dl} = -\cot \theta = \frac{dh}{dt} \times \frac{dt}{dl}$$

and hence $\dfrac{dt}{dl} = \dfrac{-\cot \theta}{dh/dt} = \dfrac{-\cot \theta}{5}$.

$$\frac{dl}{d\theta} = \frac{dl}{dt} \times \frac{dt}{d\theta} = -2L \sin \theta$$

$$dt = \frac{2L}{\text{Lead} \times \text{RPS}} \sin \theta \, d\theta$$

$$\int_0^T dt = \frac{-2L}{\text{Lead} \times \text{RPS}} \int_{10}^{60} -\sin \theta \, d\theta$$

$$T = \frac{2L}{\text{Lead} \times \text{RPS}} \times [\cos 10 - \cos 60]$$

If *Lead* is 8 mm and *T* is roughly 20 sec (1 meter height *h* at 50 mm per second), then

$$RPS = \frac{2L}{\text{Lead} \times T} \times [\cos 10 - \cos 60] = \frac{2L}{8 \times 20} \times [\cos 10 - \cos 60]$$

Therefore, the speed required from the output shaft of the gearbox $= 3.8 \times 60 = 228$ rpm. If the motor is running at 1400 rpm, the gear reduction required is $\dfrac{1400}{228} = 6.14$. This is high, assuming that a gear reduction of 5 will give 280 rpm (i.e., 4.67 *RPS*). This will give a rising time of 16 seconds.

Power required from the motor:

$$P = \left(H_y + \frac{B_y}{2}\right)\frac{dh}{dt} = \left(130 * 9.81 + \frac{93.529 * 9.81}{2}\right) * 0.05 = 86.703 \text{ W}$$

$$RPS = \frac{2L}{\text{lead} * T}(\cos 10 - \cos 60) = \frac{1150}{0.05 * 2}(\cos 10 - \cos 60) = 556$$

$$\text{Speed gear box} = RPS * 60\frac{\sec}{\min} = 556 * 60\frac{\sec}{\min} = 33452 \text{ rpm}$$

$$\text{Reduction speed} = \frac{\text{rpm from motor}}{\text{speed of the gear box}} = \frac{86.703}{33452} = 0.002$$

14.7.4 Power Screw Calculations

Thread properties

Pitch/lead (P) = 5 mm

Minor diameter = 34 mm

Major diameter = 34 + 5 = 39 mm

$$d_m = \frac{34 + 39}{2} = 36.5 \text{ mm}$$

Lead Angle (λ):

$$\tan \lambda = \frac{P}{\pi \times dm} = \frac{5}{\pi \times 36.5} = 0.0434$$

$$\lambda = \tan^{-1}(0.0434) = 2.47°$$

Thread Angle($\varnothing\varnothing$):

Assume coefficient of friction (f) = 0.15

Thread angle $\varnothing = \dfrac{29}{2} = 14.5°$

Stress calculations

Horizontal force required by the screw

$$F = -\left(Hy + \frac{By}{2}\right)\cot\theta = -400 \times 10 \times \cot(10) = 22{,}684 \text{ N}$$

Torque to raise the load (T_u):

$$T_u = \frac{F\,dm}{2}\left[\frac{\cos\varnothing \tan\lambda + f}{\cos\varnothing - f\tan\lambda}\right] = \frac{22{,}684 \times 36.5}{2}\left[\frac{\cos(14.5)\tan(0.0434) + 0.15}{\cos(14.5) - (0.15)\tan(0.0434)}\right]$$

$$= 413{,}983\left[\frac{0.1920}{0.9616}\right] = 82{,}658.8 \text{ N.mm}$$

Torsion shear stress τ_{max}

$$\tau_{max} = \frac{T_C}{J} = \frac{T_u \times 17}{\frac{\pi}{32} \times 34^4} = 10.73 \text{ MPa}$$

Axial stress σ

$$\sigma = \frac{22,684}{\pi \times 17^2} = 24.98 \text{ MPa}$$

Mohr's Circle

$\sigma_x = 24.98 \text{ MPa}, \sigma_y = 0, T_{xy} = 10.73 \text{ MPa}$

$$\sigma_{avg} = \frac{24.98 + 0}{2} = 12.49 \text{ MPa}$$

The center $C = (12.49, 0)$

The reference point $A\,(\sigma_x, T_{xy}) = (24.98, 10.73 \text{ MPa})$

Resulting maximum shear stress using Mohr's circle radius

$$R = \sqrt{\left(\frac{\sigma}{2}\right)^2 + (\tau_{max})^2} = \sqrt{\left(\frac{24.98}{2}\right)^2 + (10.73)^2} = 16.49 \text{ MPa}$$

According to TRESCA, maximum permissible shear stress

$$\frac{\textit{Yield stress}}{2} = \frac{207}{2} = 103.5 \text{ MPa}$$

Safety factor available with the least steel is $\dfrac{103.5}{16.49} = 6.3$

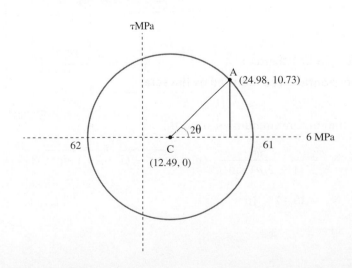

Buckling check

$$\text{Radius of gyration} = \sqrt{\frac{I}{A}} = \sqrt{\frac{\frac{\pi}{64}(17^4)}{\frac{\pi}{4}(17^2)}} = 18.06 \text{ mm}$$

$$\text{Slenderness ratio} = \frac{Length}{Radius\ of\ Gyration} = \frac{600}{18.06} = 33.22$$

Column constant for this material is 140.5
Hence, there will be no column buckling.

Bearing pressure

Assuming that there are 6 threads in contact, bearing pressure $= \dfrac{F}{n\pi dt} = \dfrac{22,684}{6 \times \pi \times 36.5 \times 2.5} =$ 13.188 MPa. Permissible value is 17 MPa. Hence the number of threads in contact should be 6 or more. The nut should be 40 mm long.

Shear stress at the root

Shear stress at the root $= \dfrac{F}{n\pi\,dmin\,t} = \dfrac{22,684}{6 \times \pi \times 34 \times 2.5} = 14.15$ MPa. This has a safety factor for the worst grade of steel $= \dfrac{207}{2 \times 14.15} = 7.3$

Final Design

The nut should be 40 mm long, having 8 threads in contact. The screw should have a 39 mm outer diameter. The length of the threaded portion is about 600 mm. Ball bearings and bearing housings must be designed to take an axial load of 22,684 N, and the safety factor may be 2.5.

The nut need not be checked, given that the nut and bolt are made of the same material.

14.7.5 Analysis of Gear Box and Shafts

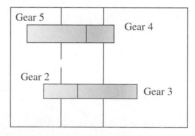

To find minimum number of teeth:

$$N_p = \frac{2k}{(1 + 2m)\ \sin^2\varnothing}(m + \sqrt{m^2 + (1 + 2m)\ \sin^2\varnothing})$$

$$N_p = 15.44 = 16$$

$$m = 4$$

Table 14.14: Gear Characteristics

Gear No.	N	D (mm)	W (rpm)	V (m/s)	Wt
2	25	56	1500	4.398	250
3	72	144	583.3	4.398	250
4	40	80	583.3	2.4434	450.18
5	100	200	233.3	2.4434	450.18

Wear on Gear 2

$$\sigma_C = Z_E \left(W_t k_o k_v k_s \frac{k_H}{d_p * b} \frac{Z_R}{Z_I} \right)^{1/2}$$

$ZE = 191 \ (MPa)^{1/2}$

$Wt = 250$

$KO = 1$

$$k_v = \frac{A + \sqrt{200V}}{A} = \frac{65.1 + \sqrt{200 * 4.398}}{65.1} = 1.455$$

$Ks = 1$

$b = 20 \ mm$ (Assumption)

$dp = 56 \ mm$

$k_H = C_{mf} = 1 + C_{mc}(C_{pf}C_{pm} + C_{ma}C_e) = 1 + 1((0.0107 * 1) + (0.15 * 1)) = 1.1607$

$$C_{pf} = \frac{b}{10 * d} - 0.025 = 0.0107$$

$Z_R = 1$

$$m_G = \frac{N_G}{N_p} = 72 / 25 = 2.88$$

$$ZI = \frac{\cos\theta \ \sin\theta}{2mN} \left(\frac{mG}{mG + 1} \right) = 0.822$$

$L2 = (12000) * (60) * (1500) = 1 * 10^9$

$ZN = 0.85$

$Zw = 1$

$Y_\theta = 1$

$Y_Z = 1$

$\sigma_c = 129.3 \ MPa$

From Table (14-14), carburized & hardened, grade 2 steel

$$n_c = \frac{1551}{11.99} = 11.9$$

Bending on Gear 2

$$\sigma = W_t k_o k_v k_s \frac{1}{b*mt} \frac{k_H k_B}{YJ}$$

$Wt = 250\ N$

$Ko = 1$

$Kv = 1.455$

$Ks = 1$

$b = 20\ \text{mm (assumption)}$

$mt = 4$

$KH = 1.1607$

$KB = 1$

$YJ = 0.36$

$\sigma = 14.65\ MPa$

From Table 14-13, flame hardened grade 1 steel

$$n = \frac{310}{39.79} = 21.1$$

Gear 3 Bending and Wear

Same as Gear 2, but

$YJ = 0.41$

$$\sigma_c = Z_E \left(W_t k_o k_v k_s \frac{k_H}{d_p*b} \frac{Z_R}{Z_I} \right)^{\frac{1}{2}} = 129.3 \text{ MPa (same for Gear 3)}$$

$$n_c = \frac{1551}{129.3} = 11.99$$

$$\sigma = W_t k_o k_v k_s \frac{1}{b*mt} \frac{k_H k_B}{YJ} = 12.8 \text{ MPa}$$

$$n = \frac{310}{12.8} = 24.2$$

Wear on Gear 4

$$\sigma_c = Z_E \left(W_t k_o k_v k_s \frac{k_H}{d_p*b} \frac{Z_R}{Z_I} \right)^{\frac{1}{2}}$$

$$ZE = 191 \,(\text{MPa})^{\frac{1}{2}}$$

$$Wt = 450.18$$

$$Ko = 1$$

$$k_v = \frac{A + \sqrt{200V}}{A} = \frac{65.1 + \sqrt{200 * 2.4434}}{65.1} = 1.339$$

$$Ks = 1$$

$$b = 20 \text{ mm (assumption)}$$

$$dp = 80 \text{ mm}$$

$$k_H = C_{mf} = 1 + C_{mc}(C_{pf}C_{pm} + C_{ma}C_e) = 1 + 1((0 * 1) + (0.15 * 1)) = 1.15$$

$$C_{pf} = \frac{b}{10 * d} - 0.025 = 0$$

$$ZR = 1$$

$$mG = \frac{N_G}{Np} = 100/40 = 2.5$$

$$ZI = \frac{\cos\theta \sin\theta}{2mN}\left(\frac{m_G}{m_G + 1}\right) = 0.0286$$

$$L4 = (12000) * (60) * (583.3) = 4.1 * 10^8$$

$$Z_N = 0.36$$

$$Zw = 1$$

$$Y_\theta = 1$$

$$Y_Z = 1$$

$$\sigma_c = 743.4 \text{ MPa}$$

From Table (14-6), carburized and hardened, Grade 2

$$n_c = \frac{1551}{743.4} = 2.08$$

Bending on Gear 4

$$\sigma = W_t k_o k_v k_s \frac{1}{b * mt} \frac{k_H k_B}{YJ}$$

$$Wt = 450.18 \, N$$

$$Ko = 1$$

$$Kv = 1.339$$

$$Ks = 1$$

$$b = 20 \text{ mm}$$

$$mt = 4$$

$$KH = 1.15$$

$$KB = 1$$

$$YJ = 0.42$$

$$\sigma = 20.63 \text{ MPa}$$

From Table 14-3, Carburized and hardened, Grade 2

$$n = \frac{448}{20.63} = 21.7$$

Gear 5 Bending and Wear

Same as Gear 4 but

$$YJ = 0.46$$

$$\sigma_c = Z_E \left(W_t k_o k_v k_s \frac{k_H}{d_p * b} \frac{Z_R}{Z_I} \right)^{\frac{1}{2}} = 743.4 \text{ MPa } (same \ for \ Gear \ 4)$$

$$n_c = \frac{1551}{743.4} = 2.08$$

$$\sigma = W_t k_o k_v k_s \frac{1}{b * mt} \frac{k_H k_B}{YJ} = 18.83 \text{ MPa}$$

$$n = \frac{448}{18.83} = 23.79$$

Summary of the Gear design:

Gear 2, Grade 1 flame-hardened, Sc = 1172 MPa and St = 310 MPa d2 = 56 mm face width = 20 mm

Gear 3, Grade 1 flame-hardened, Sc = 1172 MPa and St = 310 MPa d3 = 72 mm face width = 20 mm

Gear 4, Grade 2 Carburized and hardened Sc = 1551 MPa and St = 448 MPa d4 = 40 mm face width = 20 mm

Gear 5, Grade 2 Carburized and hardened Sc = 1551 MPa and St = 448 MPa d5 = 100 mm face width = 20 mm

Shafts Analysis

Gear 2:

N = 25

d = 56 mm

Gear 3:

N = 72

d = 144

Gear 4:

N = 40

d = 80

Gear 5:

N = 100

d = 200

Velocity and *w* Analysis:

$$V_2 = V_3 = \omega r = \frac{1500 * \pi * 56}{1000 * 60} = 4.398 \ m/s$$

$$\omega_2 N_2 = \omega_3 N_3$$

$$1500 * 25 = 72 \ \omega_3$$

$$\omega_3 = 583.33 \ \text{rpm}$$

$$V_4 = V_5 = \omega r = \frac{583.33 * \pi * 80}{1000 * 60} = 2.4434 \ m/s$$

$$\omega_4 N_4 = \omega_5 N_5$$

$$583.33 * 40 = 100 \ \omega_3$$

$$\omega_5 = 233.332 \ \text{rpm}$$

Force analysis:

$$W_{t(2,3)} = \frac{P}{V} = \frac{1.1 * 1000}{4.398} = 250 \ N$$

$$F_{r(2,3)} = w_t \ \tan(\phi) = 250 \tan(20) = 91.03 \ N$$

$$W_{t(4,5)} = \frac{P}{V} = \frac{1.1 * 1000}{2.4434} = 450.18 \ N$$

$$F_{r(4,5)} = w_t \ \tan(\phi) = 450.18 \tan(20) = 163.85 \ N$$

Shaft of gear 2:

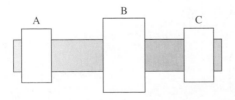

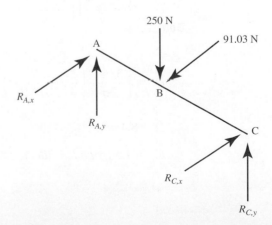

$$\sum M_A^x = 0$$

$$0 = 250(20) - R_{c,y}(40)$$

$$R_{c,y} = 125 \text{ N} = R_{A,y}$$

$$\sum M_A^y = 0$$

$$0 = 91(20) - R_{c,x}(40)$$

$$R_{c,x} = 45.5 \text{ N} = R_{A,x}$$

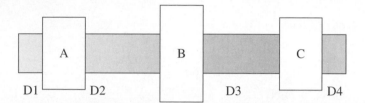

$$S'_n = S_n C_s C_R$$

Assuming material to be used as 1120 hot rolled with su = 379 MPa and sy = 207 MPa, IPR we obtain Sn from Figure 5-8 (page 175).

$$S_n = 150 \text{ MPa}$$

use C_{-s} as 0.75 as an estimate and with 0.99 reliability C_R = 0.81

$$S'_n = 91.125 \text{ MPa}$$

Calculate the torque

$$T = \frac{P}{2\pi n} = \frac{1.1 * 1000}{2 * \pi * \dfrac{1500}{60}} = 7 \ N*m$$

Calculate the diameter using safety factor of 2

$$D_1 = \left(\frac{32 * N}{\pi} \left(\left(\frac{k_f * M_a}{s_f} \right)^2 + \frac{3}{4} \left(\frac{T_m}{s_y} \right)^2 \right)^{0.5} \right)$$

$$D_1 = \left(\frac{32 * 2}{\pi} \left(\frac{3}{4} \left(\frac{7}{207 * 10^6} \right)^2 \right)^{0.5} \right)^{\frac{1}{3}}$$

$$D = 8.4 \ mm$$

$$K_f = 1.5 \ (rounded \ fillet)$$

$$M_B = \sqrt{M_x^2 + M_y^2} = 2660.47\,\text{N} * \text{mm}$$

$$D_2 = \left(\frac{32 * 2}{\pi}\left(\left(\frac{1.5 * 2660.47 * 10^{-3}}{91.125 * 10^6}\right)^2 + \frac{3}{4}\left(\frac{7}{207 * 10^6}\right)^2\right)^{0.5}\right)^{\frac{1}{3}} = 8.45\ \text{mm}$$

$$k_f = 2.5\ (sharp\ fillet)$$

$$D_3 = \left(\frac{32 * 2}{\pi}\left(\left(\frac{2.5 * 2660.47 * 10^{-3}}{91.125 * 10^6}\right)^2 + \frac{3}{4}\left(\frac{7}{207 * 10^6}\right)^2\right)^{0.5}\right)^{\frac{1}{3}} = 8.5\ \text{mm}$$

$$V = \sqrt{V_x^2 + V_y^2} = 133\,\text{N}$$

$$k_f = 2.5\ (sharp\ fillet)$$

$$D_4 = \left(\frac{2.94 * K_t * V * N}{S_n'}\right)^{\frac{1}{2}} = \left(\frac{2.94 * 2.5 * 84.98 * 2}{91.125 * 10^6}\right)^{\frac{1}{2}} = 4.63\ \text{mm}$$

Shaft of gear 5:

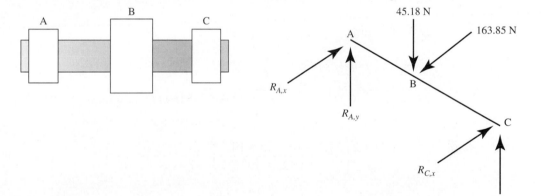

$$\sum M_A^x = 0$$

$$0 = 45.18(20) - R_{C,y}(40)$$

$$R_{C,y} = 22.59\ N = R_{A,y}$$

$$\sum M_A^y = 0$$

$$0 = 163.85(20) - R_{c,x}(40)$$

$$R_{C,x} = 81.925\ N = R_{A,x}$$

$$S_n' = S_n C_s C_R$$

Assuming material to be used as 1120 hot rolled with su = 379 MPa and sy = 207 MPa, we obtain S_n from Figure 5-8 (page 175).

S_n = 150 MPa

Use C_s as 0.75 as an estimate and with 0.99 reliability C_R = 0.81

S'_n = 91.125 MPa

Calculate the torque

$$T = \frac{P}{2\pi n} = \frac{1.1 * 1000}{2 * \pi * \frac{233.332}{60}} = 45 \text{ N} * \text{m}$$

Calculate the diameter using safety factor of 2

$$D_1 = \left(\frac{32 * N}{\pi} \left(\left(\frac{k_f * M_a}{S_f} \right)^2 + \frac{3}{4} \left(\frac{T_m}{S_y} \right)^2 \right)^{0.5} \right)^{\frac{1}{3}}$$

$$D_1 = \left(\frac{32 * 2}{\pi} \left(\frac{3}{4} \left(\frac{45}{207 * 10^6} \right)^2 \right)^{0.5} \right)^{\frac{1}{3}}$$

D = 15.6 mm

k_f = 1.5 (*rounded fillet*)

$$M_B = \sqrt{M_x^2 + M_y^2} = \sqrt{451.8^2 + 1638.5^2} = 1699.65 \text{ N} * \text{mm}$$

$$D_2 = \left(\frac{32 * 2}{\pi} \left(\left(\frac{1.5 * 1699.65 * 10^{-3}}{91.125 * 10^6} \right)^2 + \frac{3}{4} \left(\frac{45}{207 * 10^6} \right)^2 \right)^{0.5} \right)^{\frac{1}{3}} = 14.99 \text{ mm}$$

k_f = 2.5 (*sharp fillet*)

$$D_3 = \left(\frac{32 * 2}{\pi} \left(\left(\frac{2.5 * 1699.65 * 10^{-3}}{91.125 * 10^6} \right)^2 + \frac{3}{4} \left(\frac{45}{207 * 10^6} \right)^2 \right)^{0.5} \right)^{\frac{1}{3}} = 15.11 \text{ mm}$$

$$V = \sqrt{V_x^2 + V_y^2} = 84.98 \text{ N}$$

k_f = 2.5 (*sharp fillet*)

$$D_4 = \left(\frac{2.94 * K_t * V * N}{S'_n} \right)^{\frac{1}{2}} = \left(\frac{2.94 * 2.5 * 84.98 * 2}{91.125 * 10^6} \right)^{\frac{1}{2}}$$

CHAPTER 15

SELECTION OF DESIGN PROJECTS

15.1 DESIGN PROJECT RULES

This chapter introduces sample-design projects. The following rules apply to all design projects. Specific rules will also be listed for each project.

1. Students will apply a systematic design process to build the product.
2. Students will follow all of the design stages.
   a. Requirements
   b. Product concept
   c. Solution concept
   d. Embodiment design
   e. Detail design
3. Student will submit complete design documentation that includes:
   a. Design brief
   b. Customer verbatim and interpreted and prioritized requirements
   c. Requirements metric (quality function) matrix
   d. Competitive benchmarking in terms of the metrics
   e. House of quality chart
   f. Specifications with importance ratings
   g. Criteria for concept selection and their importance ratings
   h. Function diagram
   i. Concepts for consideration
   j. Pugh's matrix and decision matrix
   k. Chosen conceptual design as a sketch and scheme in words
   l. Embodiment design sequence and the calculations
   m. Complete set of production drawings
   n. Chosen analyses
   o. Complete details of the analyses
4. Students will calculate the cost and use the break-even point to find the selling price of their product.

5. Students will use the machine shop to build their design project.

6. Students will use the Gantt, CPM, or PERT methods for scheduling the different events in the design process.

7. Students will be introduced to the different manufacturing processes during the course of the design.

8. The device must be a stand-alone unit. No assistance from the operator will be allowed after actuation.

9. Any device that is available through a retailer will be disqualified. The device must be designed and constructed from the component level by team members.

10. Teams will be composed of four members, to be selected and decided on by the students.

11. Instructors can choose to allow students to evaluate each other's contribution. Typically, student evaluations of team members are kept confidential and are turned in weekly. These evaluations typically are considered for grading. It demonstrates to the instructor the value of each member's contribution to the team's work as viewed by its members.

12. Commendations and prizes may be awarded. Judging will be based on the following criteria:
    a. Performance
    b. Appearance
    c. Safety
    d. Other general engineering principles

13. Students are reminded of the general rules and honor codes of their respective institutions.

14. Students are encouraged to be creative and not restrict themselves to ideas that are currently available. This exercise should be a positive learning experience for the students; competitiveness is encouraged. However, the spirit of fairness must always prevail, and students are requested to abide by the rules and decisions of the judges and instructor.

15. The instructor reserves the right to change or add rules to fit student learning needs. Students will be informed of the changes when they occur.

15.2  ALUMINUM CAN CRUSHER

15.2.1 Objective

Design and build a device or machine that will crush aluminum cans. The device must be fully automatic such that all the operator needs to do is load cans into it. The device should switch on automatically, crush the can automatically, eject the crushed can automatically, and switch off automatically (unless more cans are loaded).

15.2.2 Specifications

- The device must have a continuous can-feed mechanism.
- Cans should be in good condition when supplied to the device (i.e., not dented, pressed, or slightly twisted).
- The can must be crushed to one-fifth of its original volume.
- The maximum dimensions of the device are not to exceed $20 \times 20 \times 10$ cm.

- Performance will be based on the number of cans crushed in one minute.
- Children must be able to operate the device safely.
- The device must be a stand-alone unit. No assistance from the operator will be allowed after actuation.
- Any device that is available through a retailer will be disqualified. The device must be designed and constructed from the component level by team members.

15.3 COIN SORTING CONTEST

15.3.1 Objective

Design and build a machine or device that will sort and separate different types of coins. A bag of assorted coins will be emptied into the device, and the contestants will be given one minute to sort and separate as many coins as possible. Coins will be disbursed in a bag containing an assortment of U.S. coins (pennies, nickels, dimes, and quarters) and other foreign coins. The device must be able to sort out and separate only the U.S. coins. The volume of coins that the bag will contain is approximately 0.3% of a cubic meter. The coins will be deposited into the device at the start time. The number of coins sorted and separated will be counted at the end of time (60 seconds). Each incorrectly separated coin will be recorded as a penalty. Each coin that is physically damaged by the device will also be recorded as a penalty. Coins that remain in the internal body of the device at the end of time will not be counted. Each round will be run independently such that each team goes one at a time and each turn is timed. Each team will compete twice on two different cycles, based on random selection. The better of the two performances will be considered in the contest.

15.3.2 Constraints

The total cost of the device should not exceed the given budget ($200). A tolerance of 10% may be allowed after discussions with the instructor. A complete breakdown of all costs must be included in the report. Receipts must be supplied. The device must be a stand-alone unit. No assistance from the operator will be allowed after actuation. The operator will be allowed to power the device either manually or by any other means. Any device that is available through a retailer will be disqualified. The device must be designed and constructed from the component level by team members. If there are any questions, the instructor should be consulted for clarification.

15.4 MODEL (TOY) SOLAR CAR

15.4.1 Objective

Design and build a model solar car that can be used as a toy by children. The primary source of energy is solar. The following are some methods to stimulate your thinking; you are not limited to these methods.

- Direct solar drive
- Battery storage (Students should note that on the day of the competition, the battery must be completely discharged. They will be given half an hour to charge the battery, using solar energy.)
- Air or steam engine (Students will be allowed to use photocells, lenses, and the like for the purpose of heating the air or for steam generation.)

15.4.2 Specifications

- The car will be required to travel 30 miles in a straight line. Extra credit will be given to cars that are remote controlled. The final competition will take place in a school parking lot.

- Performance will be based on the car that travels the distance of 30 miles in the least amount of time.

- The car must travel a distance of at least 3 miles, or it will be considered damaged. Damaged cars will get a failing grade in the performance section of the competition.

- The dimensions of the car are limited to the following maximum values: length—30 cm, width—15 cm, and height—15 cm. The car must be safe for children to use.

- Any car that is available through a retailer will be disqualified. The car must be designed and constructed from the component level by team members.

15.5 WORKSHOP TRAINING KIT

The Stirling engine kit that is used for a mechanical engineering tools class is becoming more expensive and difficult to obtain on time. The goal of using that kit was to provide students with hands-on experience using the machine shop while at the same time producing a machine that is able to convert heat into work. This project now calls for students to design another kit that can be designed and manufactured at the College of Engineering and will serve the same goal of providing hands-on experience while building a machine that is able to convert one form of energy to another. Design and build a machine shop training kit that can be used for a mechanical engineering tools class.

15.5.1 Specifications

- The kit must have teaching and training value for the students. The kit must be innovative and must utilize an engineering principle to produce work (such as the first law of thermodynamics).

- The kit must have identified training values for machine shop tools.

- The dimensions of the kit when assembled are limited to the following maximum values: length—30 cm, width—15 cm, and height—15 cm.

- The kit, once assembled, must be safe to use for students in kindergarten through grade 12.

- Any kit that is available through a retailer will be disqualified. The kit must be designed and constructed from the component level by team members.

15.6 SHOPPING CARTS

As technology advances and new materials are synthesized, certain products still are not being modified and modernized. Among those are the shopping carts used in grocery stores. As many of you may have observed, there is a tendency to conserve parking space by not designating a return cart area. Leaving carts in the parking lots may lead to serious accidents and car damage. Furthermore, many customers do not fill their carts when shopping; however, they do not like to carry baskets. Other customers like to sort products as they shop.

Design and build a new shopping cart that can be used primarily in grocery stores. The shopping cart should solve the common problems with the available carts.

15.6.1 Specifications

You must adhere to the following rules:

- Conduct surveys to measure customer needs for shopping carts.
- The shopping cart should be safe for human operation.
- The dimensions of the cart should match those of existing carts.
- The cart should be easy to operate by children of age 7 and older as well as senior citizens.
- The cart should have features that accommodate children while shopping.
- The cart should be able to accommodate both large and small items.
- Any cart that is available through a retailer will be disqualified.

15.7 MECHANICAL VENTS

Most houses have vents that open and close manually without any central control. Cities across the country advise the use of such vents to save energy. In most cases, household occupants do not use the entire house at the same time; the tendency is to use certain rooms for a long time. For example, the family room and dining room may be used heavily, while the living room and kitchen are used at certain hours of the day. To cool or heat a room, the system needs to work to cool or heat the entire house. Energy saving can be enhanced if the vents of unused rooms are closed; this will push the hot or cold air to where it is needed most while reducing the load of the air conditioning system.

Design and build a remotely operated ventilation system that will be connected to an existing duct system. The house layout will be given upon request.

15.7.1 Specifications

You must adhere to the following rules:

- The system must be safe and easily adapted to existing ventilation systems.
- The system must work for at least seven exits.
- The venting system must accommodate high and low operating temperatures.
- The system must be easy to operate by household members.
- Central units of operation are encouraged as well as remote-controlled units.
- Any system that is available through a retailer will be disqualified. The system must be designed and constructed from the component level by team members.

15.8 ALL-TERRAIN VEHICLE

The objective of this project is to design and build a model for an all-terrain vehicle (ATV). The size of the model should not exceed $15 \times 15 \times 15 \, \text{cm}^3$. The power of the model vehicle is flexible, but safety must prevail. Success of this model design may lead to a new market for ATVs, such as

- A toy for children.
- A demonstration model for object collection on other planets.
- A demonstration model for law enforcement and military purposes.

The designed model must be able to handle slopes of 45 to 60 degrees, go over rocks that are the height of the wheels, and move in mud and dry dirt.

15.9 POCKET-SIZED UMBRELLA

As technology advances, the public is searching for more convenient products to replace existing ones. In this project you are requested to design and build an umbrella that can fit in a normal pocket when folded. The following must be adhered to when designing the new umbrella:

- The umbrella must be able to shield the prospective customer from heavy rain.
- The response time to open and fold the umbrella must be reasonable and close to the response time of an existing full-size umbrella.
- The weight of the umbrella should be reasonable to fit in a pocket without damaging the pocket.

15.10 MODEL OF THERAPEUTIC WHEELCHAIR

In homes of the elderly and infirm and in medical therapy clinics, a nurse is requested to help residents or patients walk on a daily basis. Most of the residents and patients are unable to walk alone. They use standard wheelchairs most of the day. Because of cost factors, only one nurse can be assigned to each patient. The patient usually leaves the wheelchair and walks away from it with the help of the nurse. In this situation, a problem may arise if the patient needs the wheelchair urgently while he and the accompanying nurse are away from it. The nurse cannot leave the patient unattended to bring the chair, nor can she carry the patient back to the wheelchair. You are required to design and build a prototype wheelchair that will provide the necessary solution for the nurse. The model should not exceed $30 \times 30 \times 30 \text{ cm}^3$. You need to consider that the patient or resident may be walking outdoors or indoors. Usually the nurse walks the patient within a 30-meter course. In urgent situations the nurse will need the chair to be available within one minute.

15.11 DISPOSABLE BLOOD PUMP

In blood treatment applications, there is a need for a small disposable pump to reduce the interaction of patient blood with operators and equipment. The pump can be placed along with other disposable fixtures. You are required to design and build a pump with the flowing specifications:

- All blood-contacting parts are disposable.
- The pump can be mass produced to reduce cost.
- The pump must be able to achieve a variable flow rate from 0 to 100 ml/min with a volumetric accuracy of $\pm 5\%$.
- The pressure head operates between 100 mm Hg to 100 mm Hg.
- Disposable size is not to exceed $30 \times 30 \times 25 \text{ cm}^3$. The overall fixture must not to exceed $30 \times 30 \times 30 \text{ cm}^3$

15.12 NEWSPAPER VENDING MACHINE

The specific problem with existing newspaper dispensers is the ease of stealing the papers in the machine when the door is opened. Theft from existing newspaper vending machines is a source of revenue loss as well as frustration for newspaper companies today. Another limitation of the current dispensers is that they can handle only one newspaper for each dispenser, which makes them inefficient spacewise. You are required to design and build a newspaper dispenser with the following specifications:

- It can handle at least three different newspapers.
- It dispenses one newspaper at a time based on customer selection.
- Newspaper companies make most of their profit from advertisements and not from newspaper sales, so they will be unlikely to purchase machines that cost more than the current ones.
- Familiarity of the design to customers is crucial, since customers will be unlikely to purchase newspapers from designs that deviate from existing designs.
- The design size should conform to current newspaper dispenser standards, especially with regard to height.

15.13 PEACE CORPS GROUP PROJECTS

The class is assigned to a Peace Corps mission in a village of an average population of 250 somewhere on the planet where electricity is not available. The village needs the team's help to design mechanical instruments that will help the village to function.

15.13.1 Projects

The villagers require help from the team for the following:

- *A mechanism to pump water.* Their current technology consists of an open well with a bucket. For health reasons, it is best to cover the well and design a mechanical pump. The power for the pump could be provided manually or by an animal. Pumps are needed for both drinking-water sources and the irrigation system. Because of limitations in material resources, the pump must be used for both irrigation and for drinking water. (Assume the water is drinkable.) The flow rate needs to be controllable. Drinking water demand is much less than that of the irrigation system (about 1/200). The well is about 100 m deep (Don't ask how they reached that depth.)
- *A mechanism to grind wheat to produce flour.* The current technology is that each household has its own manual grinder composed of two rough, heavy disks. The disks are aligned around a central hole where the wheat is fed. This current technology is both time-consuming and a major waste of resources. It is best to set up a central location for this task that every household could use. The new mechanism must save time and could be a one-man operation.
- *A mechanism to help in seeding agricultural land.* The current technology utilizes a shovel and a digger. If you have ever planted seeds by hand, you will know that this is a labor-intensive job. Average land per household is about 1200 square meters (30 households). The system needed should be energy independent. Please do not humiliate the people by suggesting a simple attachment to a horse.

- *Develop a mechanism to collect the wheat and vegetables.* It is assumed that with the technology developed to pump the water and seed the ground, the yield will become 10 times higher than before. The task now is to develop a mechanism to collect the wheat and vegetables. Two systems are needed:
  a. One for wheat collection and packaging
  b. Another for vegetable collection and packaging (The average temperature is about 27°C during a year.)
- *A system to generate electrical energy.* This would be of interest to the village. It is very clear that one source will not be sufficient. Villagers have a lot of wood at their disposal. Wind is a more viable energy source than solar because of material.
- *A time and calendar system viewable by the whole village.* Currently time is told by the shadow length of a well-designed wood column. What is needed is a mechanism that tells the time, day, month, and year.

15.13.2 Materials

The village is remote, with limited access to material. You are allowed to have $100 worth of material shipped to you. Be wise in using the material. Remember, lumber is available by the tons.

15.13.3 Machining

Machining would be the greatest challenge. There is no machine shop available. What is available is manually driven wood-shaping equipment.

15.13.4 Deliverables

1. Teams will show three alternatives of their designs using Pro/E or another CAD tool; they will apply a systematic design process in reaching three alternative designs using CAD tools. Teams are encouraged to be creative and not to restrict themselves to ideas that are currently available.
2. Teams will show the analysis of the three alternatives as part of the evaluation process. In addition to the design objectives, the analysis plays an important part in arriving at a design decision. Details of the analysis are required.